MUSCLE PHYSIOLOGY and CARDIAC FUNCTION

LINCOLN E. FORD

*Professor of Medicine,
Physiology and Biophysics
Krannert Institute of Cardiology
Indiana University
School of Medicine*

COOPER PUBLISHING GROUP

Library of Congress Cataloging in Publication Data:

Ford, Lincoln.
Muscle Physiology and Cardiac Function

Cover Design: Michelle Cherry

Publisher: I. L. Cooper

Library of Congress Catalog Card Number: 00-103072

ISBN: 1-884125-72-7

Printed in the United States of America by Cooper Publishing Group, LLC, P.O. Box 1129, Traverse City, MI 49685.

10 9 8 7 6 5 4 3 2 1

A picture of the author as a young oarsman appears on the front cover to signal that the relationship of muscle physiology to athletics will be discussed. For this purpose, it is more relevant than a current photograph of the author that might grace the back cover. He asks "remember me as I was, not as I am."

CONTENTS

Preface   v
Acknowledgements   vii

PART I: BASIC MECHANISMS

Chapter 1:   Muscle Structure   3
Chapter 2:   Muscle Contraction   25
Chapter 3:   Crossbridges   47
Chapter 4:   Work from Chemical Reactions   77
Chapter 5:   Activation   103
Chapter 6:   Length Dependence of Activation   127
Chapter 7:   Mechanical Manifestations of Activation   147
Chapter 8:   Fatigue   166
Chapter 9:   Smooth Muscle   178
Chapter 10: Comparison of Cardiac and Skeletal Muscle   193

PART II: WHOLE BODY FUNCTION

Chapter 11: Some Consequences of Body Size   211
Chapter 12: Optimum Sizes for Some Athletics   233

PART III: CARDIAC FUNCTION

Chapter 13: Hemodynamics   263
Chapter 14: Cardiac Anatomy   290
Chapter 15: Geometric Principles Applied to the Heart   300
Chapter 16: The Frank-Starling Law of the Heart   324
Chapter 17: Inotropic Mechanisms   337
Chapter 18: Pathophysiology of Heart Failure   361
Chapter 19: Treatment of Heart Failure   381

References   397
Index   413

PREFACE

The chapters in this book began as handouts for an upper level undergraduate course. The articles suggested for additional reading at the ends of the chapters were chosen mainly because they are short, especially informative, or of historical interest, rather than to give a more comprehensive description of the material. In fact, the entire course and book are not intended to be comprehensive, but rather a collection of interesting anecdotes that have led us to our current understanding of an area in science that I am qualified to teach. The main point of this preface is to explain how it developed.

I had almost no science education until the last semester of my senior year of college, when I decided to go to medical school. Before then, I had the mistaken belief that mathematics and science had been sent to us centuries earlier from another planet, and with the exception of some arcane matters to be resolved by a band of extraterrestrials, scientific knowledge was more or less fixed. Imagine my surprise, therefore, when I enrolled in my first Biology class at the end of my college career. In spite of having no previous Biology education, I arranged to be admitted to an advanced course intended for freshmen with a strong background in high school Biology. Because the course covered the entire first year in a semester, it omitted most of the drier subjects, such as taxonomy and botany, and concentrated on the new discoveries then being made. One of the instructors was James Watson, then age 30, who had recently discovered the meaning of DNA. The next semester, as a post-graduate student, I took R.P. Levine's course in Genetics to complete the full year Biology requirement for medical school. This was a model for what I later hoped my own course would be. The lectures described the experiments that led to our understanding of genetics, beginning with Mendel and ending with the little that was then known about the genetic code.

From this experience, I learned that science was being done by ordinary mortals and that there were many discoveries remaining to be made. To a person seeking a career, it was an inspiration. I entered medical school with the ambition of becoming a scientist and never wavered from that goal. Although I often tell myself that I should grow up and get a real job, practicing medicine, I also know that unless the funding grinch intervenes, when my time is up I will find myself asking St. Peter to let me finish just one more experiment.

The main point of this history is that, as much as I am able, I have tried to convey my enthusiasm for the experiments and theories that have led to our current knowledge. In addition, I have chosen to present material that I had some prior reason to know and which fits into a larger scheme of animal function. My areas of interest are mechanics and heart failure, and for the most part I have confined myself to these areas. I have not, for example, included much biochemistry or electrophysiology of muscle, both because I know less about these areas and because they each deserve a book of their own.

Because of its student orientation, the book describes mainly the triumphs of this work, perhaps presenting some of the material in a more certain light than it deserves. I have also tried to provide sufficient background information that if a conclusion is later found to be faulty, the reasons for it will be apparent. For those readers who prefer a more detailed view of the uncertain material, I recommend books by Woledge et al. (1985) on basic muscle physiology and by Katz (1992) on cardiac physiology. Finally, I have taken the advice given Stephen Hawking (1988) in writing *A Brief History of Time,* and as much as is possible in a scientific text, have attempted to avoid mathematical equations, relying as much as possible on graphs and diagrams. For those readers interested in mathematical derivations of some of the principles, I recommend the book by McMahon (1984).

ACKNOWLEDGEMENTS

Foremost among things to be acknowledged is that I am both absentminded and a poor proofreader. Thus, gentle reader, when you encounter typographical or other errors, as surely you will, please do not assume that friends who read the typescript have let us down. The much more likely explanation is that, through oversight, I failed to act on their suggestions, or that in making the suggested correction, I committed yet another error.

I am very grateful to Drs. Susan Gilbert, Stuart Taylor, and Richard Meiss, who read every chapter and made many suggestions that improved the book enormously. Frequently, the suggestions led to long and pleasant conversations, for which I am grateful. Sir Andrew Huxley read several chapters and offered much helpful advice. I should also acknowledge that much of what I know about the history of Muscle Physiology derives from him, either from his lectures or from mealtime conversations. He very kindly corrected some mistakes in stories that I had originally written. Drs. Chiu Hui and Daryl Swartz read several chapters and made helpful suggestions. Alvin Detterline and Kevin Ho helped to compile the data on athletics. Finally, I am indebted to several classes of students who challenged me to make a better presentation.

I would especially like to thank Ms. Michelle Cherry, of Medical Illustrations Department of the Indiana University School of Medicine, who drew, and often re-drew, almost all of the figures in the book. Mr. Tom Weinzerl, associate director of the Medical Illustrations Department, also made several drawings. Apart from finding me Michelle and Tom, Mr. Craig Gossling, the director of their department, was helpful in several important ways, not the least of which was finding me a publisher, Butch Cooper. Butch has been very helpful, as well as patient in tolerating missed deadlines. I should also mention that Craig made the spectacular drawing of an athlete scaled to different sizes and shapes in Fig. 12.2, and that Tim Yates then adapted it to a computer format.

PART I

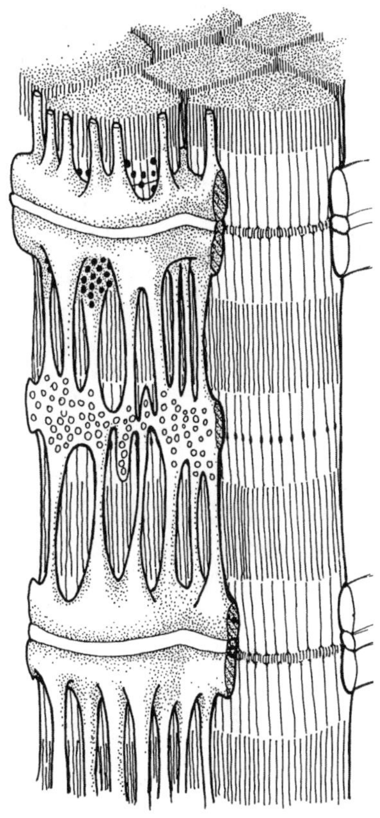

BASIC MECHANISMS

Chapter 1

MUSCLE STRUCTURE

This book is about **striated muscle**, which is divided into two categories, cardiac and skeletal, depending on its origin. **Cardiac muscle** makes up the walls of the heart and powers the circulation. **Skeletal muscle** powers almost all voluntary movement and some involuntary movement in the body. **Smooth muscle**, which is used to power most involuntary movement, does not have the regular **cross-striations** that characterize striated muscle. It is considered in only one chapter of this book, mainly because less is known about it. Much of what is known about basic contractile mechanisms comes from experiments on skeletal muscle because it is easier to study. We feel comfortable in extrapolating this knowledge to cardiac muscle because the ultra-structure of the two muscle types is the same (Fig. 1), suggesting that the same mechanisms operate in both. Although smooth muscle has similar contractile proteins and filaments, its ultrastructure is both different and less well known, so that extension of our knowledge of striated muscle mechanisms to smooth muscle seems less secure. The main purpose of this chapter is to describe the structure of striated muscle, emphasizing the ultra-structural similarities between cardiac and skeletal that justify the common description of their basic mechanisms presented in the next seven chapters.

GROSS APPEARANCE

In non-scientific terms, muscle is meat. At the macroscopic level, this meaty quality distinguishes muscular structures from

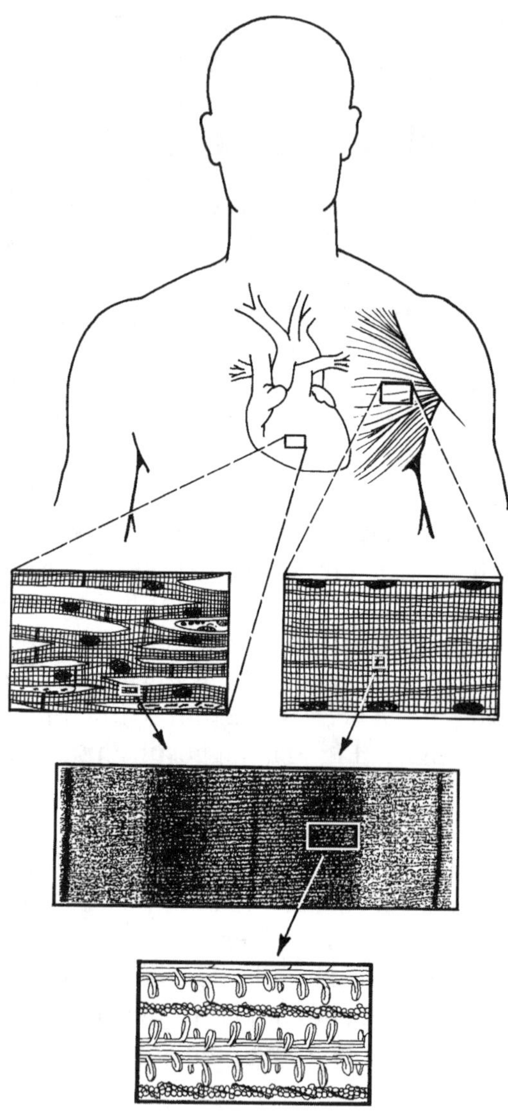

Figure 1.1. *Comparison of skeletal and cardiac muscle.*

other tissue in the body. Because the function of muscle is to pull in the shortening direction, neighboring cells in muscular structures are usually oriented in the same direction. The largest individual cells of mammalian skeletal muscle are just wide enough (about the diameter of a fine human hair) to be distinguished with the naked eye, so that the parallel orientation of the fibers is readily recognized without the aid of magnification. Mammalian cardiac muscle cells are substantially smaller in diameter. In addition, there is a need for the groups of cells to change their axial orientation through at least 90° across the heart's wall to cover all possible stresses, so that the parallel orientation of the fibers is much less apparent.

Cardiac muscle makes up about 0.5% of lean body mass and is found only in the heart; skeletal muscle comprises about 38% of lean body mass and is found throughout the body. Different skeletal muscles perform different functions, and these functional differences are represented in the color of the muscles. The red color derives from **myoglobin**, a protein similar to hemoglobin that binds oxygen and serves as a reservoir of oxygen within the muscle cells. Fast muscles, which are responsible for rapid movements, contain relatively little myoglobin and consequently are relatively light in color. (The "white meat" in the breast of domestic fowl is an extreme example.) Slow muscles, which are responsible for such activities as the maintenance of posture, contain a higher concentration of myoglobin and consequently are darker red. Cardiac muscle, which works continuously, is very dark red.

In general, the individual skeletal muscle cells, **fibers**, are all about the same length. Although many muscles have tapered ends, this shape is not always due to fibers at the edges being a different length than those at the center. In some muscles the fibers along one side originate and insert proximally in the tendons while the fibers on the other side originate and insert distally, maintaining a uniformity of fiber length (Fig. 1.2). This arrangement allows all the fibers to be the same length and yet be attached by narrow tendons. In some muscles, the tendons are longer than muscle cells,

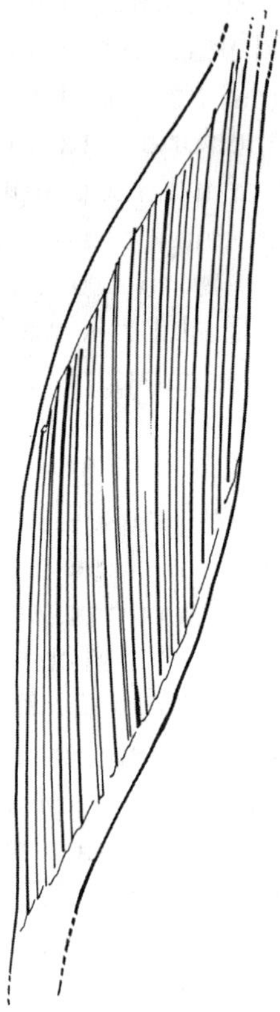

Figure 1.2. *Origins and insertions of fibers in tendons are staggered to provide constant fiber length in a muscle with tapered ends*

an arrangement that places a larger number of cells in parallel than could be fitted into the cross-section of any part of the muscle. Such muscles generate more force per unit of cross-sectional area, but shorten by a lesser fraction of overall length because the individual cells are shorter.

When fibers in a muscle are the same length, they all undergo the same **relative** length change (i.e. the same percentage length change) as the muscle length changes. They also all shorten with the same relative velocity when the muscle shortens, and they will all be at the same functional length when muscle length changes. Thus, all the fibers in the muscle will function uniformly. This uniformity of function, characteristic of individual skeletal muscles, does not obtain in the heart, and the differences in function that occur in neighboring areas of the heart are described in detail in chapters 14 and 15.

MICROSCOPIC APPEARANCE

The comparison illustrated in Fig. 1.1 shows that cardiac muscle cells are very different from skeletal muscle cells, and the functional implications of these differences are discussed in chapter 10. Individual skeletal muscle cells are very long, extending for most of the length of the muscle, and usually unbranched. They originate and insert in non-muscle tissue, either tendon or bone at both ends, and are not connected directly to their neighbors. By contrast, heart muscle cells are short and branched. They are connected to each other at the **intercalated disks**, junctions that allow both contractile force and small molecules to pass axially from one cell to another. The skeletal muscle cells have many nuclei placed just under the surface membrane and are formed by the fusion of a large number of smaller cells. The word **syncytium** is used to denote such multinucleate cells formed by the fusion of multiple cells. The individual cardiac cells have a single, more centrally placed nucleus, and the walls of the heart are formed by the incomplete fusion of the individual cells. This arrangement is called a "functional syncytium" in that the cells are functionally connected by transmembrane pores, or **channels**, that allow the passage of small molecules and ions in regions known as **gap junctions**, but retain their individual boundaries. A major function of these connections is to allow the free passage of electric current carried by ions so that there are low resistance electrical connections between neighboring cells. This arrangement is sometimes call an **electrical syncytium**.

Skeletal muscle cells are generally wider, 30-60 μm diameter, as compared with 5-15 μm diameter for cardiac cells. Both types of cells are substantially longer than they are wide, but the length/width ratio of cardiac muscle cells, about 5-10, is very much smaller than that of skeletal muscle, which in extreme cases can be more than 10,000.

Striated muscle is distinguished microscopically by having regular bands, striations, that traverse the fiber axis. The contractile elements are grouped together in **myofibrils** that run longitudinally in the cells. These myofibrils are difficult to resolve in the light microscope but are readily recognized in electronmicrographs. In general, a larger fraction of the cross-sectional area is occupied by these contractile myofibrils in skeletal muscle than in cardiac muscle. For the most part, the cellular cross-section not occupied by myofibrils or nuclei are filled either by mitochondria, which are barely resolvable by light microscopy, or by the sarcoplasmic reticulum, which is too small to be seen with the light microscope. Slow muscles, which depend heavily on oxidative metabolism, have a greater abundance of mitochondria. Fast muscles, which are required to cycle calcium through sarcoplasmic reticulum more rapidly, have a greater abundance of sarcoplasmic reticulum. Since the sarcoplasmic reticulum occupies less volume than the mitochondria, faster muscles generally have a greater cross-sectional area available for the contractile myofibrils. Cardiac muscle, which can be regarded as the slowest form of striated muscle, has the least amount of its cross-sectional area devoted to the myofibrils.

ULTRASTRUCTURE

Sarcomeres

The appearances of the two types of striated muscle become progressively more similar as the magnification of their images increases (Fig. 1.1). The cross-striations of both types of muscle result from the repetition of **sarcomeres** arranged in series along the length of the cell. Fig. 1.3 illustrates the structures that give the sarcomeres their characteristic appearance. The sarcomeres are separated from each other by dense **Z-lines**. At the center of the sarcomere is an **anisotropic** structure called the **A-band**. By anisotropic is meant that the structure exhibits birefringence, i.e. it has a higher index of refraction for light polarized in one plane than the other. Electronmicroscopy reveals that this birefringence is caused by a regular arrangement of **thick (myosin) filaments** oriented parallel to the fiber axis. The length of the filaments, and of the A-band, is about 1.6 μm in all types of mammalian striated muscle. Both the A-band and the individual filament lengths remain constant when overall muscle length changes, at least in mammalian striated muscle. Some invertebrates have muscles with very different lengths of A-bands, and in some, the A-band

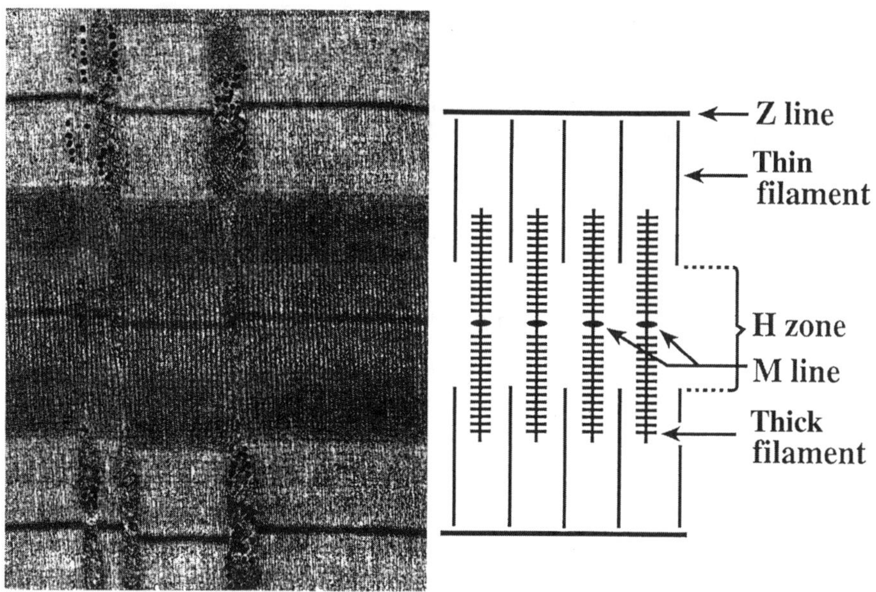

Figure 1.3. *Sarcomere structure.*

length appears to shorten during activation, but the physiological significance of this shortening has not been discovered.

The space between A-bands is called the "I-band" because it is **isotropic**, i.e. it does not exhibit birefringence. Its length varies exactly with sarcomere length as the muscle shortens. The Z-line bisects the I-band. Electronmicroscopy shows that **thin (actin) filaments** arise in the Z-line and extend through the I-band to end in the A-band. The constancy of A-band length and the direct variation of I-band length with sarcomere length was taken as strong evidence that the thick and thin filaments slide past each other when the muscle changes length.

In the center of the A-band is the **H-zone**, which is devoid of thin filaments. The ends of this zone are defined by the ends of the thin filaments, and its length varies directly with sarcomere length. In the center of the H-zone is a thin line, called the **M-line**, which is now known to be the locus of connections between neighboring thick filaments.

Under physiological conditions, sarcomere lengths of range between about 2.0 and 3.0 µm in skeletal muscle and between about 1.7 and 2.3 µm in cardiac muscle. Thus, the amount of thin filament overlap with thick filaments varies substantially over the physiological range of lengths.

Filament lattice structure

Thin filaments are composed of individual **actin** molecules joined together in paired strands twisted into helices (Fig. 1.4 A). The axial distance between individual actin molecules along each strand is 5.4 nm and the helix undergoes a half twist every 36.5 nm.

Thick filaments have regular side projections repeating every 14.3 nm (Fig. 1.4 B), also arranged in a helix that repeats every 42.9 nm. These projections are the force generating **crossbridges** that produce the sliding force between the two types of filaments. Force is generated by the cyclical attachment and detachment of crossbridges to the actin molecules of the thin filaments, as ex-

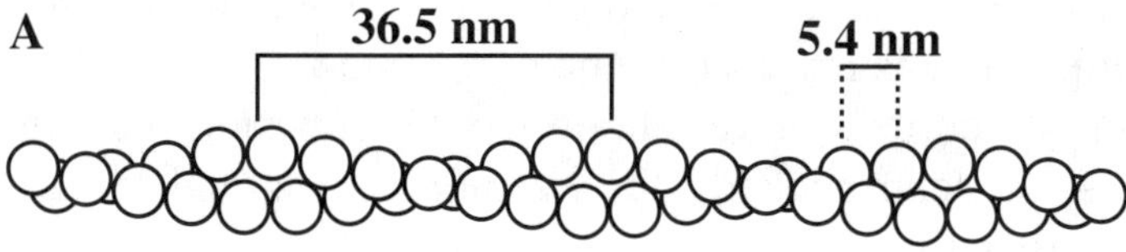

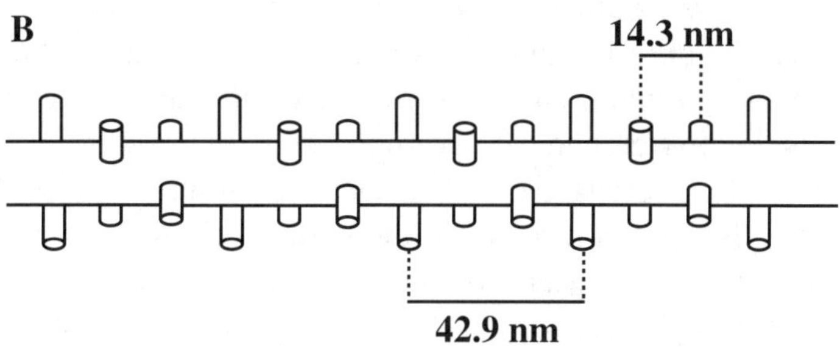

Figure 1.4. *Myofilaments: A) thin filament; B) thick filament. From H.E. Huxley (1969), with permission.*

plained in Chapter 3. It should also be said that while Fig 1.4 B, taken from a 1969 publication, shows two projections, one on each side of the thick filament at each 14.3 nm repeat, called a **crown**, there are now known to three projections per crown.

The absence of any small common denominator relating the 14.3 nm repeat of the myosin crossbridges to the 5.5 nm repeat of the actin monomers insures that the cycling of neighboring cross-bridges arrayed along a filament will not be synchronized as the filaments slide. The lack of synchronization insures smooth force generation as the individual cross-bridges cycle between the most and least optimal positions relative the their attachment sites.

The central 0.2 μm of the thick filaments are devoid of side projections. This **bare zone** in the center of the filaments creates a lighter appearance at the center of the A-bands of fixed and stained muscle (Fig. 1.3). The individual thick filaments exhibit different polarity at each end, such that the thin filaments are drawn from either side toward the center of the A-band. This thick filament polarity is reversed in the bare zone.

In cross-sections of vertebrate striated muscle, thick and thin filaments are seen packed into a regular hexagonal lattice, with twice as many thin filaments and two thin filaments between each pair of thick filaments (Fig. 1.5). This lattice structure insures that force generating cross-bridges operate in a nearly uniform environment. The environment is not entirely uniform, however, because lattice volume remains constant as the sarcomere shortens, so that cross-sectional area varies inversely with sarcomere length. When muscle is stretched, for example, sarcomere lengthening is accompanied by decreased distances between filaments. Similarly, osmotic swelling or compression of whole cells alters inter-filament distances. These alterations in lattice spacing are accommodated by flexible "hinge regions" in the myosin molecules, as explained under the heading *molecular structure*, below.

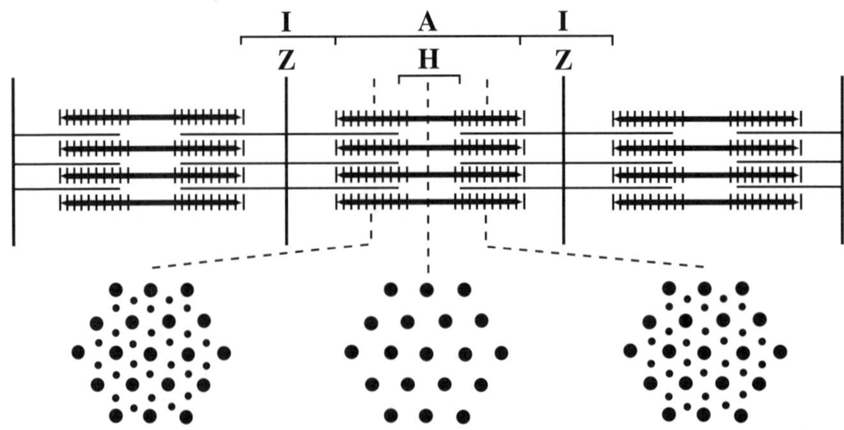

Figure 1.5. *Myofilament lattice in longitudinal section and cross section. From H.E. Huxley (1969), with permission.*

The repetitiousness of the myofilament lattice produces very regular X-ray diffraction patterns, much like a crystal of some uniform chemical compound. These X-ray diffraction patterns have been used extensively to study the structure of muscle, and these studies have provided a great deal of information about the structure-function relationships in muscle. In spite of the regularity of structure, neighboring cross-bridges along a filament experience very different interactions with actin for two reasons. First, as mentioned, the spacings between myosin heads and between actin attachment sites are not related by a small common factor, insuring that neighboring attached crossbridges will always have different orientations to their actin attachment sites. Second, the twist of the actin helix causes attachment sites on actin to face the myosin filaments favorably for only short axial distances. Thus, there are short **target zones** on the actin filaments where the crossbridges from one thick filament can attach. In regions between target zones the crossbridges do not attach as readily, if at all. One consequence of this arrangement is that filament sliding causes some cross-bridges to move into more favorable positions for attachment and others to move away from such positions. This alteration of more and less favorable orientation for cross-bridge attachment and force generation causes further heterogeneity of cross-bridge activity and smoother contractile movement. In addition, the unfavorable attachment regions between target zones on a thin filament suggests that only a fraction of the total number of crossbridges are able to attach at one time. These considerations add an uncertainty to the calculations, made in the next two chapters, of the total number of myosin molecules that participate in contraction at any given instant.

A great deal is known about the contractile protein lattice of striated muscle because it can be studied by X-ray diffraction in the living state. It is an excellent tissue for using this technique to study molecular movements during physiological activity. In general, the filament lattice structures of mammalian cardiac and skeletal muscle are virtually identical. In addition, the contractile proteins, actin and myosin, are very similar, and in at least one case, derive from the same gene in both muscle types. These considerations justify direct extension of knowledge of contractile mechanisms elucidated in skeletal muscle to cardiac muscle.

Internal membranes

In addition to the surface membrane and mitochondria, striated muscle contains two sets of internal membranes relevant to its function (Fig 1.6). One of these, the transverse tubule, **t-tubule**, is an inward extension of the surface membrane transversely across the fiber. In skeletal muscle these inward extensions form a network of tubules, called the **t-system,** that traverse the cell diameter. Although the diameter of these tubules is very small, their lumina are open to the extracellular space. Large molecules that cannot cross membranous barriers will diffuse quickly into the tubules soon after they are placed in the extracellular space. The purpose of these tubules is to carry the electrical impulse from the surface membrane to the interior of the cell, providing a rapid spread of activation throughout the cell's cross-section. In cardiac muscle cells, which are much thinner, there is more heterogeneity in the invagination of the surface membrane. Some cell types, such as those of the left ventricle, have a t-system similar to that of skeletal muscle. Other cell types, such as those of the atria, have deep clefts lined by surface membrane.

A second set of internal membranes, called the **sarcoplasmic reticulum**, forms closed vesicles that are not continuous with the extracellular space. These vesicles resemble fenestrated, flattened sacs that run longitudinally through the sarcomeres. They surround myofilament bundles, called **myofibrils.** At the ends of the sarcomeres these flattened sacs bulge to form the **terminal cisterns** that are closely applied to the t-tubules. The structure comprised of a single t-tubule with a terminal cistern attached to either side is called a **triad**. The t-tubules and sarcoplasmic reticulum are joined together by regions of specialized membranes. As ex-

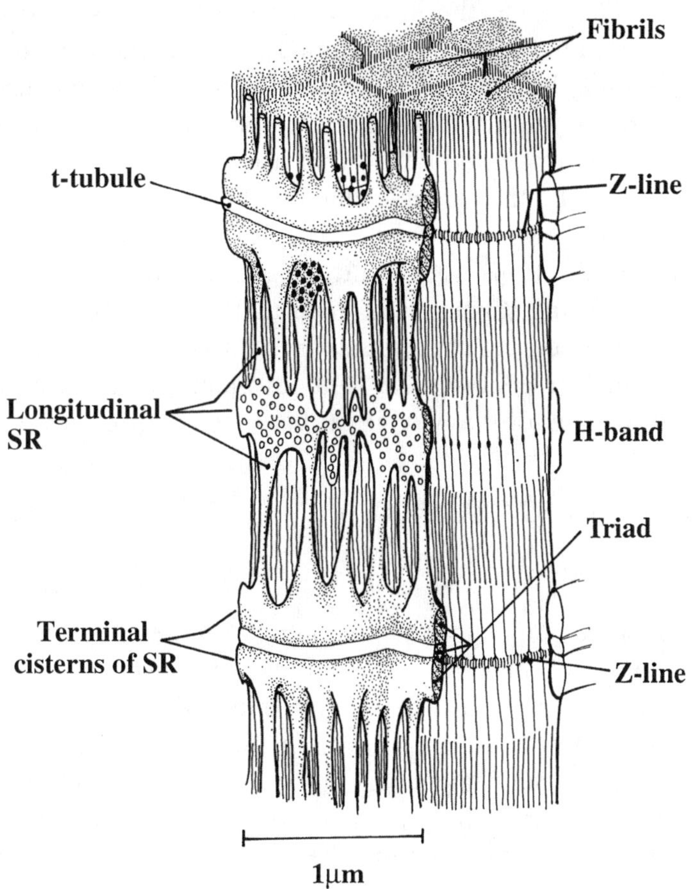

Figure 1.6. *Sarcoplasmic reticulum (SR) and t-tubules of frog muscle. Redrawn from Peachey (1965), with permission.*

plained in Chapter 5, the influence of the inward spread of the electrical impulse at the sarcolemma is transmitted to the sarcoplasmic reticulum at these junctions. This influence causes the sarcoplasmic reticulum to release calcium, that in turn activates the contractile apparatus. Relaxation occurs when this calcium is re-accumulated by the sarcoplasmic reticulum.

A cautionary word related to species differences is required here. The descriptions above and the picture in Fig. 1.6 are for frog muscle. Many species, and in particular mammals but others as well, have two sets of t-tubules in each I-band. Chapter 5 describes a small diversion in the study of muscle activation that resulted from a lack of knowledge of this species difference.

MOLECULAR STRUCTURE

Thick filaments

The myosin molecules that make up thick filaments are composed of three pairs of proteins. The largest of these, called **heavy chains**, are large proteins that are joined into dimers (Fig 1.7). The individual heavy chains have a molecular weight of about 500 kilodaltons and each heavy chain has associated with it two different **light chains**. The heavy chains are divided into a globular **head region** and a **rod portion**. The dimers of myosin are formed so that the rod portions are intertwined and the two heads project from the same end. Brief enzymatic digestion of myosin reproducibly yields two fragments of the molecule called heavy meromyosin (**HMM**) and light meromyosin (**LMM**). The light meromyosin comprises much of the rod portion of the intact molecule, while heavy meromyosin comprises the remainder of the rod portion and the head region. Further enzymatic digestion of HMM yields two fragments called sub-fragment 1 (**S-1**) and sub-fragment 2 (**S-2**). S-1 is the crossbridge **head**. It contains the apparatus required for contraction in that it attaches to actin filaments and hydrolyses ATP.

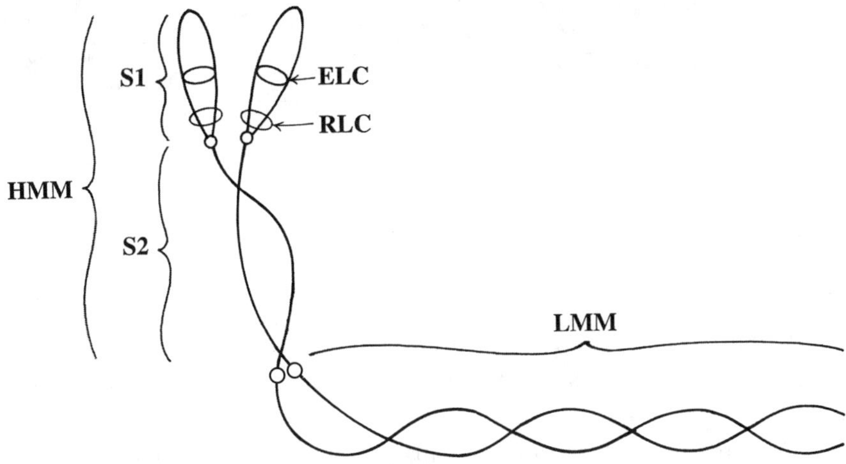

Figure 1.7. *Myosin molecule containing two heavy chains, each having two light chains (ELC and RLC). Circles indicate hinge regions that are susceptible to the enzymatic cleavage that yields the three fragments indicated.*

Myosin molecules are arranged in thick filaments so that their rod portions are incorporated into the body of the filament (Fig. 1.8). In each half-filament the molecules are oriented so that the end with the S-1 heads point toward the end of the filament and the tails point toward the center. The tails of the center-most molecules abut where the polarity of the molecules reverses in the center of the filament. This abutment of the tails accounts for the absence of crossbridge projections in the bare zone at the center of the filament. It is also one of the properties of the contractile apparatus that gives muscle contraction directionality. Activated S-1 heads drive thin filaments toward the myosin tails. The reversal of polarity at the center of the thick filaments provides that thin filaments are driven toward the center of the sarcomere from both ends.

The observation that enzymatic digestion of myosin yields reproducible fragments indicates that the cleavage sites in the protein are more exposed than the rest of the molecule. The likely explanation for this is that most of the molecule exists as a coiled **alpha-**

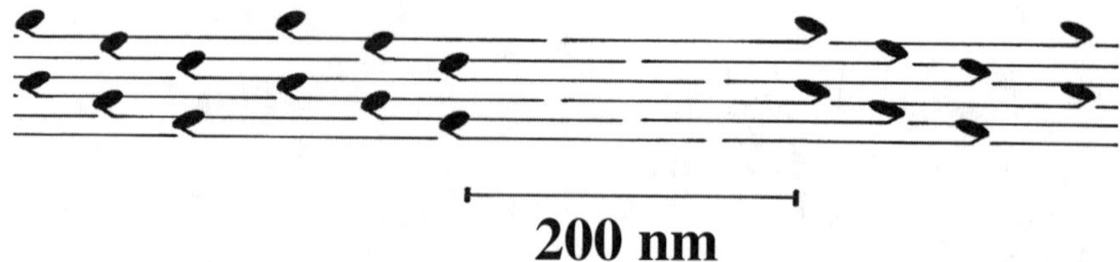

Figure 1.8. *Abutment of myosin tails of thick filaments produces a reversal of polarity at the center of the filament.*

helix that is both rigid and resistant to enzymatic digestion. The reproducible cleavage sites occur where there is no helix, and as a consequence, where the molecule is flexible. These flexible parts are known as **hinge regions** because they allow the otherwise rigid molecule to bend at these points. The hinge between the LMM and HMM portions of the molecule allows the HMM portion of the myosin to swing away from the filament backbone so that the S-1 heads can reach the thin filaments. The hinge between S-1 and S-2 gives the head freedom to move as it binds to a thin filament. Together, the two hinges allow crossbridges to function with some independence of the spacing between thick and thin filaments.

Each myosin S-1 head has two different light chains attached. One is called the **essential light chain** because it cannot removed from myosin without disrupting myosin function. The other is called the **regulatory light chain** because its phosphorylation promotes activation in muscles where activation is controlled by myosin rather than actin.

During sustained activation, regulatory light chains can become phosphorylated in response to physiological stimuli, but the functional significance of this phosphorylation, or indeed of the light chains themselves, has so far eluded definition.

Thin filaments

Actin is a protein which can exist in two forms, globular or **g-actin** and filamentous or **f-actin**. filamentous actin is simply g-actin that has polymerized into strands twisted into paired helices (Fig 1.4 A). These helices make up the backbone of the thin filaments and provide the attachment sites for the myosin cross-bridges. Thin filaments reconstituted from purified actin will stimulate myosin molecules to hydrolyze ATP in an unregulated manner. In contrast to this, thin filaments that have been isolated without purification, called **native thin filaments**, will stimulate myosin to hydrolyze actin only in the presence of calcium. This **calcium sensitivity** is conferred by a protein complex, **troponin-tropomyosin**, that is present in the native but not the reconstituted filaments.

Tropomyosin is a long molecule that lies along the thin filament, near the groove in the helix (Fig 1.9). There is one tropomyosin on each side on the filament and each tropomyosin molecule spans seven g-actin monomers in one strand of the helix. Troponin is a complex of three proteins that are attached to each tropomyosin molecule and confer the calcium sensitivity. One of these three, **troponin-C**, binds calcium. Another, **troponin-T**, binds the complex to tropomyosin, while the third, **troponin-I**, alters the calcium sensitivity as it becomes phosphorylated and dephosphorylated in response to hormonal manipulation.

Structural studies show that tropomyosin shifts its position on the thin filament when calcium is released into the myofilament space. This shift suggests that activation is regulated by the steric hinderance of myosin binding. In the relaxed state, tropomyosin covers crossbridge binding sites on the thin filaments. Calcium

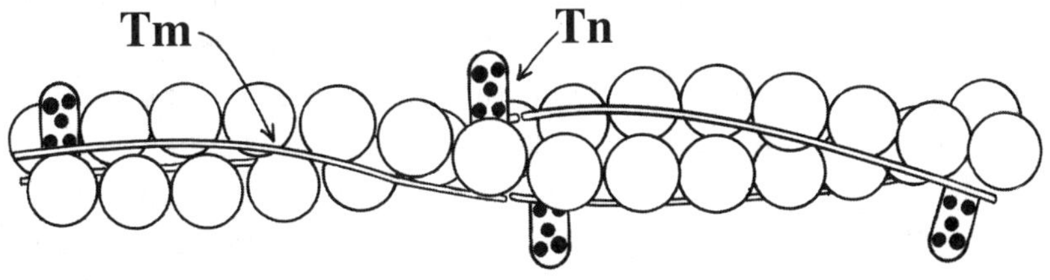

Figure 1.9. *Native thin filaments containing the f-actin helix, tropomyosin (Tm) and the troponin (Tn) complex.*

binding to troponin produces activation by moving tropomyosin aside, thus exposing the crossbridge attachment site.

Molecular similarities and differences among motile cells

A major point of this chapter is that the contractile mechanisms elucidated for skeletal muscle and described in the next chapter are probably operative in cardiac muscle as well. In addition, the same mechanisms are likely to be responsible for movement in other types of motile cell. This belief in the similarity of the contractile mechanisms arises from the finding that the contractile proteins in the different cell types are very similar.

Similar types of actin are found in many different cells that move, including all forms of muscle, as well as other motile cells. The prevalence of actin suggests that thin filaments are involved in many types of cell movement. One regulatory protein of the thin filaments, troponin, seems to be peculiar to striated muscle and has not yet been found in smooth muscle or motile cells. This observation suggests that the regulation of contraction may differ among cell types, even though the basic contractile mechanisms are the same. Differences in regulatory proteins of cardiac and skeletal muscle appear to be minor in comparison with the differences between striated muscle and other cell types.

Myosin-like molecules are found in non-muscle cells that move. These molecules have an ATPase head region similar to that of myosin but differ in their rod portions. These molecules do not form filaments as readily, suggesting that the molecular mechanisms of cell motility muscular contraction are similar, but that the aggregation of myosin into thick filaments varies among the cell types, perhaps to accommodate plastic changes of shape.

There are also small differences in the myosin types found in different types of skeletal muscle. These different types, called **iso-forms**, are distinguished functionally on the basis of the speed with which they hydrolyze ATP. Slow muscles, often responsible for maintaining posture, contain slow myosin, while fast muscles contain fast myosin. Many muscles contain mixtures of both myosin types and these muscles shorten at intermediate velocities. Other isoforms include a very fast myosin responsible for eye movement and different cardiac isoforms that are among the slowest isoforms. In at least one type of small mammal the fast isoform of cardiac myosin is electrophoretically indistinguishable from the slowest isoform of skeletal muscle. Again, this observation suggests that the contractile mechanisms in the two types of muscle are very similar.

PARALLEL ELASTIC ELEMENTS

An unstimulated muscle resists stretch with a passive force that increases progressively with lengthening. Passive length-tension relationships typical of skeletal and cardiac muscle are shown by the lower curves in Fig. 1.10, where the upper curves show the superimposed active length-force relationship. The structures responsible for this resistance to stretch are called the **parallel elastic elements** because they are functionally in parallel with the contractile elements.

There are several components of these parallel elastic elements. The solid passive curve in Fig. 1.10 shows the passive length-tension relationship for a whole skeletal muscle. The mus-

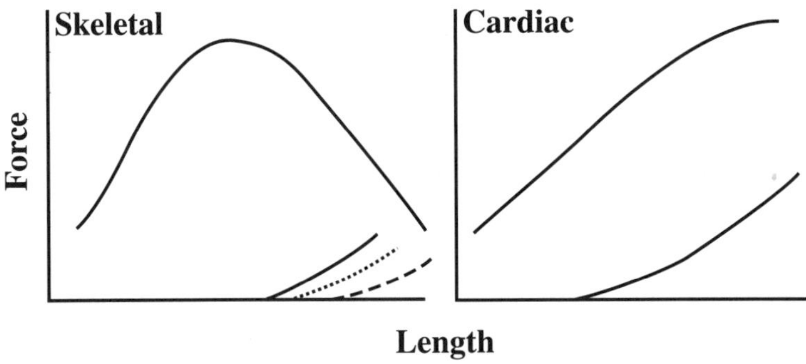

Figure 1.10. *Length-tension relations for skeletal and cardiac muscle.*

cle tension is expressed as a percentage of the maximum force that the muscle can develop. If all but one cell is dissected away, the remaining parallel elastic elements are more compliant; both the slope of the curve is less steep and the length at which force appears is longer (dotted curve in Figure 1.10). This observation suggests that some of the passive resistance to stretch is provided by the connective tissue in the extracellular space of the muscle. A further increase in compliance can be achieved by dissecting away the surface membrane of the cell. Thus, some of the parallel elastic elements appear to reside in structures, probably more connective tissue, that is firmly associated with the surface membrane. The **skinned fiber** that remains after removal of the surface membrane continues to resist stretch, albeit with less force (dashed curve, Figure 1.10). In addition, the filaments do not diffuse away when the membrane is removed. These findings suggest the presence of intracellular structures that hold the sarcomere together and resist stretch.

The structures, both intracellular and extracellular, that provide muscle with its passive properties are much less well studied than those responsible for the active properties. Recent evidence suggests that there are filaments made of a protein called **titin** that link the Z-line with the M-lines at the center of the A-band. The exact points of attachment of these titin filaments have not been fully elucidated, so that their precise function is not fully understood. The reason that they have not been as extensively studied is that they are not stained when tissue is prepared for electron-microscopy in the usual manner. For a long time their existence was inferred only from indirect evidence, such as the presence of distinct cleavage planes in the I-bands of fibers that were severely stretched. Only recently have the filaments been demonstrated by exposing muscle tissue to anti-titin antibodies during preparation for electronmicroscopy. The importance of these passive structures is that they maintain lattice integrity and the uniformity of sarcomere architecture.

As shown in Figure 1.10, the parallel elastic elements in cardiac muscle are shorter than those of skeletal muscle. When the muscle is stretched much beyond its optimum length for active force development, passive tension rises to very high levels and the muscle is usually damaged irreversibly. The heart requires that its muscle have these less compliant passive structures because it is not protected from over-stretch by the skeleton in the same way as skeletal muscle. As explained in Chapter 10, dilation of the heart beyond the point where its muscle develops maximum force could have disastrous consequences for the circulation. The exact structures responsible for the stiff parallel elastic properties of cardiac muscle are even less well understood than in skeletal muscle because preparations of the smaller cardiac cell adequate for such studies have not yet been developed. Thus, the dissection of the passive elastic properties into intracellular, membrane-associated, and whole muscle components, as illustrated by Fig 1.10 for skeletal muscle, has not yet been done for cardiac muscle.

ROBUSTNESS OF THE CONTRACTILE MECHANISMS

As mentioned, the filament lattice of intact muscle is so regular that it has been studied extensively with X-ray diffraction techniques developed initially for elucidating the structure of crystals. Although these studies have provided a great deal of information about the relationships between structure and function in muscle, it should be emphasized that much of the regularity of structure is not needed for contraction. For example, when a single skeletal muscle cell is "skinned," i.e. has its membrane dissected away, the equatorial X-ray reflections, caused by the regular spacing between thick and thin filaments, often disappear unless extreme care is taken to avoid damaging the lattice during dissection. In spite of this disruption, the cells produce their usual force and shorten with the expected velocity when activated. Similarly, when skinned fibers are activated, the optical diffraction pattern caused by the regularity of sarcomere spacing often disappears with no change in the contractile performance of the muscle. Thus, the molecular mechanisms underlying the contractile processes do not depend on the regularity of the filament lattice. While the regularities of structure have provided invaluable tools for the investigation of structure-function relations, the mechanisms are sufficiently robust that disruption of this regularity does not greatly alter contractile function.

SUGGESTED READING

H.E. HUXLEY (1969) The mechanism of muscular contraction. *Science* **164**: 1356–1366.

Chapter 2

MUSCLE CONTRACTION

The purpose of this chapter is to describe the physiological repertoire of muscle. Before beginning this description, the term "contraction" will be defined and units of measurements discussed.

When a muscle is stimulated electrically with its ends held firmly, force is generated in the shortening direction. If the stimulus is not repeated, force rises to a peak and then falls (Fig. 2.1 A). This brief response to a single stimulus is called a **twitch**. If instead the muscle is stimulated repeatedly at a sufficiently high rate, force rises to a higher plateau level where it remains until the stimulus train ends (Fig 2.1 B). This sustained response is called a **tetanus**. Under physiological conditions twitches are the only contractions produced by cardiac muscle. A sustained tetanus would have the disastrous effect of interrupting the rhythmic contraction-relaxation cycles needed for the pumping action of the heart. Conversely, skeletal muscle usually produces tetani of varying duration, required by its various physiological functions.

When the ends of a muscle are held at the same length it is said to be **isometric** (same length), and the types of response produced by electrical stimuli are also termed isometric (e.g an isometric twitch or an isometric tetanus). If the ends of a muscle are not held at a constant length and the load on the muscle is less than the isometric force, the muscle shortens. The ability of an activated muscle to shorten and/or to produce force in the shortening direction is termed **contraction**. The term is used irrespective of loading. For example, the term "isometric contraction" is used even though there is no reduction in muscle length. Similarly, an activated muscle that is being stretched by a load greater than its

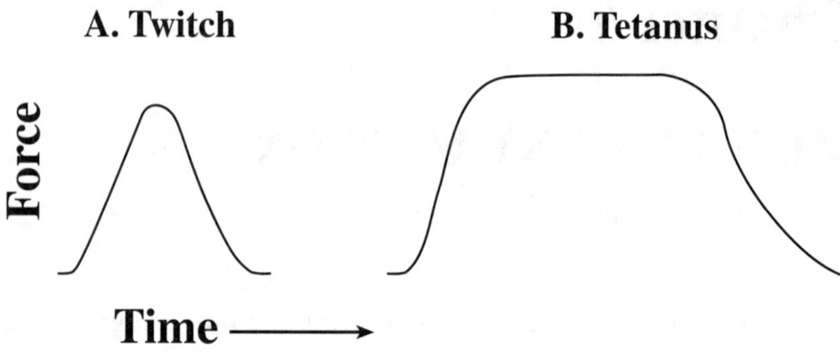

Figure 2.1. *Isometric contractions: A) twitch; B) tetanus.*

isometric force is also said to be in the seemingly contradictory state of "lengthening contraction."

The capabilities defined by the term "contraction," shortening and force generation in the shortening direction, define the entire repertoire of a muscle's mechanical ability. It cannot, for example, actively lengthen itself or produce other movements. Such activities as projection of the tongue are accomplished through the anatomical arrangements of shortening muscles. Other dynamic physical properties of the muscle are heat production, which accompanies contraction, and the electrical activity of the membranes. Most of the remainder of the chapter defines the inter-relations between the dynamic physical properties, force generation, muscle length, shortening velocity and heat production. But first some of the units used to define these properties are presented.

PHYSICAL DIMENSIONS

Force

It is convenient to express measurements in terms of the units used to calibrate the instruments. A transducer used to measure muscle force is frequently calibrated by hanging weights from the muscle attachment with the transducer oriented so that its measurement axis is vertical. Thus, it is common, at least in the laboratory, to express the force generated by a muscle in terms of **grams** (g). For the purposes of many calculations, however, it is both more convenient and more rigorous to convert grams to **newtons**, the **Systeme Internationale** (*SI*) force unit. At a time when SI units were less familiar, Andrew Huxley was known to point out that the conversion factor is approximately the weight of one apple, i.e., 102 grams per newton. To be useful, raw force must be related to the size of the muscle. This is done by dividing the force by the muscle cross-sectional area, yielding the units of pressure, N/m^2. The symbol P is often used to denote muscle force, perhaps because it is expressed in pressure units. It is common to express these pressure units as **pascals** (Pa). A convenient mnemonic, again due to Huxley, is that the pascal might also be a measure of the quality of universities; it is the number of newtons per square meter ($Pa=N/m^2$). Skeletal muscle typically generates about 150 to 400 kPa depending upon the experimental conditions.

A caveat is required regarding the cross-sectional area used to normalize force. Because muscle volume remains constant when length changes, cross-sectional area varies inversely with the length change. If accurate comparisons are to be made between different muscles, cross-sectional areas should be measured at the same functional length. The need for this uniformity arises because force is generated by individual filaments. Although the number of filaments in the cross-section does not change, both the distance between them and the muscle diameter vary inversely with the square root of the length, so as to maintain constant volume.

Length

Length measurements are similarly related in their final analysis to filament arrangements. While it has been said that the basic anatomic unit of muscle is the sarcomere, it is more accurate to say that the basic unit of the filament lattice is the half-sarcomere, which contains one Z-line and one set of thin filaments with the associated set of half thick filaments oriented in the same direction. More importantly, when length changes of the overall muscle are divided by the number of half-sarcomeres in series, the result is expressed in terms of the sliding distance experienced by the individual thick and thin filaments. This sliding distance represents the length change experienced by the individual crossbridges.

While it is useful to calculate length changes in terms of half-sarcomeres, length itself is usually expressed in sarcomeres, both because the phrase "sarcomere length" is more convenient than "half-sarcomere length" and because this is the quantity measured.

It is nearly impossible to measure sarcomere length in whole muscle, so that length is often expressed as a fraction of some standard length, such as the length where maximum force is developed (e.g. Fig. 2.2), or the **body length**, defined as the maximum length of the muscle in *situ*. These standard lengths are frequently denoted L_0 for skeletal muscle. The reference length for cardiac muscle is called L_{max}, both because it is the length where maximum force is developed and because it is the longest length where the muscle can stretched without damage. It should be emphasized

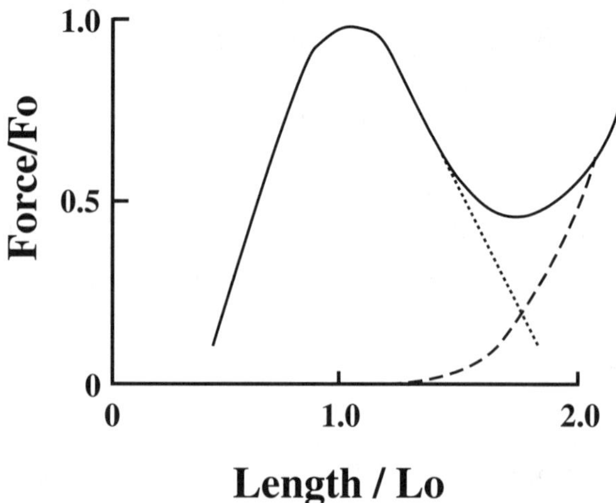

Figure 2.2. *Force-length relation in whole muscle.*

that the sarcomere length at these functional lengths may vary among preparations or in the same preparation under different conditions. One of the distinct advantages of single fibers studies is the ready determination of sarcomere length.

Cross-sectional area

Whole muscle cross-sectional area is often estimated by dividing muscle weight by specific gravity, 1.05, to obtain volume, and dividing this volume by the length (A=Wt/[1.05xL]). Single fiber cross-sectional area is often estimated either by measuring a single diameter and assuming a circular cross-section or by measuring two orthogonal diameters and assuming an ellipse. All estimates of cross-sectional area are subject to substantial error, and reliable results require both experience and great care.

Work and heat

The work done by a muscle is equal to its force times the distance shortened (work=P·dL). When force is expressed in newtons and length change in meters, the work is given in **joules** (J=N·m). Heat and chemical energy are also expressed in joules, so that the total energy output from the muscle can be determined as the heat plus the work (E=heat+work) and this total energy liberation can in turn be related to the chemical metabolism in the same terms.

Power

It is sometimes convenient to express mechanical activity in terms of work rate, power (PV), equivalent to the product of force times velocity (PV = P·V = dw/dt = P·dL/dt). The unit of power is the watt, equal to an energy rate of 1 joule per second (W=J/s). This is the same unit used to calculate electrical power. It is sometimes useful to know that a person with a basal metabolic rate of 2,067 kCal/day has an average basal expenditure of 100 watts.

Normalization

To compare forces in different muscles, force is usually normalized to cross-sectional area, but for many purposes it is more convenient and perfectly satisfactory to normalize it to its maximum force, F_o, developed under some standard condition (e.g. Fig. 2.2.)

When force is normalized to cross-sectional area and length change to a standard length, the product of the two denominators $(A{\cdot}L)$ is muscle volume. If volume is multiplied by the specific gravity of 1.05, the muscle work is then be expressed as work per gram (J/g). This normalization allows work to be compared with other energetic measurements that are also normalized to weight, such as heat, oxygen consumption, and chemical energy.

Efficiency

Dimensionless efficiency ratios can be determined when all the same units are used. For example, it was assumed for a time that all energy not realized as work was liberated as heat, and muscle efficiency was then calculated as the work divided by the total energy (e=work/[work+heat]). It was later recognized that this is not the same as the thermodynamic efficiency because the free energy available from a reaction is determined, in part, by the concentrations of the reactants and by entropy changes. It is, however, a measure of the chemomechanical efficiency because the denominator, heat plus work, represents the total energy available from the relevant chemical reaction. Thus, it is possible to use the heat and work measurements to determine the extent of a chemical reaction, once the reaction is known. As explained in Chapter 4, the immediate substrate for contraction is ATP, but this is replenished immediately by the transfer of phosphate from creatine phosphate. Unless the transfer is inhibited, the immediate measurements of heat plus work indicate the amount of creatine phosphate hydrolysis. If measurements are made after a delay, the relevant reactions are those that replenish the creatine phosphate.

Energetics of exercise

An example of the caloric equivalent of exercise is derived here to show how these calculations can be used. The **calorie (cal)** is the amount of heat required to raise the temperature of 1 gm of water by 1°C. It is not an SI unit but is converted to joules by the factor 4.18 J/Cal. The calorie counted by dieticians is actually a **kilocalorie (kcal)**, equivalent to 1000 cal or 4.18 kJ. The work equivalent of one kcal is thus 4,180 Nm and would be expended in lifting a 102 g apple 4,180 meters. Assuming the efficiency of muscle to be 0.5, the caloric equivalent of 1 kCal could be realized in lifting the apple only 2,090 m. To calculate the fat loss resulting from lifting the apple, it is only necessary to know that the metabolism of 1 gm of fat produces 9 kCal. Thus, if the energy in fat were extracted with perfect efficiency, lifting an apple by 2,090 m might consume as little as 110 mg of fat.

To put this work in still more familiar terms, consider not one apple but 100 apples weighing 10.2 kg, 22.4 lb, which would have to be lifted 1/100 the height, 20.9 meters, to consume one kCal. Furthermore, the work need not be done either in lifting or in one continuous motion. It might, for example, be done with repetitively smaller motions on an ergometer set to 22.4 lb of resistance. If the distance moved in a single repetition were 0.5 m, 41 repetitions would be required to consume one kCal.

In making these calculations, a muscle efficiency of 0.5 was assumed. The total body efficiency is substantially less, about 0.1 to 0.25, for three reasons: 1) the chemical reactions that immediately fuel contraction, ATP and creatine phosphate hydrolysis, are later reversed by other reactions which consume energy; 2) additional work is required to service the muscle, as for example with additional blood flow from the heart and additional respiratory effort; and 3) there are mechanical inefficiencies of the body movements, as for example in the need to provide postural support or in providing the return stroke of the ergometer, when no measurable work is done. Thus, number of repetitions required to burn one calorie would be about 10 to 20 rather than 41.

ISOMETRIC CONTRACTION

Relationships between muscle length and isometric force are plotted in Fig. 2.2, which shows that there are three separate forces to consider: 1) the passive force (lower, dashed curve); 2) the active force (middle, dotted curve); and 3) the sum of these two, the total force (upper, solid curve). For the present discussion, only the active force will be considered because it is the extra force added by the activated contractile machinery.

As shown in Fig. 2.2, developed force depends strongly on muscle length. The reasons for this dependence are described in Chapters 3 and 6. For the present discussion, it is only important to recognize that the dependence exists and that care must be taken during experiments to insure that the variations in length do not confound the interpretations.

SHORTENING CONTRACTIONS

There are many ways to load a non-isometric muscle. The most common experimental practices is to keep the load constant, i.e. to maintain the muscle **isotonic**. When the isotonic load on a tetanized skeletal muscle is less than the isometric force, the muscle shortens at a nearly constant velocity. When the force is suddenly reduced from the isometric level to an isotonic load, there is a sudden, rapid shortening, attributed to recoil of the series elastic elements, followed by a rapid approach to steady velocity (Fig. 2.3. There are brief **velocity transients** during which velocity is first higher and then lower than the final value, but these end before the half-sarcomeres have shortened by 15 nm. After these transients the velocity is nearly constant.

The qualification "nearly" is used here, however, because in most muscle preparations velocity declines progressively with shortening, so that the shortening records are curved slightly. While this curvature is well recognized, some early workers described shortening as being constant, and the reasons for this are

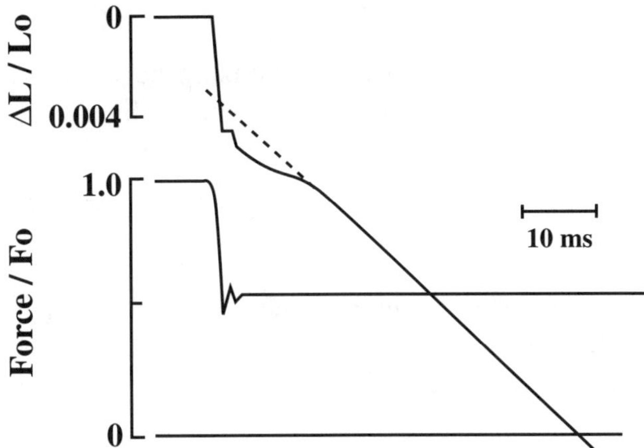

Figure 2.3. *Quick release to an isotonic load. Adapted from Podolsky (1960) and Civan and Podolsky (1966).*

not known. On the one hand, it is possible that their preparations did genuinely shorten with isotonic velocities that remained constant. Alternatively, it is relatively easy to overlook a small curvature in a record, even when the slope has declined by 30 to 50%. A final explanation is that the early records were obtained under conditions where fresh crossbridges were recruited as the shortening progressed, either because it was occurring during the rise of activation or because there was increasing filament overlap.

The discoveries of the early velocity transients and the progressive decline of velocity during sustained contraction raise the issue of the proper method for measuring velocity. While this would seem to be a trivial matter, easily resolved by experiment, it has also been a source of several substantial controversies. The ideal conditions for measuring velocity are probably during a sustained tetanus, shortly after a quick release to an isotonic load, and after the velocity transients have died away. The muscle can then be considered to be in a quasi-steady state. In the descriptions that follow, the measurements are assumed to have been made under those conditions unless otherwise stated.

Force-velocity relationships

There is an inverse, curvilinear relationship between the isotonic force and the steady-state velocity, as shown by the plot in Fig. 2.4. The equation most commonly used to define the force-velocity data is a rectangular hyperbola that is asymptotic to a line below the force axis and to a line to the left of the velocity axis. The distance below the force axis is conventionally designated by the constant **b** and the distance to the left of the velocity axis by the constant **a**. The equation used to describe the relationship between force, **P**, and Velocity, **V**, is thus:

$$(P+a)\cdot(V+b) = c \qquad\qquad (2.1)$$

where c is a constant. This equation was developed to describe muscle heat measurements that were subsequently retracted by their discoverer. It has continued to be used widely because it gives a good empirical description of the force-velocity properties of many types of muscle operating under many different conditions.

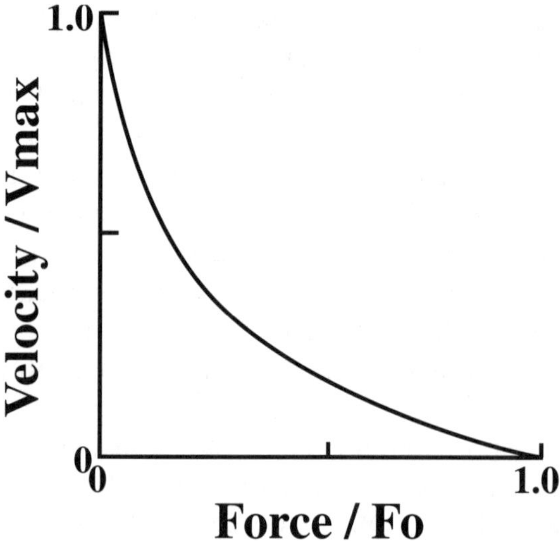

Figure 2.4. *Force-velocity relationship.*

HEAT PRODUCTION

The most elementary human experience will reveal that muscle contraction produces heat; shivering is a common method of keeping warm. Anyone who has experienced cold climates will appreciate that muscle contractions which include shortening, as during vigorous physical activity, produce more heat than simple isometric contractions, as occur during shivering. Heat production during contraction of isolated muscle was first demonstrated by Helmholtz in the nineteenth century. The belief that these thermodynamic measurements would yield insight into contractile mechanisms resulted in rigorous studies of heat production during contraction. In 1923 and 1924 W.O. Fenn demonstrated that more heat is produced during shortening, when a muscle does work, than during isometric contraction, when it does no work. This **Fenn Effect** has important implications for theories of contraction, as described in Chapter 4. All of the chemical energy liberated during contraction appears either as heat or as work. The Fenn Effect thus implies that total energy liberation increases with the work load.

The potential importance of the Fenn Effect resulted in a great effort to measure accurately the heat produced by contracting muscle. In 1938 A.V. Hill published quantitative measurements showing that the extra heat, i.e. the amount of heat produced in addition to the isometric amount, was directly proportional to the distance shortened. He later retracted this scheme, but his older experiments are worth further description here because they contributed to the development of the original crossbridge models and his later data contributed to its evolution.

Superimposed heat, force, and length records from four different contractions are shown in Fig. 2.5 A to illustrate how Hill made his measurements. The muscle was held isometric in all four contractions until a steady force level was achieved and the rate of heat liberation (i.e. the slope of the heat record) was nearly constant. (The heat rate at the onset of contraction was higher than in

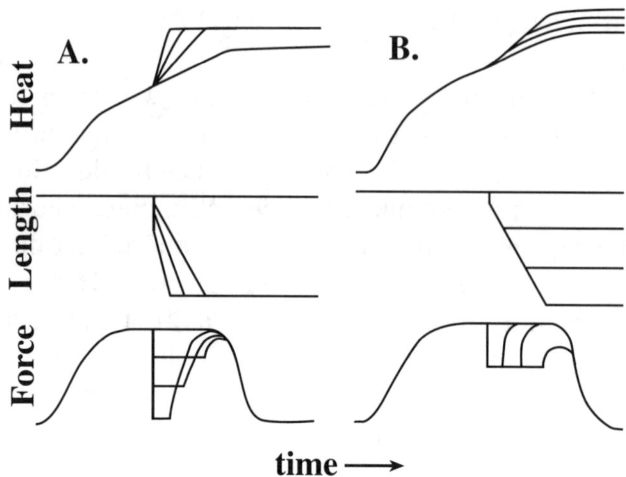

Figure 2.5. *Heat (upper), length (middle), and force (lower) records superimposed for 4 separate contractions. Left-hand panels (A) show the same shortening distance at different speeds. Right-hand panels (B) show different shortening distance at the same speed. Adapted from Hill (1938).*

the steady state, so that waiting for a steady state made the measurements less complicated.) In one contraction the muscle was held constant throughout. In the other three it was allowed to shorten the same distance beginning at the same time but at different rates. The total amount of heat liberated was measured at the end of each contraction. As shown in these records, the extra heat of shortening was the same in all three contractions, indicating that the extra heat was independent of the shortening speed. Furthermore, the heat appeared to come out nearly simultaneously with the length change, but this point is more difficult to establish, as described below.

In a separate set of four contractions (Fig. 2.5 B) the effects of three distances of shortening are compared with the isometric. In this experiment the extra heat varied linearly with the distance shortened. If the extra heat varied with the distance shortened and not with the rate of shortening, the rate of heat production would vary linearly with the rate of shortening.

The rate of extra heat liberation was thus found empirically to be equal to a constant, "a", times the velocity ($dh/dt = a{\cdot}V$). The rate of mechanical energy liberation, power, is equal to the product of force times velocity ($dw/dt = P{\cdot}V$). Hill found that the total rate of energy liberation (heat rate plus work rate) was highest at maximum velocity, where isotonic force was zero, and that it declined linearly to zero as force increased to the isometric level (designated Po). It was thus described by the equation

$$(P+a){\cdot}V = (Po-P){\cdot}b \qquad (2.2)$$

which is graphed as the solid line in Fig. 2.6. This equation can be re-arranged to

$$(P+a){\cdot}(V+b) = (Po+a){\cdot}b \qquad (2.3)$$

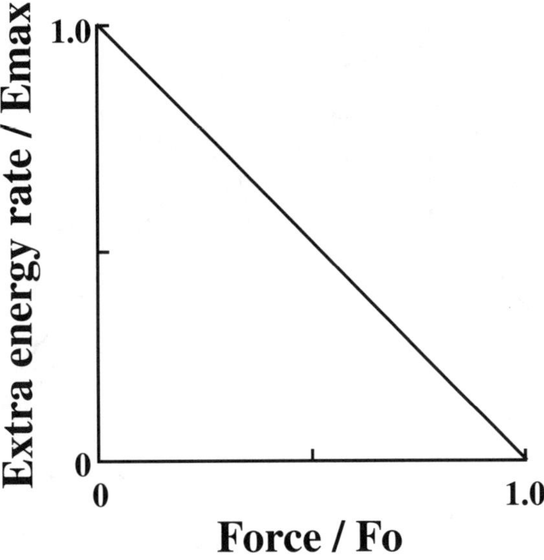

Figure 2.6. *Total energy rate defined as the work rate plus rate of shortening heat rate, plotted as a function of the isotonic force. Adapted from the data of Hill (1938)*

which is the same as equation 2.1 when "c" is replaced by the constant term "(Po+a)·b". In this scheme, "a" is the proportionality constant relating extra heat rate to velocity, and "b" is the slope of the line relating total rate of energy liberation to force.

Equation 2.3 gives excellent empirical descriptions of the force-velocity curves in many muscle preparations but the constants a and b are no longer considered relevant to heat production because Hill later retracted the scheme. In 1964 he published experiments showing that the extra heat for a given amount of shortening depended on velocity; less heat was produced at lower loads and higher velocities. The difference between Hill's 1938 and 1964 are summarized in Fig. 2.7.

The question arises as to how Hill could have made such a mistake in at least one of his sets of experiments. This question has not been answered fully. In 1964, he pointed out that a decline

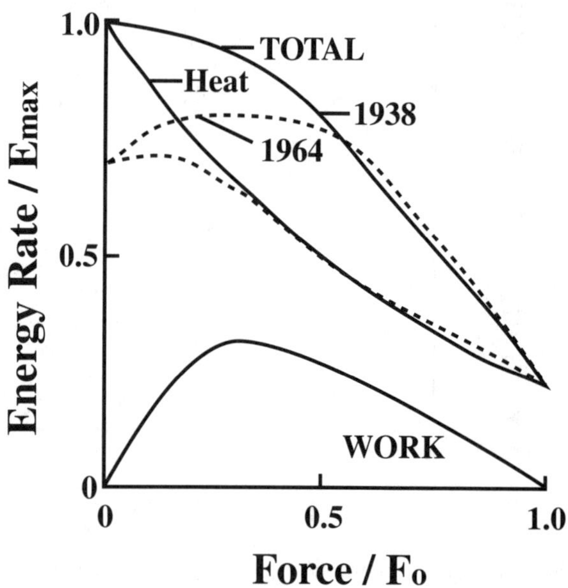

Figure 2.7. *Contractile parameters derived from Hill's 1938 and 1964 data.*

in extra heat at the highest velocities was seen in his 1938 experiments but he had overlooked this decline in favor of the simpler, linear scheme he originally proposed. Another important factor is that it is extremely difficult to determine the exact time course of heat liberation because of lags in the measuring apparatus. In 1938 Hill measured heat at the end of stimulation. In 1964 he measured it soon after shortening ended, just as force recovered to near its isometric level. Subsequent heat experiments (Fig.2.8) have shown that heat due to shortening continues to rise for a long time after shortening ends, so that the time of measurement is critical.

Delay in shortening heat production

Fig. 2.8 shows the results of an experiment that measured the difference in shortening heat in paired contractions where both contractions had the same velocity but different extents of shortening. Timing and length were adjusted so that shortening ended at the same length and time in both contractions (Fig. 2.8 A). The only differences between the two contractions in a pair were, therefore, initial starting length and the distance shortened. The extra heat produced by the longer shortening is thus measured as the difference in the two records after shortening ends (Fig. 2.8 B).

This method has the advantage that processes associated with the onset of shortening and with force recovery after shortening were the same in both contractions, so that the differences in heat production in the two contractions were due entirely to differences in the extent of shortening. It should also be emphasized that this type of analysis would have been impossibly tedious in Hill's day, before the advent of digital recording techniques.

As shown in Fig. 2.8 B, the extra shortening heat rises for a long time after the shortening ends, and the rise is greater for more rapid shortening. The point to be made from these observations is that shortening heat measured at the end of shortening shows a strong dependence on velocity, while that measured after force has recovered shows little dependence on velocity (Fig. 2.8 C).

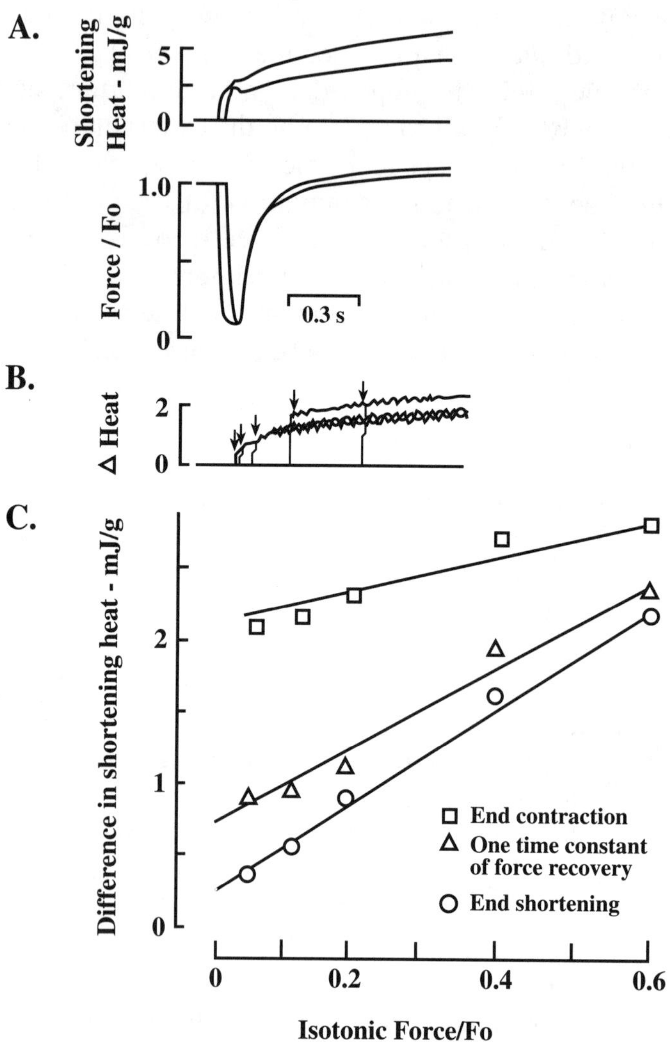

Figure 2.8. *Heat measured in paired contractions having two extents of shortening at the same speed. Timing and length were adjusted so that shortening ended at the same time and length in both. A) A pair of force and heat records. B) Difference in heat records, measured from the end of shortening, for 5 paired contractions. C) Heat difference measured at the times indicated in the inset. Adapted from Ford and Gilbert (1987), with permission.*

This seemingly arcane issue is belabored here because the 1938 data, but not the 1964 data, indicate that chemomechanical efficiency achieves its peak at substantially lower velocities than power, a conclusion also found for athletic endeavors in Chapter 12.

INSTANTANEOUS PARAMETERS

Velocity, power, heat, and efficiency calculated from Hill's 1938 data (eq. 2.3), are plotted against force in Fig. 2.9 A. Mechanical power, i.e. the rate of work production, is defined as the product of force times velocity, and the total energy rate is the sum of the heat rate plus the work rate.

The chemomechanical efficiency of muscle, defined as the mechanical power divided by the total energy rate, is shown in the lower graph. The peaks of the power and efficiency curves show that efficiency reaches its peak at a higher force, and lower velocity, than power. This difference arises because the total energy, used as the denominator for efficiency, has a negative slope in the region of these peaks. This is not true for the 1964 data, graphed in Fig. 2.7.

A barrier to intuitive understanding of mechanical physics is explained by the right-hand ends of the heat and total energy curves in 2.7 and 2.9 A. When the muscle is isometric, and therefore generating no mechanical work, it still produces heat and uses chemical energy. The process therefore creates the sensation of doing work, even though no external work is done by the muscles. Beginning students of physics often have difficulty in understanding that isometric force production does not accomplish any work.

INTEGRATIVE PARAMETERS

The data graphed on the ordinate of Fig. 2.9 A are the instantaneous output of a muscle. These data can have a very different relationship to the load than the total work done and distance

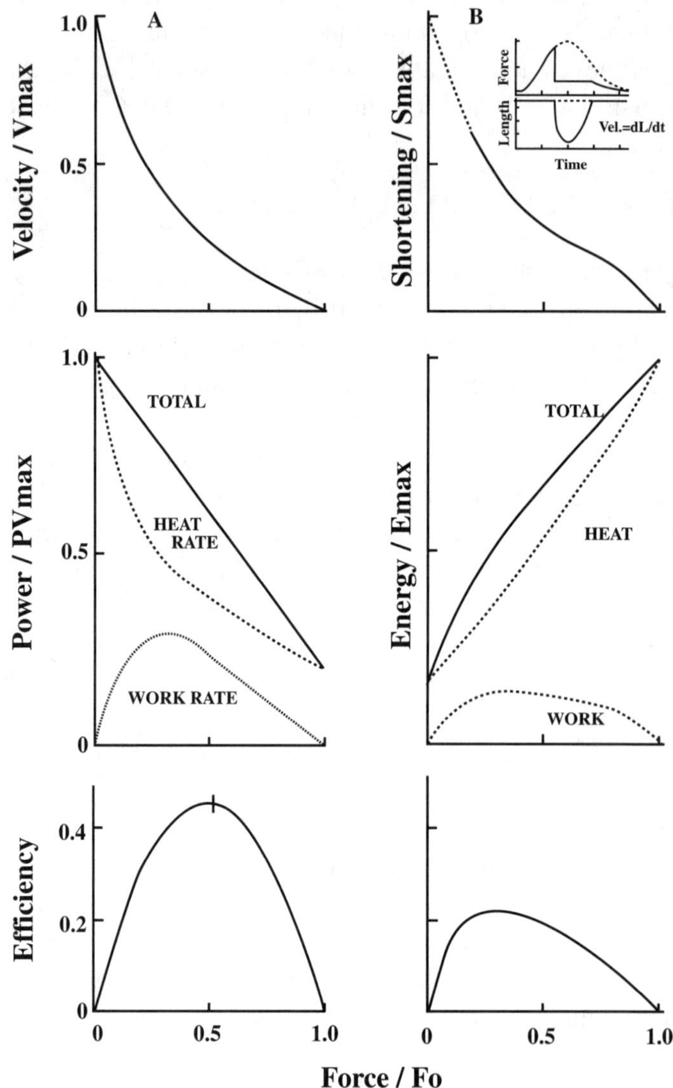

Figure 2.9. *Relationship between physical parameters in muscle. A) instantaneous rates in tetanic contractions of skeletal muscle (Hill, 1938). B) values integrated over a twitch in cardiac muscle (Gibbs and Gibson, 1969). The inset in the upper panel of (B) shows that shortening abbreviates the twitch. Adapted from Ford (1981).*

shortened, at least for twitch contractions of heart muscle. An example is shown in Fig. 2.9 B. Instead of velocity, the distance shortened during an afterloaded contractions is plotted. The distance shortened is relatively higher than the velocity at low loads, so that the peak in the force-work curve is broader than the peak of the force-power curve. More importantly, the heat and total energy output curves have positive slopes over the entire force range. The cause of these differences is shown by the force and length records in the inset of Fig. 2.9 B. The act of shortening inactivates the muscle and abbreviates the duration of contraction. This inactivation greatly reduces energy output at low loads, while the positive slope causes peak efficiency to occur at lower loads and higher velocities than peak work. Finally, efficiency for the integrated twitch is lower than for instantaneous power in a maximally activated tetanus, as expected from the greater energy cost of initiating and relaxing the twitch.

CHEMOMECHANICAL vs THERMODYNAMIC EFFICIENCY

Calculations of efficiency as the ratio W/(H+W) (Fig. 2.9) by Hill brought vigorous objections from his younger colleagues because this formulation fails to account for the entropic changes that must occur. An adequate explanation is beyond the scope of this book, but their objection is based on the principle that some heat produced by chemical reactions cannot be converted to work, even in an ideal system. Energy is lost irreversibly as the entropy increases. The amount of the entropy change depends upon the concentrations of reactants and products, which are not specified in the experimental results. Thus, thermodynamic efficiency, defined as the fraction of work extracted from the total available energy is substantially higher than that plotted in Fig. 2.9 by an unknown amount.

On the other hand, the denominator in the ratio (H+W), called enthalpy, gives an accurate measure of the extent of chemical reactions. Thus, the ratios plotted estimate the fraction of work extracted from the chemical reactions, i.e. the chemomechanical efficiency. To the extent that the concentrations of chemical reactants in the muscle are known, the entropy changes can be estimated and the thermodynamic efficiency calculated. The possibility of obtaining an accurate estimate may have been one reason why Hill was undaunted by his colleagues' objections. A more cogent reason is likely to have been that the important conclusion to be drawn from these curves is not the absolute value of efficiency, but the shapes of the curves. Efficiency peaks at a substantially slower velocity than power in the 1938 data but not in the 1964 data, and this relationship is not likely to be altered by corrections for entropy. In 1949, before the 1964 experiments, Hill published a seminal paper on the relationship between muscle physiology and athletic performance. As discussed in Chapter 11, that paper makes a significant point of the differences in the velocities at which peak power and peak efficiency are achieved.

RECOVERY REACTIONS

A complete understanding of muscle energetics is further complicated by recovery reactions. While Hill's 1938 experiments suggested that the extra heat of shortening was released simultaneously with the length change, he had previously shown that there was substantial heat liberated long after the muscle had relaxed. This recovery heat is attributed to reactions that restore the muscle to its baseline state. The delay in the appearance of shortening heat shown in Fig. 2.8 is intermediate between the relatively brief lags of Hill's apparatus and the recovery heat seen much later. Its cause is not known, but one possibility is that isometric cross-bridges store some energy which is consumed during rapid shortening. Such a consumption of stored energy might account for the velocity slowing that occurs during isotonic shortening. The slowing might occur, for example, if bridges become inactive after they give up the stored energy and are not available to help carry the load until their energy stores have been repleted. The energy storage might occur if the bridges released the energy derived from hydrolysis of a single ATP molecule in several cycles, rather than in one, as is discussed at the end of Chapter 4.

A. V. HILL

No history of the study of muscle would be complete without a description of A.V. Hill (1886–1977). This book has a responsibility to point out how some of the more momentous of Hill's conclusions were incorrect. This is unfortunate because his contributions to muscle physiology, and to science in general, were enormous. A measure of both his stature and his predilection to bold conclusions is shown by his receiving the 1922 Nobel Prize for a theory of contraction that was proved wrong a short time later. Although his larger theories were later shown to be incorrect, many of his methods are still used today. He excelled at mathematical descriptions. The equations presented above for describing the force-velocity curves are but one example of a "Hill equation." There is another for describing diffusion with reaction, as well as "Hill coefficients" used to assess cooperative phenomena, first worked out to describe oxygen binding to hemoglobin.

By all accounts, Hill was a man of enormous intellect and great presence. At times he could be overbearing, and even frightening, but always interesting. At one time he represented the British University Constituency as a member of Parliament, and perhaps his greatest contribution to science was not in the laboratory but in his position as Biological Secretary of the Royal Society during World War II. World War I British policy dictated that all young men should contribute equally to the battlefield effort, irrespective of education or ability. This policy led to the indiscriminate use of educated young men in battle, with disastrous consequences for the succeeding generation of academe. A much different policy was therefore developed for the second world war, largely due to Hill's efforts. He was personally responsible for placing young scientists in military laboratories where they made more substantial contributions than they might have made in combat and where they received valuable practical experience. The great success of British science immediately following the war was due mainly to the return of these young persons to peacetime work.

Hill also had a great personal influence in the development of young scientists' careers. Wallace Fenn, who became one of America's greatest physiologists, discovered the Fenn effect while working as a post-doctoral fellow literally in Hill's basement. In the acknowledgments of his first paper he thanked Mrs. Hill for the use of her basement, which was apparently the most vibration-free environment he could find for his sensitive instruments. In his 1985 memoir, published in the Journal of Physiology, Bernard Katz recalls escaping from Nazi Germany with considerable anxiety, having nothing but his new medical degree and A.V. Hill's address. At the end of a brief meeting with Hill he had what was to become one of the most successful careers in modern science. In describing how his own career had paralleled Hill's, Andrew Huxley relates that on arriving at the Gunnery Division of The Admiralty to do theoretical work, his first task was to read the book written on the subject by Hill during the First World War.

Apart from his personal intervention for Katz, Hill was a leading figure in the Academic Assistance Board (later the Society for the Protection of Science and Learning) that helped Jewish and other dissident academics escape from Germany and find work.

One of Hill's less prescient observations, published in 1951, was that "the ultimate mechanism of muscular motion seems to lie far outside the limit of optical resolution." This statement was intended to justify the general disregard that muscle physiologists had shown for the sarcomeric structure of striated muscle. While the statement is undoubtedly true at the molecular level, the next chapter will show that the re-discovery of these structures led to one of the most fascinating stories in modern Biology.

Chapter 3

CROSSBRIDGES

A revolution in the study of muscle occurred in 1954 with the simultaneous publication in *Nature* of two short papers describing light microscope observations of sarcomeres in muscles undergoing length changes. One was by A.F. Huxley and R. Niedergerke and the other by H.E. Huxley and Jean Hanson. Both papers showed that the length of the A-band did not change while the I-band length changed in parallel with sarcomere length. Both pair of workers interpreted their results as indicating that length changes in muscle were accompanied by sliding of two sets of filaments past each other, without a change in the filament length. This **sliding filament hypothesis** led to whole new classes of theories to explain contraction.

A.F. Huxley, grandson of T.H. Huxley and half-brother of Aldous and Julian Huxley, had with Alan Hodgkin discovered the ionic basis of the action potential in nerve a few years earlier. Although he had spent six years in the war effort work, by age 35 he had discovered how nerve worked and was looking for another area of study. He therefore turned his attention to muscle. By his own admission, this choice was dictated, in part, by a desire to indulge his hobby in optics. Some earlier reports had suggested that there might be changes in the optical density of the A-bands during contraction, and his initial intention was to investigate these changes. To do this he needed a microscope capable of showing faithfully the ends of the A-bands in a thick, contracting fiber. This need gave him the opportunity to build a new type of interference microscope. Once filament sliding was discovered, however, his attention quickly turned to the mechanisms that drive the sliding, and his subsequent work, described in this chapter, has focused on these mechanisms.

H.E. Huxley, no relation of A.F., has pioneered the study of muscle structure using several different techniques, including electronmicroscopy and X-ray diffraction. Much of what is described in Chapter 1 is the result of his work. To resolve the sarcomere structures in unfixed tissue, he and Jean Hanson used the newly developed phase contrast microscope. In the preceding year H.E. Huxley had published an electronmicroscope study of fixed tissue showing the presence of the two types of filaments and their arrangements in the A and I bands.

A.F. Huxley has subsequently made an extensive survey of the early literature on the microscopy of muscle and found that by the end of the nineteenth century there was sufficient information available to conclude that there were filaments in muscle and that filament sliding occurred. Such a conclusion was not reached, however, in part because the interpretation of the images became a source of controversy and in part because attention was diverted to thermodynamic and biochemical studies before any significance could be attached to the sarcomere structure.

The sliding filament proposal was at first vigorously resisted in some sections of the scientific community. Some resistance arose, naturally enough, from those who had a vested interest in other theories. The folding of proteins into tertiary structures was becoming the subject of study, and theories of contraction deriving from "conformational changes" of proteins were being developed. The proponents of these theories were encouraged when H.E. Huxley found longitudinal filaments in the sarcomeres because the folding of such filaments would lead to muscle shortening. They were less than encouraged when the filaments were found to slide rather than fold. An even larger source of resistance came from those of a chemical bent. They had difficulty in understanding how chemical reactions occurring over a few angstroms could drive the sliding of filaments separated by 10-20 nm. Most were won over by the crossbridge hypothesis.

In their Nature paper describing the sliding filaments, A.F. Huxley and Niedergerke suggested that the filaments might be driven to slide in the shortening direction by discrete interactions at specific points between the thick and thin filaments. Immediately after the publication of this paper, A.F. Huxley developed a comprehensive mathematical model of such discrete interactions, later called crossbridges. By the time the theory was published, in 1957, H.E. Huxley had described the side projections from the thick filaments that are now recognized as the crossbridges. As described below, the original theory was able to account for all of the contractile behavior of muscle known at the time. This behavior was explained by the attachment of bridges in a high-force state. Subsequent experiments led A.F. Huxley and R.M. Simmons to modify the original theory in 1971 to account for force generation after attachment. In 1973, A.F. Huxley published an additional modification to account for the change in Hill's shortening heat data. Before discussing these theories, a monumental experimental observation will be described.

SARCOMERE LENGTH-FORCE CURVE

Fig. 3.1 summarizes the conclusions of Gordon, A.F. Huxley, and Julian (1966) showing that crossbridges are the likely force generators in muscle. Fig. 3.1 A plots isometric force as a function of sarcomere length in a frog skeletal muscle fiber maximally stimulated at a temperature slightly above freezing. Sarcomere length in a central segment of the fiber was measured electronically using a "spot follower" that detected the positions of pieces of metal foil attached to the fiber at either end of the central segment. The electronic signal from this follower was then used to control a servo-motor that maintained the length of the central segment exactly isometric. This system eliminated from consideration both the ends of the fibers, which often had a different sarcomere length than the central segment, and changes in length of structures, such as tendons, that were in series with the central

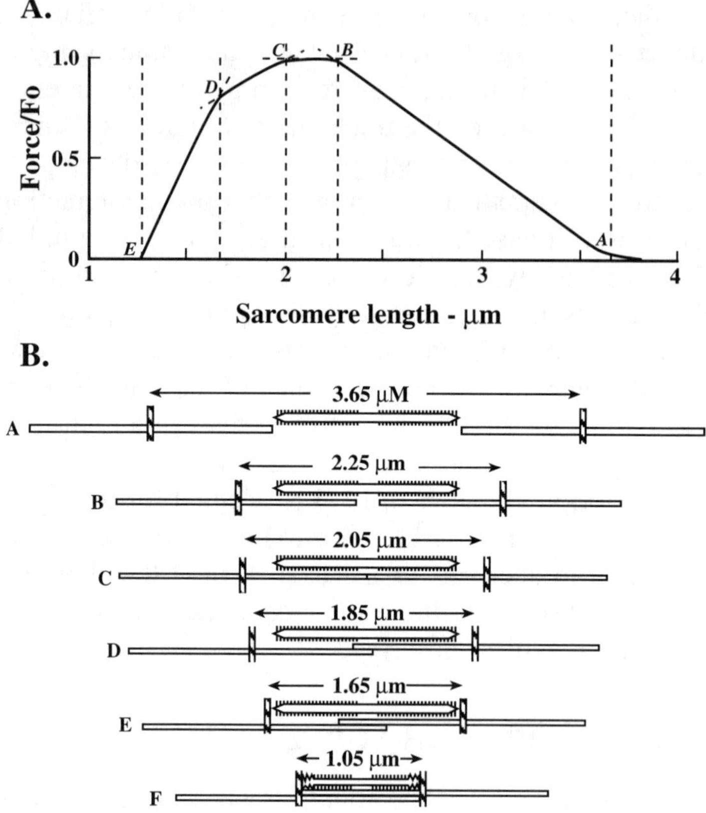

Figure 3.1. *A) Force-sarcomere length relation in single frog fibers. B) Filament relations at each corner of the curve. Redrawn from Gordon, Huxley, and Julian (1966), with permission.*

segment. Even more important than the sophistication of the apparatus was the authors' care in either eliminating or accounting for
variations in sarcomere length that develop in fibers during sustained tetani. The work is an elegant example of a conceptually
simple but technically demanding experiment that leads to an
unimpeachable conclusion.

An important feature of the data graphed in Fig. 3.1 A is that it
is a summary from a large number of contractions in many mus-

cles, with the isometric force from each contraction plotted as one point at a single length. The main point to be made from the plot is that the length-force "curve" is composed of a series of straight lines joining distinct corners. The corners occur at critical sarcomere lengths shown schematically in Fig. 3.1 B.

Force decreases linearly with sarcomere length above 2.2 μm, reaching zero at 3.65 μm, the length where overlap of thick and thin filaments ceases. This observation indicates that filament overlap is required for force production.

Force is constant at its maximum plateau between 2.0 and 2.2 μm sarcomere length. This plateau region corresponds to the lengths over which the ends of the thin filaments cross the bare zone devoid of cross-bridges in the center of the thick filaments. Changing sarcomere length changes filament overlap but does not change the number of crossbridges available to interact with thin filaments. The observation that force is proportional to thin filament overlap with crossbridges, and not just with the thick filaments, strongly suggests that crossbridges are responsible force generation.

Length-force relationships for whole muscle, such as that shown in Fig. 2.2 in the last chapter, do not usually show the straight lines and distinct corners seen here. A likely reason is that the individual fibers reach the critical sarcomere lengths at slightly different muscle lengths, thus obscuring the sharp changes of slope and obscuring the corners. In addition, unless the dissections are impeccable, the individual sarcomeres along a single muscle fiber often show some variation in length which also tends to confound the interpretation of the results.

Force declines in two stages from its maximum plateau as sarcomere length is decreased below 2.0 um, the length at which the ends of the thin filaments just touch in the center of the sarcomere. The slope of the length-force relationship is much steeper at sarcomere lengths below 1.65 μm, the length at which the thick filaments touch the Z-lines. There are several mechanisms responsible, and these are described in Chapter 6.

THE ORIGINAL CROSSBRIDGE THEORY

Because muscle contraction is smooth and continuous, it is easy to expect a tight link between the driving chemical reactions and the force generating movements. In fact, such tight linkages generally do not occur. Chemical reactions, including those generating contraction, take place by a series of discrete, stochastic steps. Some of the success of A.F. Huxley's theories derives from his recognition of the quantal nature of these processes.

The original theory posited that force and shortening were driven by the cyclical attachment and detachment of crossbridges. The unattached crossbridge, shown schematically in Fig. 3.2 A, was imagined able to move randomly parallel to the fiber axis as the result of Brownian motion. The bridge was tethered to the myosin filament by a spring, and force in the spring varied linearly with the axial deviation, designated x, from the mid-position, where force was zero (Fig. 3.2 B). Force was generated between the thick and thin filaments when the crossbridge was attached to an actin site in something other than its rest position. As shown in Fig 3.2 B, force was designated as positive when it is was exerted in the shortening direction. At any instant, the force in an individual attached bridge could have any value, positive or negative, determined by the relative positions of the actin attachment site and the crossbridge rest position.

The spring was an essential part of the model; it allowed the crossbridges to act independently of each other to generate force, perform work, and hydrolyze ATP. Filament sliding was not needed for bridges to go through their cycle and liberate energy.

The total force generated between the filaments was determined as the sum of the forces in the individual bridges. The net force exerted in the shortening direction resulted from the higher probability that an attached bridge would be strained in the positive force direction. This probability was provided rate constants for attachment and detachment of crossbridges that varied with bridge position. Fig 3.2 C plots the **rate functions** for attachment

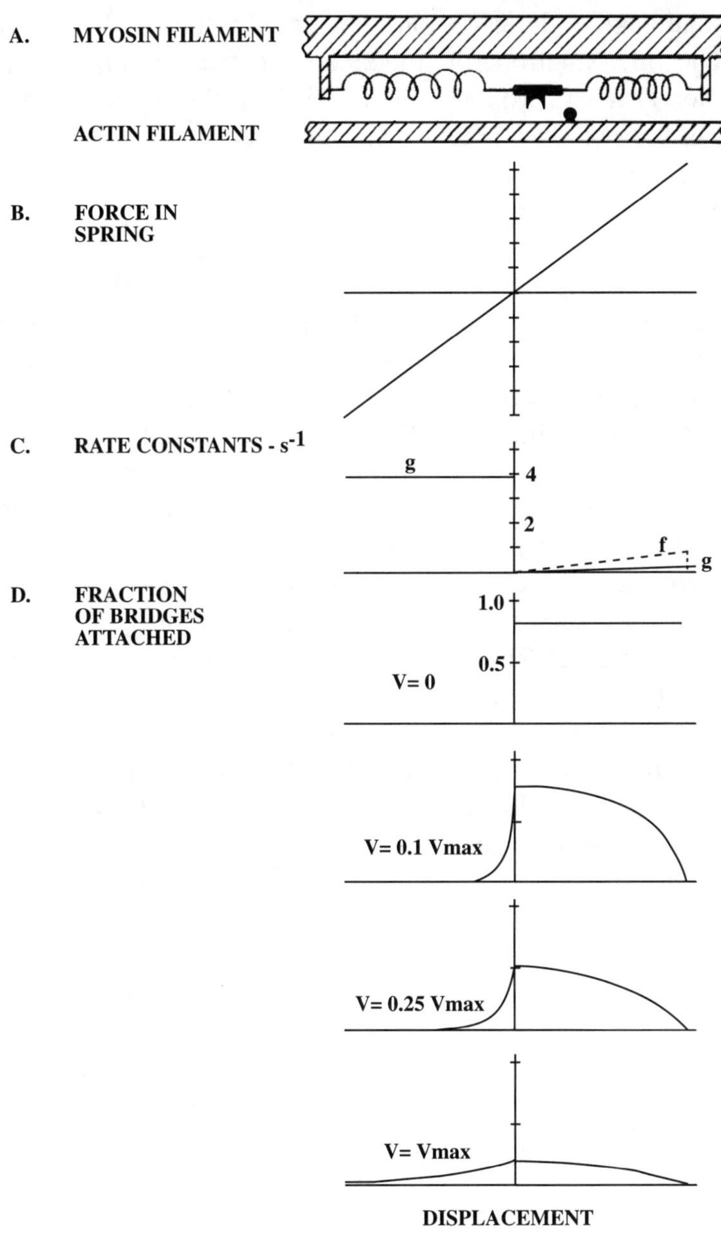

Figure 3.2. *Two-state crossbridge model. Adapted from A.F. Huxley (1957), with permission.*

and detachment, i.e. the variation of rate constants with bridge position. The rate function for attachment, **f**, was assumed to increase linearly from zero as the bridge was displaced from its mid position in the positive force generating direction and to be zero when the bridge was in a negative force generating position. The rate function for detachment, **g,** was assumed to be large and constant over the range of negative force generating positions and to be a constant fraction of the attachment function in the positive force generating range. These simple, linear functions were chosen mainly because they gave an excellent description of the data but also because they led to explicit algebraic solutions.

The model is defined completely by three simple properties, the spring constant defined by the linear relationship in Fig. 3.2 B and the two rate functions for attachment and detachment plotted in Fig. 3.2 C. It is a **two-state** model because the bridges can exist in only two states, attached or detached. At each bridge position, i.e. for each value of x, the fraction of attached bridges was designated n and the fraction unattached was therefore 1−n. For each value of x, the rate of bridge attachment was determined as the product of the value of the rate constant, f, for that position multiplied by the fraction of bridges unattached in that position (attachment rate = f·[1−n]). Similarly, the rate of detachment was determined as the product of g multiplied by the fraction of bridges attached, (detachment rate = g·n). The effect of filament sliding on the distributions of attached bridges is shown in Fig. 3.2 D-G.

When the muscle is isometric and the filaments are stationary, the bridges detach from the same positions where they attach. Since the attachment rate is zero over the entire negative force generating range, there are no "negatively strained" bridges in the isometric state. Furthermore, since the ratio of detachment to attachment rate is constant for each position in the positive force-generating range, the fraction of attached bridges is constant over this range (Fig. 3.2 D).

Filament sliding alters the distribution in several ways. Since all bridges detach from a lower force-generating state than they attach, the average force per bridge is reduced. The higher the rate of shortening, the greater the reduction in this average force. Some bridges that attach in a positive force-generating state are brought into the negative force-generating state before they can detach. While in this state they resist shortening by generating a negative force. This effect increases with velocity, and there is no external force at maximum velocity because there is then an exact balance between negative and positive forces (Fig. 3.2 G). The number of attached bridges is also reduced at higher velocity because shortening brings the attached bridges into the region where they detach more rapidly. Thus, filament sliding has two effects that reduce force; it decreases both the number of attached bridges and the average force per attached bridge.

The effect of sliding velocity on the total force produced by the attached bridges is plotted against velocity as the circles in Fig 3.3, with Hill's rectangular hyperbola superimposed. As shown,

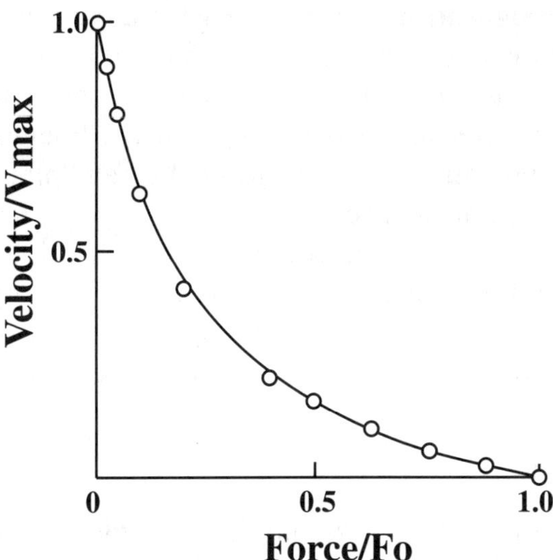

Figure 3.3. *Force-velocity relations calculated from the Hill equation (solid curve) and the original crossbridge model (circles). Redrawm from A.F. Huxley (1957), with permission.*

there is very close agreement between the calculations and the empiric measurements, even though the equations used to describe them were generated by very different functions. A major triumph of the cross-bridge model was that it could explain the constant steady-state velocities found in the muscle without requiring that the muscle have a viscous resistance to shortening. Earlier muscle physiologists were at a loss to explain the constant isotonic velocity because a constant force usually produces acceleration unless it is resisted by a viscosity, and there was no evidence of the required viscosity in the passive muscle.

Another triumph of the original cross-bridge model was that it explained quantitatively the energetics of contraction. It did this by assuming one ATP molecule is hydrolyzed in each crossbridge cycle. There is some crossbridge cycling, and some ATP hydrolysis, in the isometric state because the crossbridges can detach from the same position in which they attached in a force-generating state. The cycling rate is increased during shortening because the bridges are brought to negative force-generating positions where they are more likely to detach. The more rapid the shortening, the greater the increase in cycling rate. By the proper choice of rate functions, the relationship between total energy liberation and isotonic force found empirically by Hill in 1938 and plotted in figure 3.4 was matched almost exactly.

In addition to explaining the force-velocity relations and the energetics of contraction, the original crossbridge theory accounted for the relatively slow rise of force during an isometric tetanus. The significance of this explanation is described in Chapter 7.

TRANSIENTS

The crossbridge theory was not just a single contractile mechanism, but a class of theories, each with its own set of rate functions and other properties. The rate functions chosen in the original theory gave a good match to most of the physiological phenomena known at the time, but it was also acknowledged that

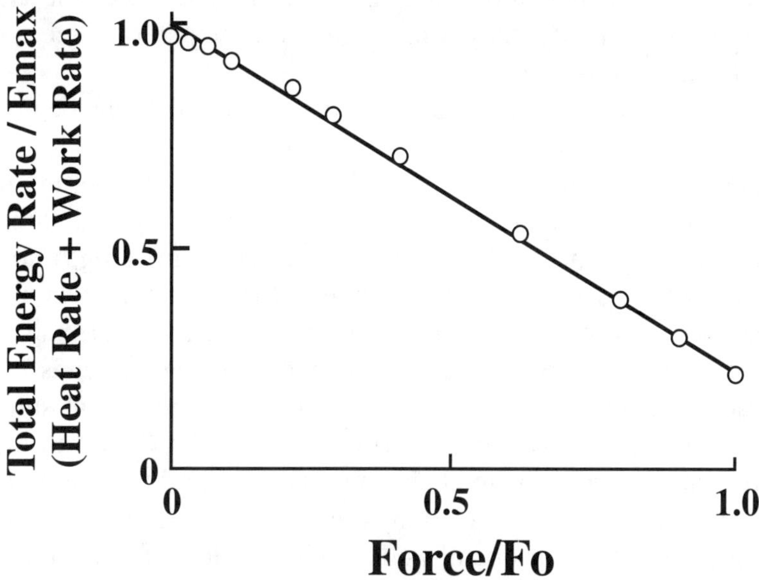

Figure 3.4. *Force-energy relations calculated from Hill's 1938 data (solid line) and the original crossbridge theory model (circles). Redrawn from A.F. Huxley (1957), with permission.*

another set of functions might have given equally good matches. What was needed was not more theory but new data. One class of data not yet observed were **transients**, deviations from steady behavior when a system passes from one steady state to another.

Apparently, the possibility of some type of transient behavior was considered before it was observed. In a study of force-velocity properties published in 1958, Jewell and Wilkie stated that they did not see any transients, even when they made the transition from isometric to isotonic state within one millisecond. In this reference to the unobserved transients they stated that these transients were expected from A.F. Huxley's crossbridge theory and cite a page in his paper as the source for their expectation. Even the most knowledgeable reader of that page would have difficulty in discerning any anticipation of transients, but this reference indicates that the expectation existed. Thus, when they were described by Podolsky in 1960 they were already the subject of some speculation and discussion.

The two mechanical properties of a muscle that can be manipulated are force and length. To simplify interpretation, one of these is usually held constant while changes in the other are measured. A presumed advantage of isotonic shortening is that the series elastic elements are maintained at constant length so that overall muscle velocity reflects the velocity of the sarcomeres. Since the sarcomeres are in series, they all experience the same force. Thus, the force and velocity measured at the ends of the muscle accurately reflect the changes in the individual sarcomeres. With these considerations in mind, Podolsky carefully measured the length and velocity changes that occur during and immediately following the transition from the isometric to the isotonic state. An example of one of Podolsky's transients is shown in Fig. 2.4 of the last chapter, and a similar example from Huxley and Simmons (1973) is shown in Fig. 3.5 A, here. As shown in both figs, there is an initial rapid velocity immediately following the step to the isotonic load, and this rapid, early shortening is not easily distinguished from the recoil of the series elastic elements that occurs in response to the change in force. The most conspicuous part of these velocity transients is a brief period when velocity slows substantially, and may even become negative for a short time, so that the muscle lengthens a small amount. After a period of time required for the muscle to shorten by about 15 nm/half-sarcomere, these conspicuous transients disappear, and shortening progresses at a steady velocity. Careful observation shows, however, that small damped oscillations may be superimposed on the length record for an additional one or two cycles.

Although recognized as important to a full understanding of muscle function, the **velocity transients** could not be explained until the **tension transients** were discovered. In 1971, Huxley and Simmons described a distinctive pattern of force recovery when a single muscle fiber was suddenly stepped from one isometric sar-

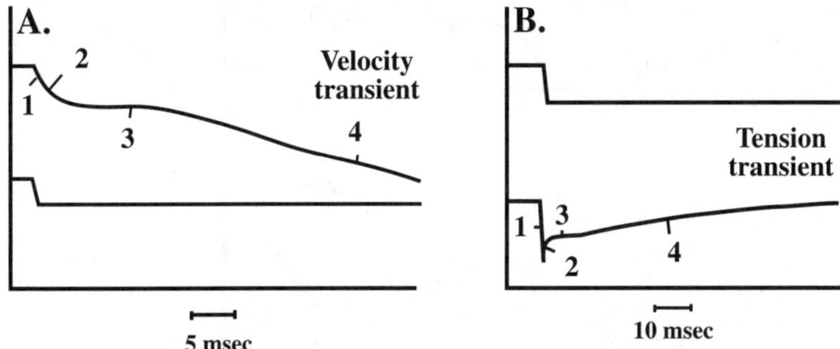

Figure 3.5. *Comparison of (A) velocity transients and (B) force transients. Numbers indicate the phases. In each panel, upper trace is length and lower trace force. Redrawn from Huxley and Simmons (1973) with permission.*

comere length to another. To eliminate the confounding effects of the changes in series elastic element length, they used the spot follower to measure and control the central sarcomeres in the single fiber. Huxley and Simmons divided the force transients into four distinct phases (Fig 3.5 B), each of which could be explained by separate cross-bridge mechanisms, and each of which had a rough correlation with a phase in the velocity transient (Fig. 3.5 A). The importance of their interpretation is that it further explained how the cross-bridges could act independently of each other, and how force was generated by stochastic processes.

Phase 1 is a force change simultaneous with and nearly proportional to the length change. The extent of the force change for a given length change is a measure of the stiffness of the preparation. The half-sarcomere stiffness can be inferred from plots of the extreme tension reached at the end of the step, designated T_1, against the length change per half-sarcomere in Fig. 3.6.

Phase 2 is a period of partial, rapid tension recovery immediately following the length change. In velocity transients it appears as a period of shortening that is faster than the steady level and slower than the initial phase 1 elastic recoil. An advantage of tension transients is the clear separation between Phases 1 and 2. The

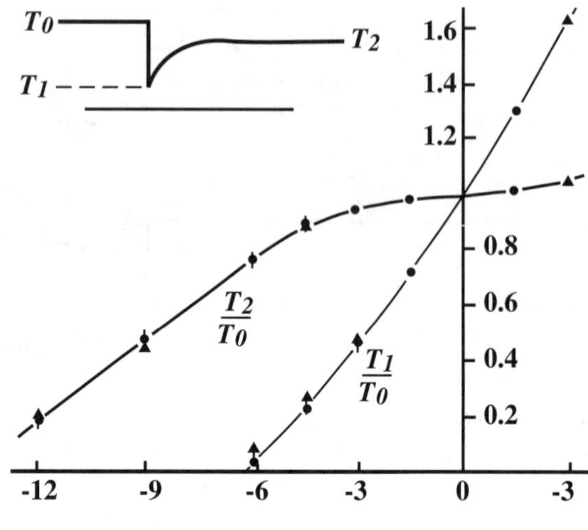

Length change (nm/half- sarcomere)

Figure 3.6. *Extreme tension reached at the end of the step (T_1) and the interme-diate level to which tension recovers during Phase 2 (T_2) vs. the length change.*

force change reverses sign rather than simply slowing. The level to which force recovers at the end of Phase 2, $\mathbf{T}_2$, is plotted against the half-sarcomere length change in Fig. 3.6.

Phase 3 is a distinct pause or inflection in force recovery. Sometimes there is a **delayed fall** of force after the initial recovery following a small release. This is the most conspicuous phase of the velocity transients.

Phase 4 is a later, slow recovery to the final isometric value due to redistribution of crossbridge attachments. The correlate of Phase 4 in the velocity transients is the steady shortening that pre-vails after the initial transients have died away.

INTERPRETATION OF THE TRANSIENTS

The tension transients were interpreted in terms of a funda-mentally different bridge, shown schematically in Fig. 3.7. During

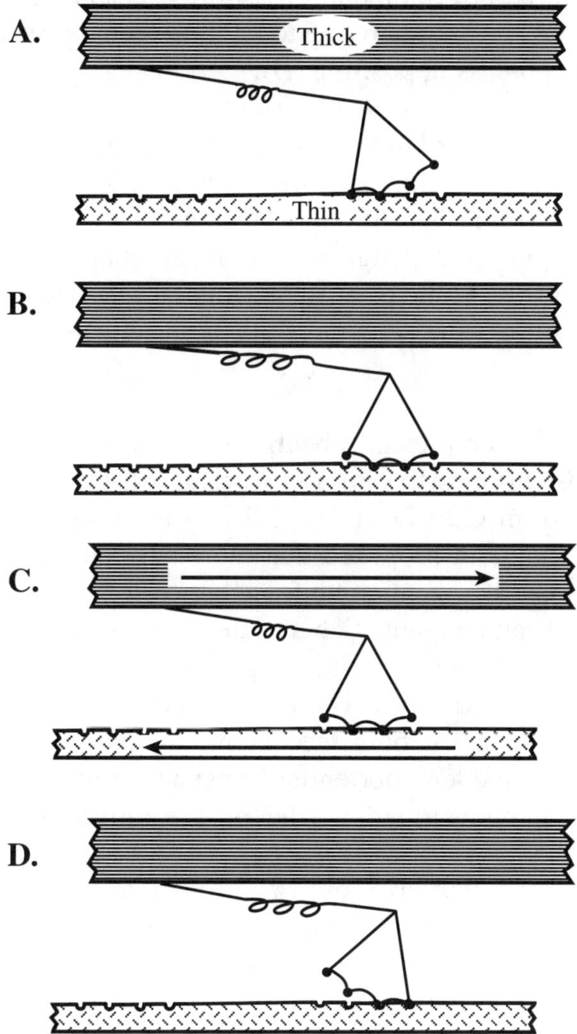

Figure 3.7. *Huxley-Simmons crossbridge showing the discrete movements between attached positions.*

isometric force development, crossbridge heads attach in a low force state (A) and some rotate to (B), stretching a spring and beginning the process in position (B).

Phase 1 was attributed to an immediate change in the length of the crossbridge spring. For bridges in position (B), some of the strain in the spring is relieved. The slope of the T_1 curve in Fig. 3.6, the ratio of the instantaneous force change to the length change, is the sarcomere stiffness. Since each crossbridge has a compliant spring, the sarcomere stiffness reflects the number of attached bridges.

Phase 2 is interpreted as being caused by movement of the attached cross-bridges to partially re-establish the isometric force (transition from C to D in Fig. 3.7). In the original crossbridge model, Huxley had proposed that the bridges attached in a pre-stretched condition, generating force between the filaments immediately upon attachment. Others later suggested that the cross-bridges move to generate force after they attach (Reedy et al., 1965; H.E. Huxley, 1969). The Huxley and Simmons experiments provided evidence of both the extent and the dynamics of this movement. A sudden shortening transiently relieved spring tension, which redeveloped as the bridge moved further to re-stretched the spring. Since the extent of force redevelopment depended both on step size and range of crossbridge motion, the range of motion could be estimated from the T_2 curve in Fig. 3.6.

For small length changes, the phase 2 tension recovery is almost complete, as expected if most of the bridges can move sufficiently to re-develop their original force. It is less complete for larger steps because some bridges reach the end of their range at length changes shorter than the step. For the very largest steps, all of the bridges reach the end of their range at length changes shorter than the step, so that the slope of the far end of the T_2 curve simply reflects the stiffness of the bridges. For these large steps, the T_2 curve should parallel the T_1 curve but be displaced by the average distance the bridges have moved from their isometric positions to the end of their range.

Phase 3 is not well understood. It almost certainly results from delayed detachment of bridges from high force positions before they re-distribute their attachments to sites appropriate to the new length, but there are several mechanisms that might account for it, and they are not mutually exclusive.

Phase 4 is attributed to the detachment of bridges from their initial isometric positions and reattachment at positions appropriate to their new length. In velocity transients, Phase 4 is the period of constant shortening and results from the steady crossbridge cycling.

T_1 and T_2 *curves* are used to estimate ranges of crossbridge movement and the fractions of attached bridges in high and low force states. To understand how this is done, consider that bridges attach in an average zero-force position and then move to stretch their springs and develop force. The filament sliding required to reduce average force to zero after all attached bridges have moved to the end of their range should match the force-generating movement. Thus, the zero force intercept of the T_2 curve, about 13 nm, indicates the average range of motion of an attached bridge.

The zero force intercept of the T_1 curve is reached before the crossbridges move. It occurs at shortening distances less than the intercept of the T_2 curve because low-force bridges that have not undergone the power stroke are driven to negative forces by the shortening. If the spring constant for the crossbridge is the same at both negative and positive forces, the ratio of the two intercepts will indicate the fraction of the attached bridges that have undergone their power stroke and moved to high force states.

The area under the T_2 curve has the dimensions of work, i.e. force times length change. It represents the maximum work that could be done by all of the isometrically attached bridges if they were to move a load through their full range of attached motion. A substantial disparity is found when this integral is compared with the free energy available from the hydrolysis of one ATP per crossbridge. The area under the T_2 curve is about 2 J per liter of muscle. Assuming the free energy of ATP hydrolysis to be 56 kJ/mole and 280 μM crossbridge heads in muscle, the total free energy from hydrolyzing one ATP molecule per bridge is 16 J per liter of muscle, about 8 times the integral of the T_2 curve. This 8-fold factor might just be accounted for by some corrections that must be applied. These include the efficiency of the energy transfer, the fraction of bridges attached in the isometric state, and the uncertainty about the two heads of myosin. If the efficiency were 50%, only one of the two heads could attach at any instant, and only half the myosin dimers participated, then the factor of 8 would be explained exactly. Another very different explanation, also considered by Huxley and Simmons, is that the crossbridges undergo more than one power stroke and can produce more than one Phase 2 recovery per ATP molecule hydrolyzed. Again, this is an area that awaits further clarification.

THERMODYNAMIC MODEL

It would have been a substantial advance if Huxley and Simmons had simply interpreted the Phase 2 force recovery as being due to movement of attached crossbridges, and used the T_1 and T_2 curves to indicate the range of crossbridge movement. But they did substantially more. They made the further observation that the speed of Phase 2 force recovery varied with the size and direction of the length step, being most rapid for large releases and slowest for stretches. They then postulated a thermodynamic model to explained this behavior. Bridge movement was imagined to occur in discrete, stochastic steps between different, well-defined, attached positions. The probability of movement between positions, and therefore the speed of force recovery, depended on a balance between the fixed energy available from ATP hydrolysis and a variable activation energy required to make the movement. This activation energy varied because it included the energy required to

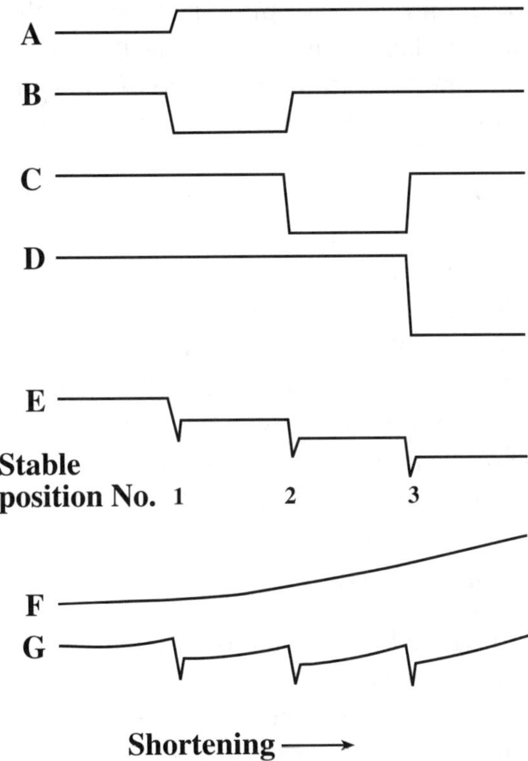

Figure 3.8. *Thermodynamic model of crossbridge movement. Redrawn from Huxley and Simmons (1973), with permission.*

stretch the spring when the bridge moved in the shortening direction, and this energy was increased when the spring was stretched.

The bridge was imagined to progress down an energy cascade as it moved in the shortening direction (Fig 3.8). Bridge attachment was made firm by double attachment, with a relatively broad energy trough for each of the single attachment points (Fig. 3.8 A-D). Firm, two-point attachment resulted in a narrow energy well that was the sum of the energies for the single attachments, so that the bridge paused in each double attachment well as it moved (Fig. 3.8). In the absence of the spring (Fig. 3.8 E), the activation energy to be overcome was simply equivalent to the depth of the trough and was higher for movements in the lengthening than in the shortening direction. In the presence of the spring, the energy of stretching the spring (Fig. 3.8 F) was added to the energies associated with each attached position (Fig. 3.8 G) so that the activation energy for shortening includes the energy in the spring. This additional energy was small for large releases because the strain in the spring was relieved, and very little energy was required to restretch it. Phase 2 force recovery was fast under these conditions. For small releases there was more force, and consequently more energy stored in the spring, so that a higher activation energy was required for bridge movement, resulting in slower force recovery. The activation energy for stretches was highest because it required moving back up the energy cascade.

Comparison of models

The major difference between the original crossbridge theory and the Huxley-Simmons scheme is that the original theory was intended to explain steady state behavior while the later scheme was intended to account for transients. The Huxley-Simmons model did not include crossbridge attachment or detachment and would not explain Phases 3 and 4 of the transients. A complete model will thus require that the two models be combined to include both crossbridge movement and attachment and detachment. The next chapter explains that a complete model will also require that chemical transitions be included. But it should also be emphasized that the original model can still explain steady state-behavior as long as the rate functions, f and g, reflect other transitions.

Although not needed to explain steady-state behavior, the addition of a force-producing power stroke to the original model made it more intellectually satisfying because it specified more precisely how the energy of ATP hydrolysis was used to drive contraction. In the original model, muscle force was presumed to arise from attachment of bridges after they had been moved to favorable positions by brownian motion. McClare (1971) suggested that this was a "Maxwell's demon" that derived its motive force from the thermal energy of the surrounding medium. He was not correct because the original model specified that ATP hydrolysis was needed to detach the crossbridges, so that the energy initially obtained from brownian motion was restored by the hydrolysis of one ATP molecule per cycle, but the precise mechanism of energy transfer was unspecified.

The recognition that there might be multiple attached states with at least one power stroke between them made it possible to consider other transitions between attached states. The first of these was proposed by Huxley to account for Hill's 1964 heat data.

TWO STAGE ATTACHMENT

As described above, the original cross-bridge theory accounted for Hill's early (1938) heat data by allowing bridge cycling rates to increase directly with shortening velocity. Hill's later (1964) heat data indicated that the extra heat of shortening declined at high velocities. (See Fig. 2.7 of last chapter for a comparison of the two sets of data.) Since mechanical work is also less at higher velocities, Hill's later finding implied that the total energy liberation rate was lower at maximum than at intermediate velocities. If the rate of total energy liberation were proportional the rate of ATP hydrolysis, and this hydrolysis rate was in turn proportional to the rate of crossbridge cycling, Hill's later data would imply a lower cycling rate at high velocities.

In 1973 Huxley proposed a relatively minor modification to his theory that would account for these lower cycling rates at higher velocities. The Huxley-Simmons theory specified that the crossbridges attached in a low-force state and then rotated to produce higher force. In the modification Huxley proposed that the low force attachment occurred in two stages. The first was a state from which the bridge could detach without ATP hydrolysis or force generation if the filament sliding brought it to an unfavorable position. The second was a state in which the bridge was committed to ATP hydrolysis and force-generating movement. The transition from the first to the second state was more likely to occur when the bridge remained in the first state. At higher filament sliding velocities, the amount of time the bridges remained in this state was reduced, so that fewer bridges would make the transition. In this modified theory, the initial attachment rate increased at high velocities, just as in the original theory. ATP hydrolysis was diminished, however, because bridges were detached before they moved to the second state.

The two stage attachment was immediately seen by some as the explanation of a number of phenomena. It became equated with a weak binding state seen in biochemical studies of actomyosin. Whether the first of the two stages of attachment is the same as the weak binding state remains to be proven. It is clear that the addition of another state to the theory provides another degree of freedom to explain a variety of physiological phenomena. The ability to accommodate a substantial change in the experimental data with a minor modification of the model illustrates the robustness of this class of theories.

TESTS OF CROSSBRIDGE MECHANISMS

The presence of the spring in the crossbridge is expected to cause sarcomere stiffness to vary directly with the number of attached crossbridges. Thus, changes in stiffness can be used to assay changes in the number of attached bridges. This assay for attached bridges allows several aspects of the theory to be tested. Some of the more important of these tests will be described here, but first the nature of stiffness and compliance will be reviewed.

Stiffness is defined as the change in force that results from a given length change, i.e. the force change divided by the length change. Compliance is the inverse of stiffness, the length change divided by the force change. The choice of the parameter to be expressed, stiffness or compliance, depends on the way that results are to be used. When two compliant elements are added in series, their compliances sum linearly. That is, the length change required to obtain a given force change is the sum of the length change in the two elements, while the force change is the same in both. Inversely, when two compliant elements are added in parallel, their stiffnesses sum linearly; the force change produced by a given length change is the sum of the force change in the two elements, which each experience the same length change. When assaying for bridges attached in parallel, stiffness is the parameter to be measured. When assessing structures in series, the results are analyzed in terms of compliance.

Non-crossbridge sarcomere compliance

If stiffness is to be used as an assay for attached crossbridges, it must first be determined how much it varies with changes in crossbridge number and how much other structures contribute to it. The results of the first such experiments are shown in Fig. 3.9.

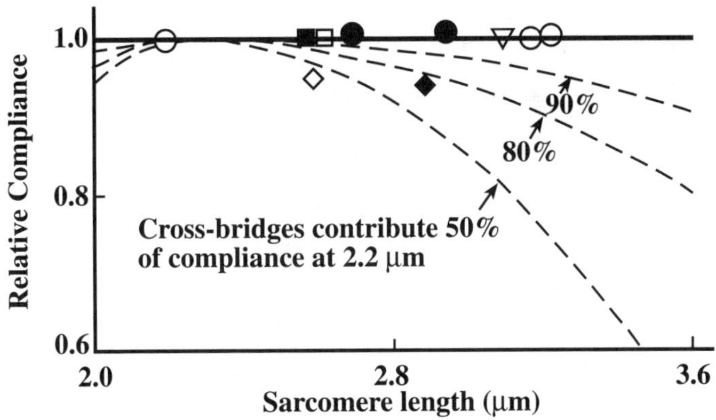

Figure 3.9. *Relative compliance, expressed as the amount of length change required to bring isometric force to zero vs sarcomere length. The interrupted curves indicate the results expected if only the indicated percentage of the compliance were in the crossbridges. Different symbols represent different experiments. Redrawn from Ford et al. (1981) with permission.*

The numbers of bridges were varied by varying the overlap of the thick and thin filaments, as Gordon et al. (1966) had done earlier to show that crossbridges produce force (Fig. 3.1). Because the compliance of other structures in series were being sought, the results were expressed in terms of the compliance. The interrupted lines in Fig. 3.9 indicate the result that would have been obtained if the filaments had contributed the amount of compliance indicated. The results were interpreted as showing that "at least 80%, and probably well over 90% of the instantaneous [sarcomere] compliance" was in the crossbridges themselves. Furthermore, since the relative compliance, i.e. compliance normalized to force, remained constant at different amounts of overlap and different forces, the results suggested that stiffness varied in proportion to the number of attached bridges.

While the data in Fig. 3.9 appeared very convincing, it should be mentioned that subsequent workers have suggested that there is substantially more compliance in other structures, particularly the thin filaments. In addition, recent x-ray diffraction studies have been interpreted as indicating a slight length change in the thin filaments, both during the rise of tetanic force and following the fall of force caused by sudden shortening. The reasons for the disparity in the results is not yet known. It is possible, for example, that filament length changes are damped and therefore not apparent when measurements are sufficiently "instantaneous." It is also possible that changes in the lateral spacing between filaments at different sarcomere lengths had an effect that compensated for changes in filament compliance. Since it is possible that the non-crossbridge compliance varies among preparations, the more recent findings dictate that the exact relationship must be measured under each condition where sarcomere stiffness is used as an assay for attached crossbridges.

It should also be emphasized that the suggestion that some sarcomere compliance is in non-crossbridge structures does not invalidate the use of stiffness as an assay for attached crossbridges. It simply means that a more complex relationship than a strict proportionality must be used to deduce changes in crossbridge number from changes in sarcomere compliance.

Feet off the ground

As mentioned earlier, a major triumph of the crossbridge theory was its explanation of the force-velocity properties of muscle. Steady shortening has two effects that reduce force; it reduces the average force per bridge, and it reduces the number of attached bridges. This aspect of the theory is sometimes referred to as the "feet off the ground" hypothesis because of Huxley's cartoon representing the thick filament as a tug-of-war team (Fig. 3.10), which has fewer feet on the ground and generates less force per foot when the filaments are moving.

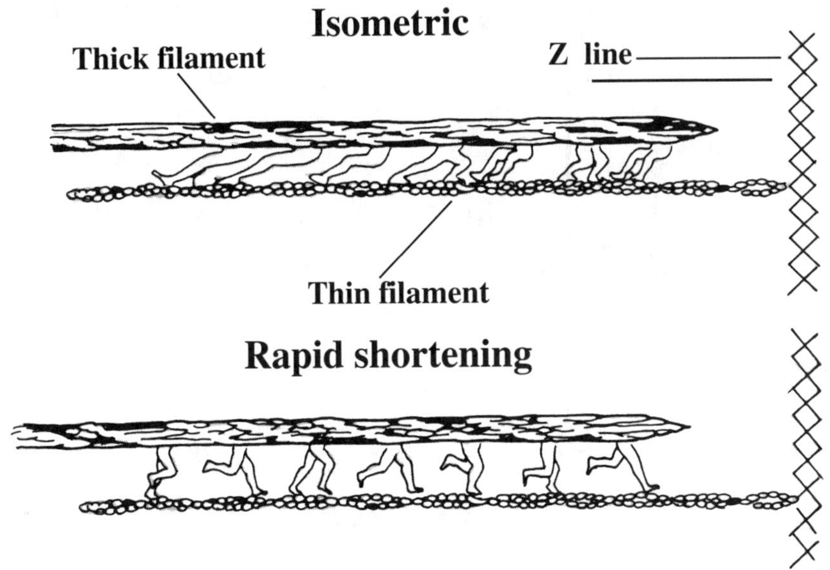

Figure 3.10. *Isometric vs. rapidly shortening muscle. Redrawn from Huxley (1969) with permission.*

Proof for this aspect of the theory is shown in Fig. 3.11, where stiffness is plotted against isotonic force. Stiffness declines as isotonic force decreases, indicating fewer bridges attached during shortening. On the other hand, stiffness does not decline in proportion to force, so that there is still substantial stiffness at zero load, indicating that a substantial fraction of bridges are attached when the muscle is shortening at maximum velocity. This is the result predicted by the original crossbridge theory and illustrated in Fig. 3.2 G. A substanatial number of bridges are attached at maximum velocity, but attached bridges pulling in the shortening direction are exactly balanced by bridges that have pulled through their useful range but remain attached to resist further shortening.

Low-force attached states

The final aspect of the Huxley-Simmons theory to be discussed is that bridges are presumed to attach in a low force state and then

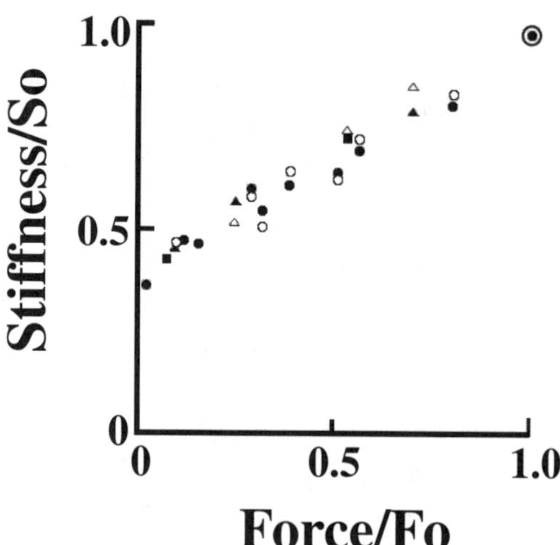

Figure 3.11. *Decline of stiffness associated with isotonic shortening. Different symbols represent different experiments. Redrawn from Ford et al. (1985) with permission.*

move or rotate to stretch their springs and generate force. If this is so, stiffness should rise in advance of force. A lag of 11 to 16 ms between force and stiffness was first described by Cecchi et al. (1982) and later by Ford et al. (1986).

MECHANICAL ADDITIONS TO THE THEORY

Head vs. tail

While it is known that myosin S-1 heads contain the binding sites for both ATP and thin filaments, there has until recently been some question as to whether the tail region had a role other than as a simple mechanical link. This question was answered with elegant simplicity by Spudich and his colleagues (Toyoshima et al.,1987). They developed an ***in vitro* motility assay** in which actin filaments floating in a solution are propelled by myosin molecules stuck to glass at the bottom of the chamber. The thin filaments are labelled with a fluorescent dye to make them visible with the light microscope. When a filament touches a myosin molecule, it is moved in a characteristic direction. Only the S-1 heads were needed to obtain the movement seen with the complete molecule.

Spudich and others have recently developed an even more elegant but much less simple technique for measuring the mechanical ability of single myosin molecules. They first developed laser **light traps** to hold and independently position two tiny beads. An actin filament is suspended between the beads in a way that it can be straightened and moved next to a single myosin molecule fixed to a pedestal. Force on the filament is indicated by an increase in the signal required to maintain the position of the beads at the ends of the filaments (Finer et al., 1994). They are now using this technique to investigate the effects of peptide additions and deletions in the myosin. One such experiment, related to a head vs. tail issue, will be described here.

Huxley and Simmons (1971) originally drew the crossbridge spring in the S-2 link between the head and thick filament for convenience in simplifying the drawing (see Fig. 3.6). In a later review, Huxley (1974) presented the drawing in Fig. 3.12 to make the point that both the spring and that the site of crossbridge motion could be anywhere. Spudich and colleagues showed that it was not in the S-2 link, where it had been drawn originally; the S-1 head stiffness was not altered by omitting the S-2 link. More recent evidence from Spudich's laboratory suggests that the compliance is in the head itself, possibly associated with flexion of the head shown in the left hand diagrams of Fig. 3.12.

Multiple power strokes

As described above, Huxley and Simmons (1971) concluded that it seemed likely that crossbridges move by more than one power stroke while attached. Recent experiments by Lombardi et al. (1992) suggest that the bridges may go through more than one complete cycle of attachment, power stroke, and detachment per ATP molecule hydrolyzed. The biochemical implications of this suggestion are described in the next chapter, and only the mechanical data will be presented here.

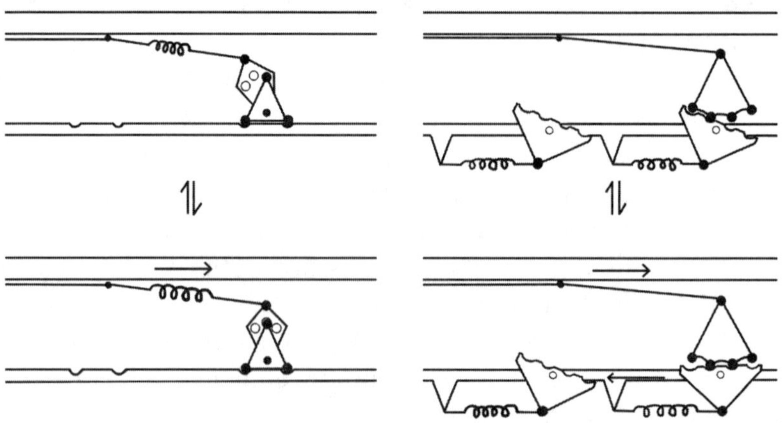

Figure 3.12. *Alternative crossbridge models. Redrawn from Huxley (1974), with permission.*

The duration of a single cycle of ATP hydrolysis during shortening can be calculated as the ATPase rate divided by the number of myosin molecules available to perform the hydrolysis. Lombardi et al. observed the tension transients produced by several steps applied in rapid sequence. When the interval between the individual steps was very short, the tension at the end of several steps was no different from the tension observed after a single large step of the same overall magnitude. As the interval between steps was increased, the tension recovery was greater, indicating an increase in the ability to generate new transients. They further found that this ability grew at a rate that was substantially faster than the ATPase rate. From this they concluded that there can be more than one mechanical cycle per ATP molecule hydrolyzed.

EVOLUTION OF THE CROSSBRIDGE THEORY

The discussion above emphasizes that crossbridges are not defined by a single theory but by a class of theories united by a common mechanism in which force between filaments is generated by a mechanical link that is cyclically made and broken. They differ from each other both in the types of states for a cycling bridge and in the rate functions for passing between states. Different states have been postulated by various investigators to account for specific experimental results. When properly conceived, these additional states add additional degrees of freedom, so that the enlarged theory will continue to account for much of the earlier data. For example, the proposal that crossbridge movement after attachment accounted for the transients did not diminish the ability of the theory to account for the earlier steady state data. This modification also enabled the further addition of the two stage attachment to accommodate Hill's more recent heat data without diminishing its ability to account for much of the other data. As described in the next chapter, biochemical experiments reveal a moderate number of transitions between chemical states. Much current research in muscle contraction is directed toward determining the nature and order of these states.

SUGGESTED READING

HANSON, J. and H.E. HUXLEY. (1953) Structural basis of the cross striations in muscle. *Nature* **172**: 530–532

HUXLEY, H.E. and J. HANSON. (1954) Changes in the cross-striations of muscle during contraction and stretch and their structural interpretation. *Nature* **173**: 973–976.

HUXLEY, A.F. (1974) Muscular contraction (Review lecture) *J. Physiol.* **243**: 1–43

Chapter 4

WORK FROM CHEMICAL REACTIONS

Muscle has long fascinated scientists because of its ability to generate work at a constant temperature. Perhaps the earliest credible theory of contraction was developed from the work of Schwann in 1835[1]. He noted that muscle developed the maximum force at its normal body length and that less force was developed at shorter lengths. The positive slopes of the length-force relation of muscle resembled the positive force-extension curve of a spring. It was therefore concluded that stimulation caused muscle to change from a relatively compliant spring to a tightly stretched spring, as if activation had initiated an "all-or-none" reaction that cocked the spring. In this theory, the energy invested in the spring would either be liberated as work, if the muscle were allowed to shorten, or dissipated as heat, if it were allowed to relax without shortening. The all-or-none aspect of the theory therefore predicted a fixed energy liberation during a contraction, so that if shortening and external work were performed, less heat would be released. This "cocked spring" or "new elastic body" theory prevailed for almost 80 years until it was disproved by the Fenn effect. Fenn showed that more heat, not less heat, was produced when the muscle shortened. In addition to disproving the cocked spring theory, Fenn showed that muscle could adjust its energy release to accommodate its load.

---

[1]See Needham (1971) p.53 for a reference to this work and its subsequent interpretation by Weber.

While the Fenn effect disproved the theory of one large cocked spring, it is entirely consistent with multiple small springs that are cyclically stretched and released many times during a contraction. The crossbridge theory described in the preceding chapter suggests that chemical reactions leave the myosin heads in a state where they can undergo a movement that cocks a spring. The purpose of the present chapter is to explain what is known about the relationship between the chemical reactions and the force generating movements of the crossbridges. Our knowledge about this comes from two very different types of experiments which technical limitations have prevented from joining until relatively recently. The first of these, biochemical studies of homogenized muscle, has the advantage of providing access to individual reactions and the major disadvantage of eliminating the primary object of interest, contraction. The second, correlation of mechanical events with chemical assays, leaves the function intact but does not permit easy dissection of separate reactions, especially not with a high time resolution . The third technique, which weds the first two, uses naked myofibrils where mechanical function can be studied in different chemical environments. Because the literature in these areas is large, this chapter describes only the most pertinent information derived from each type of experiment.

BIOCHEMICAL STUDIES

The earliest biochemical experiments on muscle were simply studies of the changes that occurred with time in the "brie" obtained by mincing a muscle. Fiske and Subbarow (1927) and Eggleton and Eggleton (1927) found that phosphate was produced from an organic compound identified by Fiske and Subbarow as creatine phosphate. In 1929 Lohmann described **adenosine triphosphate (ATP)** in muscle and in 1932 Meyerhoff and Lohmann described the transfer of phosphate in biochemical reactions. It is now known that it is ATP and not creatine phosphate that is hydrolyzed to **adenosine diphosphate (ADP)** by myosin and actin, and that ATP is replenished immediately by the transfer of phosphate from creatine phosphate. This phosphate transfer serves to keep the concentrations of ATP and ADP nearly constant in muscle. In addition to maintaining a nearly constant environment for the myofilaments, it avoids a potentially catastrophic increase in activation. As explained in Chapters 5 and 7, both ADP accumulation and ATP depletion tend to increase the activation of the actomyosin by a cooperative mechanism. If ADP were not rephosphorylated immediately, the combined ATP depletion and ADP accumulation could have disastrous consequences because activation would be sustained, worsening these imbalances. An interesting feature of this system is that muscle contains a high concentration of **creatine kinase** (abbreviated **CK** or **CPK**), the enzyme which catalyzes the phosphate transfer. The enzyme is used as a serum marker for muscle damage, particularly of heart muscle during heart attacks.

The energy for rephosphorylating creatine and ADP derives from other biochemical reactions. In skeletal muscle, the primary energy source is **glycogen**, long chains of glucose. In cardiac muscle the energy derives mainly from short chain fatty acids and lactate. Most of the energy from both metabolic pathways is derived aerobically from the **tricarboxylic acid "Krebs" cycle** operating in conjunction with the mitochondria. An important source of energy in emergency situations, however, is the initial anaerobic production of lactic acid from glycogen. Although the energy from this source is only about 6% of the total available from the complete aerobic catabolism of glycogen, it is sufficient to sustain muscle contraction in sprints, albeit at the cost of lactic acid accumulation.

The immediate substrate for muscle contraction is magnesium-ATP. Frequently, the magnesium is omitted from the abbreviation, but is assumed to be there, unless otherwise specified. Fig. 4.1 shows its chemical structure and the hydrolysis reaction which yields the reaction products, ADP and phosphate. This hydrolysis provides the energy for muscle contraction as well as for other biological processes. While it is sometimes said that the terminal phosphate bond is a "high energy bond," the experiments to be de-

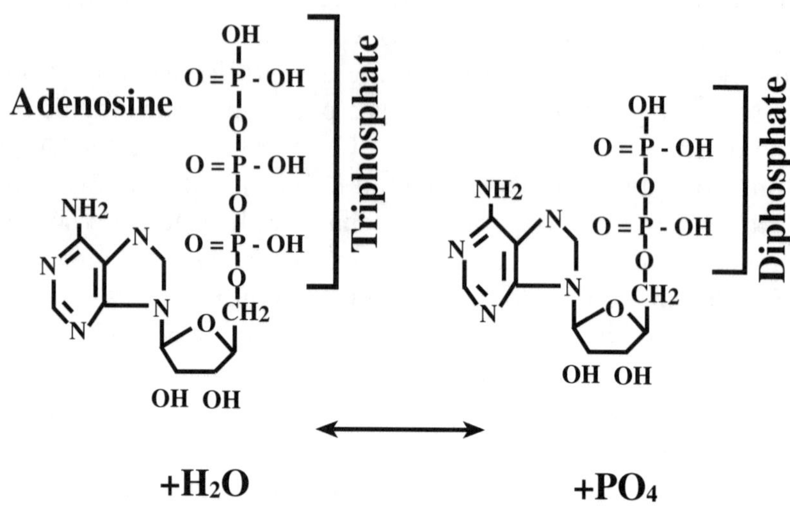

Figure 4.1. *Hydrolysis of ATP to ADP and phosphate.*

scribed here show that the energy is not stored in the bond, but in
the hydration energy of the reaction products.

When myosin and actin are mixed, they hydrolyze ATP if it is
available and bind tightly to each other if it is not. When ATP is
present, the actin and myosin are partially associated. The ATP hydrolysis occurs only if the actin is assembled into filaments (as f-
actin). The binding and the reactions described here will not occur
if actin is in its monomeric form (as g-actin). By contrast, myosin
will participate in these reactions not only when it is in a
monomeric form, but when it has been cleaved into sub-frag-
ments. Specifically, it is the S-1 head, the part of the molecule that
projects from the thick filaments, that binds to actin and hy-
drolyzes ATP. Myosin, or the S-1 fragment, will hydrolyze ATP in
the absence of actin, but at a rate which is very much slower than
the actin-myosin mixtures, called **actomyosin.**

Several hours after an animal dies, its muscles become very
stiff, a state called **rigor mortis**. It is now known that this is due to
the tight binding of myosin to actin that occurs in the absence of
ATP. The state of myosin binding to actin in the absence of ATP,

called **rigor**, is a very useful phenomenon for the study of muscle because it marks a specific state in the crossbridge cycle.

Lymn and Taylor's experiments

A classic example of the use of the rigor state can be found in the experiments of Lymn and Taylor (1971). They mixed actin and myosin without ATP to produce rigor complexes and then added ATP. Using optical techniques to measure binding they found that the ATP bound very rapidly to the complex and then the myosin dissociated from the actin at a rate faster than they could measure. The rate of ATP binding was about 500/s, and the dissociation rate was greater than 1000/s. By contrast, the steady state ATPase rate was very much slower, about 0.9/s. Furthermore, ATP appeared to remain bound to the myosin when it detached, since ATP binds tightly to isolated myosin and not to actin.

To learn what happened to the ATP bound to myosin, Lymn and Taylor mixed ATP with myosin alone, and after waiting varying periods, injected the mixture into an acid to denature the protein and release the bound ATP, ADP, and phosphate. They found: 1) only about half the ATP was hydrolyzed to ADP and inorganic phosphate, even after a long delay; 2) the initial hydrolysis rate was about 150/s; and 3) neither the ATP nor the reaction products were released from myosin until it was denatured. In a third set of experiments they mixed the myosin-ATP complex with actin and found that there was a burst release of some phosphate from the protein complex. Taken together, these experiments suggested the "four state" scheme shown in Fig. 4.2.

This scheme posits that myosin (M) cyclically attaches to and detaches from actin (A). Because the rigor complex (AM) binds ATP so rapidly and then dissociates, the binding and dissociation must occur at the very end of the attached phase of the cycle, after all other reactions have occurred. The very rapid dissociation further suggests that this step is almost irreversible, so that the initial association of actin and myosin must be due to binding in a different state that is separated from the final state by a relatively slow reaction. The observation that ATP is hydrolyzed on the unattached myosin further indicates at least two detached states, M.ATP and M.Pr, where Pr represents the reaction products, ADP and phosphate. The observation that neither ADP nor phosphate was released until the myosin re-bound to actin further suggested that there were several additional reactions between the two attached states shown in Fig. 4.2. The Lymn and Taylor work was published in the same year as the Huxley and Simmons description of the tension transients described in the preceding chapter. Both studies suggest that a power stroke occurs when the myosin crossbridge is attached to the actin filaments.

Energetic considerations

The finding that ATP and reaction products were present in approximately equal proportions on the unattached myosin suggests

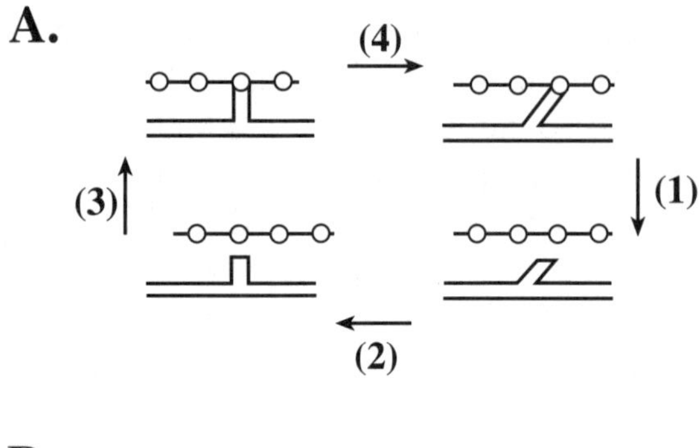

Figure 4.2. *Four-state model of Lymn and Taylor (1971). A) Mechanical transitions. B) Chemical reactions. Redrawn with permission*

that there is almost no energy loss in the immediate hydrolysis of ATP. This conclusion comes from the considerations that the rate of hydrolysis equals the rate of reassociation of the products. The similar concentrations of reactants and products indicates similar forward and backward rate constants, and therefore, similar activation energy barriers for the two reactions. A substantial energy drop with hydrolysis would produce a larger activation energy for the backward reaction and a higher proportion of products, which did not exist.

This conclusion about the low energy of the bond is supported by later experiments of Bagshaw et al (1975). They observed a rapid exchange of radioactively labelled oxygen on the terminal phosphate of ATP when it was bound to myosin. Phosphate has three oxygen atoms bound when attached to ADP and acquires a fourth from water when the terminal phosphate bond is hydrolyzed (Fig. 4.1). When this bond is made and then broken, an oxygen atom on the terminal phosphate is exchanged with the water in solution. If the inorganic phosphate is free to rotate when attached to myosin, then each cycle of making and breaking the terminal bond will exchange 1/4 of the oxygen atoms. The rapid exchange of oxygen found by Bagshaw et al. therefore suggested multiple making and breaking of the terminal phosphate bond of ATP bound to myosin. The rapid bond formation further indicated a low bond energy. Since energy is released when ATP is hydrolyzed to ADP and phosphate, these experiments support the concept that the energy is not in the bond itself, but in the energy of hydration of phosphate when it is released from myosin. The low bond energy has two implications for the present discussion, one mechanistic and one semantic.

The semantic issue arises from the custom of calling the terminal phosphate bond a "high energy bond." Barbara Banks has led a one-person crusade, reaching as far as high school text books, to scourge this term from the scientific glossary when applied to ATP. The experiments just described indicate that the energy is not in the bond itself. In addition, the bond energy, about

55 kJ/mole is relatively small, about 14% of the energy of a carbon-carbon bond. While there is nothing wrong with a term when everyone understands what it means, Dr. Banks makes the point that this particular form of words may lead to a misunderstanding of the underlying mechanisms. Some further energetic measurements indicate how she may have been correct.

The finding that the energy derived from ATP hydrolysis did not come from the bond itself led White and Taylor to measure the energy release associated with other reactions in the crossbridge cycle. Most reactions they could measure produced very little energy release, but they were unable to measure the energy associated with phosphate release. This finding leads to the conclusion that most of the free energy drop occurs in close association with phosphate release from actomyosin.[2] A major point to be made from this consideration is that phosphate release leaves the crossbridge with a store of energy that can be used to stretch its internal spring. The reaction schemes described next suggest that phosphate release in some way gates the mechanical energy release. It seems likely that the energy of hydrating the phosphate is in some way imparted to the crossbridge.

Reaction schemes

The foregoing descriptions suggest that force and shortening in muscle are generated by a multistate crossbridge cycle in which mechanical transitions (power stroke, attachment and detachment of bridges) alternate with chemical reactions (ATP binding, hydrolysis, and product release). While the distinction between chemical and mechanical transitions may be artificial, it is made here on the basis of the types of experiments used to study them. Mechanical transitions are assessed on the basis of mechanical measurements, such as force and stiffness; chemical reactions are assessed from changes in chemical concentrations. A scheme for

---

[2]See Huxley (1980, p. 93) for a description of this unpublished conclusion.

such a multistate model incorporating the known transitions (Fig. 4.3) was presented by Huxley (1980) in his Sherrington Lectures. Two important features of the scheme should be emphasized. The first is that each transition occurs as a separate stochastic event. Thus, the finding of a large free energy drop associated with phosphate release does not imply that phosphate release and the power stroke are synonymous, or that they occur together, but rather that they are closely related in the cycle. The second important feature of the model is that the serial nature of the transitions ensures an orderly process for energy release. Attachment of crossbridges to actin allows the phosphate to be released. This release in turn

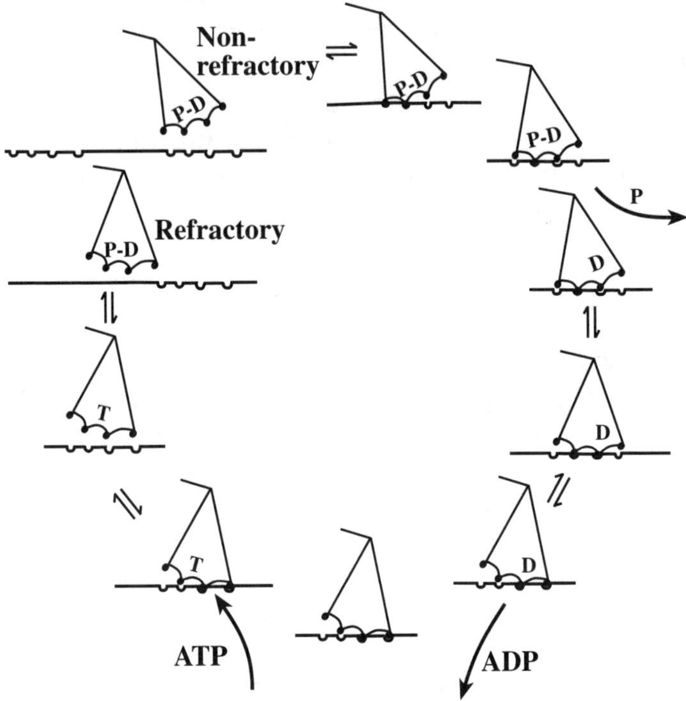

Figure 4.3. *Huxley's (1980) multistate crossbridge cycle. Mechanical transitions (arrows) alternate with chemical transitions, indicated by changes in letters. A=ATP, D=ADP, P=phosphate. Redrawn with permission.*

gates the power strokes, which must occur before the ADP can be exchanged for a new ATP.

An objection that some chemists raise to a scheme such as that shown in Fig. 4.3 is that it fails to consider a number of possible transitions. For example, they might assert, correctly, that for every attached state there should a matching detached state, even if it is almost never occupied. A possible resolution of this discussion is to imagine a scheme such as that shown in Fig. 4.4 where every attached state has a matching detached state. These intermediate detached states are realistic in that they will be occupied if bridges are forcibly detached from intermediate attached states by rapid filament sliding. In this scheme, the path taken by the detached bridges is usually not the same taken by attached bridges. In the absence of forcible detachment there are only two detached states to consider. When forcible detachment occurs, the bridges reattach quickly to actin sites further along the thin filament and do not take the paths indicated by the interrupted lines unless they are pulled out of reach of an attachment site.

Alternative states

A final point to be made about schemes such as those shown above is that there is always the possibility of a bridge taking an alternative route to another state in the cycle. When this occurs, additional states must be postulated. If, for example, phosphate release normally occurs before the power stroke, as shown Fig. 4.4, the

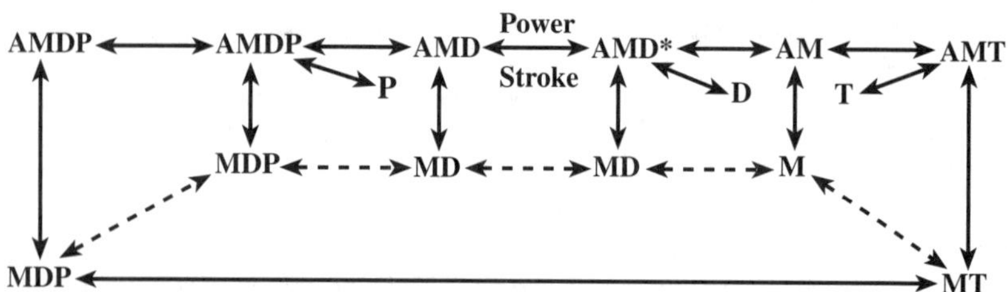

Figure 4.4. *Simplified crossbridge cycle that provides a detached state for every attached state.*

postulate that it may also occur after this transition requires an additional high-force state with attached phosphate. This consideration raises two issues. The first is which states and transitions are consistent with the available data. The second is whether the bridges can be made to occupy states not normally in the cycle. The goal of much of the current research in muscle is directed at determining the exact sequence and rates of transitions in the cycle, with the object of learning the orderly process that converts chemical energy into physical work.

EXPERIMENTS ON WHOLE MUSCLE

The study of whole muscle requires that the reactions of interest be isolated. Heat measurements were pursued long after all the relevant thermodynamic conclusions could be extracted because they could be used as surrogates for chemical measurements and had a greater accuracy and better time resolution than any of the available chemical methods. At the same time, it was recognized that the direct chemical measurements were needed to correlate with the heat experiments. The major difficulty with such measurements is that the muscle must be destroyed in preparation for the assay, so that only one chemical determination can be made on each muscle. The variability in chemical content between muscles requires that many muscles be assayed when a high degree of resolution is required. In addition, recovery reactions immediately restore the ATP and later replenish the creatine phosphate, so that poisons must be used to inhibit the recovery reactions. In spite of these limitations, some useful conclusions regarding muscle energetics have been derived from such studies.

One of the earliest of these studies, by Kushmerick and Davies (1969), was among the most comprehensive. These authors measured the relationship between mechanical work and phosphate production, reflecting ATP hydrolysis, in the presence of dinitrofluorobenzine used to inhibit creatine kinase. To maximize phosphate production in single contractions, they activated the muscles at long lengths and allowed them to shorten to their optimum length before freezing them for chemical assays. Their results relating the work done to the amount of ATP hydrolyzed are shown in Fig. 4.5. As indicated, there is a maximum in the work and efficiency curves in the region of intermediate velocities. An interesting feature of these experiments is that when estimated efficiency was increased by 33% to account for the energy spent on activation (interrupted curve in Fig. 4.5), the estimated maximum efficiency of the myofilaments approached 100%. While such a high efficiency seems unlikely, and while subsequent experiments have suggested an efficiency close to 50%, the high value found in this seminal study should be kept in mind.

Subsequent studies have refined the Kushmerick and Davies work with more rapid freezing techniques, shorter distances of shortening, and possibly better chemical determinations, but with the exception of the high efficiency, the results are about the same.

The reason for making an issue of the high efficiency here is the long shortening distance employed in the initial study. When a

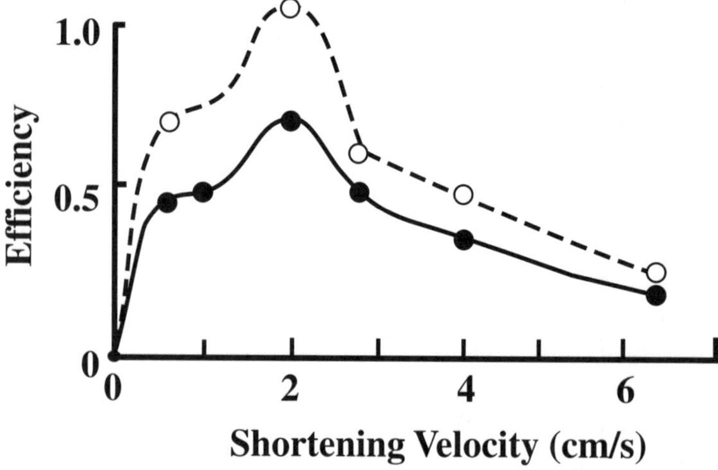

Figure 4.5. *Kushmerick and Davies (1969) measurements of the ratio of work to energy of ATP hydrolysis. They increased the ratio by 33% (interrupted curve) to account for the ATP hydrolyzed by non-mechanical reactions. Adapted with permission.*

muscle begins shortening from such a long length, there is almost no overlap of the crossbridges with the thin filaments until the shortening occurs. Thus, the energy for the work is extracted from bridges that are fully charged with ATP. This may be a very different circumstance than when the work is done over a shorter distance nearer full filament overlap. Under these conditions, bridges will have been active before the shortening and work began. If it is assumed that a bridge undergoes a complete cycle of attachment and detachment and returns to the same energetic state each time one ATP molecule is hydrolyzed, there should be no energetic difference between these experiments. If, on the other hand, the bridges undergo partial cycles, liberating the hydrolysis energy over multiple cycles of attachment and detachment, it is possible to achieve differences in the apparent efficiencies, depending on energetic differences in the bridges at the beginning and end of shortening. Such "partial cycles" could explain the higher in efficiencies measured by Kushmerick and Davies compared with the subsequent works.

Spectrophotometric measurements

The inability to make multiple direct assays of chemical changes in the same muscle has led to the development of other methods for assessing chemical changes. The earliest and most extensively studied surrogate measurement has been heat, which has the disadvantage of being non-specific. There are now available much more specific physical techniques for measuring chemical changes in the living state. One of the most promising is nuclear magnetic resonance (**NMR**) spectroscopy. This technique uses a high density magnetic field to align the nuclei of some atoms and then a radio frequency pulse to perturb the alignment. As the nuclei return to their unperturbed alignment, they emit radio frequency energy at a wavelength characteristic of the immediate chemical environment of the nucleus. The wavelength is highly specific for each chemical form of the nucleus, and so the technique is capable of resolving all the different forms of a particular atom, such as phosphorous. For example, the technique can distinguish the three phosphate atoms in ATP, the two phosphate atoms on ADP, and the ionization state of inorganic phosphate. The only problem with the technique is that the signals are very weak. To obtain an adequate signal-to-noise ratio, it is necessary to collect and average signals over an extended period of many seconds or minutes. For measurements of steady intracellular concentrations this requirement does not create a large difficulty. By contrast, resolution of rapid concentration changes may requires that the muscle be sampled for very brief periods, in the order of milliseconds. In addition, it is necessary to wait several seconds between radio frequency perturbations. Thus, a measurement of the phosphate release during the first millisecond after the onset of shortening, for example, might require averaging the responses to several thousand identical contractions. As a consequence, the results of such experiments are not yet available. While these limitations have so far impeded the generation of results about the chemical transitions in the crossbridge cycle, there have been some important issues resolved by NMR spectroscopy in muscle. The use of NMR to study the changes in concentrations of various phosphorous compounds during hypoxia and fatigue will be described in Chapter 8. One of the most relevant measures is a precise estimate of intracellular pH.

For years a controversy continued over estimates of intracellular pH. The measurements centered around two values. One was asserted by a group that believed the hydrogen ions to be freely diffusible and thus distributed according to the membrane potential and its Donnan equilibrium. These measurements showed the value to be a full pH unit below the extracellular value. Another group found the pH to be approximately the same as the extracellular fluid, requiring that hydrogen be pumped out of the cell against an electrochemical gradient. A precise measurement was determined as a by-product of the NMR spectroscopy of intracellular phosphate. The technique gives an accurate estimate of the ratio of the monobasic and dibasic forms of inorganic phosphate, which is in turn determined by pH. The measurements show that the resting intracellular pH is only slightly less than the extracellular value, and that it rarely falls below about 6.8, at least in isolated muscle where the extracellular pH is maintained constant. Thus, a mechanism exists for removing protons from the intracellular fluid and maintaining a high degree of pH homeostasis for the contractile units.

NAKED MYOFIBRILS

The first preparation of fibers without membranes was described in 1947 by Albert Szent-Györgyi, who had the laboratory equivalent of a green thumb. He had earlier won the Nobel prize for the isolation of vitamin C, a task which had eluded several generations of chemists. He placed whole muscles in a solution containing 50% glycerol and 50% physiological saline and then kept them at -20°C for several weeks. The low temperature inhibited both the proteolytic enzymes and growth of microorganisms. We now know that the glycerol has two effects. The first is to prevent freezing and the formation of crystals that damage cells when they are frozen. The second is to dissolve the membranes, so that chemicals placed extracellularly could readily diffuse into the myofilaments. Thus, when the muscles were removed from the freezer and the glycerol, they could be made to contract by the direct application of ATP. At the time, the activating mechanisms of muscle, described in the next chapter, were unknown, as were the roles of ATP and the surface membranes, but the value of such a preparation was readily recognized. The 1954 study of Huxley and Hansen demonstrating the constancy of the A-band and leading to the sliding filament hypothesis were done using myofibrils from glycerinated muscle. The reason for mentioning Szent-Györgyi's green thumb is that he always expressed surprise as his successes. When asked by a young admirer how he had happened on the 50/50 mixture of saline and glycerol, later found to be optimum, he replied "My boy, I don't even know why I used glycerol!"

A second type of preparation was published by Natori (1954). He first isolated segments of single fibers under oil and then mechanically dissected away the outer membranes, leaving both the myofibrils and the internal membranes intact. This became an excellent preparation for studying the activating system, as will be described in the next chapter, and the term **skinned fiber** was coined to describe it. At first, drops of solution were placed on the fibers in oil and the response observed under the microscope, but in 1967 Hellam and Podolsky published a method for gripping the ends of fiber segments, both so that the fibers could be moved between solutions and so that force generation could be measured. This was quickly recognized as a powerful method of studying the influence of chemical environment on the contractile proteins, and many variations on the methods have evolved. Because of the prominence of the experiments using the preparation, fibers that have had their membranes dissolved are sometimes called "chemically skinned," in spite of the priority of Szent-Györgyi's method, even when the dissolution is caused by glycerol. Other authors sometimes refer to their chemically skinned preparations as "permeabilized" to emphasize that some membrane structure and function are left behind.

When a skinned fiber is placed in an aqueous environment, all of its diffusible intracellular constituents are replaced by the chemicals in the bathing medium. The relevance of the results obtained with these preparations to physiological function therefore demands a knowledge of normal intracellular environment. Issues such as the physiological intracellular pH can become important because this can be set to any value by the experimental conditions. In addition, methods must be used to avoid substantial diffusion gradients for reactants. For example, most experiments now use an ATP regenerating system consisting of creatine phosphate and creatine kinase to avoid substantial radial gradients for ATP and ADP. In addition, retention of fixed charges on proteins causes the development of an osmotic differential at the cell surface, so that removal of the surface membrane causes the filament lattice to swell. Godt and Maughan (1976) first showed that skinned fibers can be re-compressed by adding to the bathing solutions large, osmotically active polymers that cannot diffuse into the lattice. Goldman (1987) later showed that the maximum velocity was about 50% higher if the fibers were not so compressed. He argued that this resulted from the increased distance between filaments allowing negatively strained crossbridges to buckle rather than resist shortening. He further showed that his results could be modelled quantitatively with Huxley's (1957) original equations when the absence of adequate polymer concentration caused crossbridge stiffness at negative strain to be less than at positive strain.

The goal of most of the work on the contractile behavior in skinned fibers can be summarized by saying that it is intended to discover the exact series of transitions in the crossbridge cycle and the rate constants or functions for these transitions. Naked myofibrils are ideal for this work because the effects of variations in the chemical environment on physical transitions can be measured directly. The mechanical parameters assayed are force, velocity, and stiffness, as well as various forms of transient responses. Changes in stiffness are used to assay changes in the number of attached bridges; changes in the force/stiffness ratio signal alterations in the average force per attached bridge; and changes in velocity are taken to reflect variations in cycling rate. Some examples supporting the multistate scheme in Fig. 4.4 follow.

ATP-ADP exchange on high force states

In skinned rabbit psoas muscle fibers at 0°C, raising the ADP concentration to 5 mM or lowering the ATP concentration from 10 to 2 mM have similar effects on force, velocity and stiffness. Stiffness is increased by 10-20%, isometric force is increased by about 25% and maximum shortening velocity is reduced by about 35%. The reduction in velocity indicates a reduction in crossbridge cycling rate, the increased stiffness indicates an increase in the number of attached bridges, and the increase in the force/stiffness ratio indicates more average force per bridge. Taken together, these results suggest that the two interventions detain bridges in high force states, after they have undergone the power stroke. This is expected from Lymn and Taylor's earlier studies of the interaction of ATP with proteins in solution. They showed that ATP binding was followed by such rapid dissociation of actin and myosin that it was likely to be the last transition in the attached phase of the cycle. Furthermore, since ATP is hydrolyzed to ADP, it is expected that the two will occupy the same binding site and that ADP must be released before ATP can bind. Further insights into the exchange can be obtained from the interactive effect of ADP and ATP on velocity shown in Fig. 4.6.

Maximum shortening velocity increases with increasing ATP concentration, and the relationship between the two can be linearized by plotting the inverse of velocity against the inverse of substrate concentration, as in Fig. 4.6. In addition, ADP decreases velocity in a manner that depends on the ATP concentration. For each ADP level, the velocity increases with ATP concentration, and all velocities extrapolate to the same absolute maximum. By analogy with enzyme kinetics, the plots in Fig. 4.6 suggest two

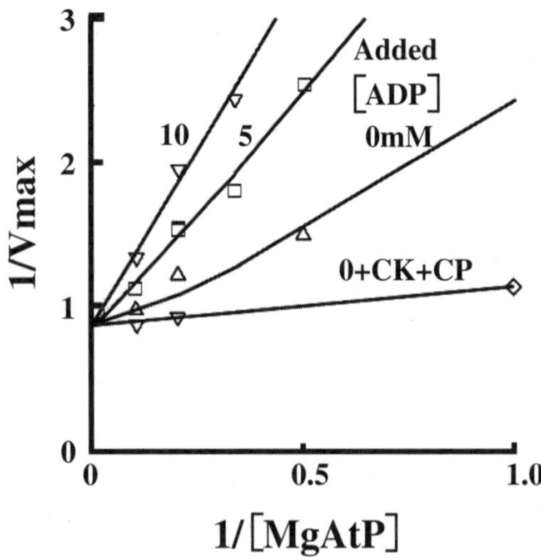

Figure 4.6. *Interactive effects of ADP and ATP on maximum shortening velocity. From Seow and Ford (1997), with permission.*

conclusions: first, there is a reversible exchange of ATP and ADP, and second, the exchange occurs without a step intervening between the release of ADP and the binding of ATP. If there were such a step, it would not be possible to overcome completely the effects of intrafiber ADP with high concentrations of ATP.

Together with the Lymn and Taylor (1971) experiments on isolated proteins, the data in Fig 4.6 suggest that the ADP-ATP exchange occurs on a high force state, immediately before myosin detachment. These experiments thus support the order of the final attached transitions shown in Fig. 4.4.

Low-force transitions

There are several low force transitions that occur before the power stroke. Two that will be explained here are one inhibited by high ionic strength and another inhibited by phosphate. Raising ionic strength inhibits contraction. In rabbit skinned fibers, raising the level from 125 mM to 360 mM depressed isometric force by 50% and stiffness by 25%, so that the force/stiffness ratio is decreased by 33%, without any change in shortening velocity. The decreased stiffness indicates fewer attached bridges, and the decreased force/stiffness ratio indicates a lower average force per attached bridge. Together these findings indicate that increasing ionic strength causes an accumulation of bridges in low force states without otherwise impeding crossbridge cycling.

The nature of this state is further indicated by the alterations in the transient tension responses to step changes in length shown in Fig. 4.7, which are analyzed according to the procedures shown there. The responses to three sizes of step are shown in Fig. 4.7 A. The computerized records on the left are printed at a slow speed to show the entire period of recording, while the records on the right are printed at an increased speed to resolve the rapid events associated with the step. At the end of the recording period, a large release is applied to make the muscle go slack and define the zero force level. Note that the response to the stretch is recorded for a longer period than the response to the releases, and that the largest release is slowed to avoid making the muscle go slack during the recording period, when sarcomere length is being servo controlled.

The records obtained in the presence of 125 mM and 360 mM ionic strength are superimposed in Fig. 4.7 A. The fibers developed half as much force at the high ionic strength, but the records have been scaled to make their isometric forces coincide in the superimposed plots. The rationale for this scaling and superimposition is the consideration that when the force is reduced because bridges are detained in a low-force state, all the developed force will be produced by non-detained bridges. Thus, the responses of these "normal" bridges will scale in proportion to the isometric force and the difference between the records will reflect the responses of the non-detained bridges. To a first approximation, therefore, the difference records (Fig. 4.7 B), reflect the responses

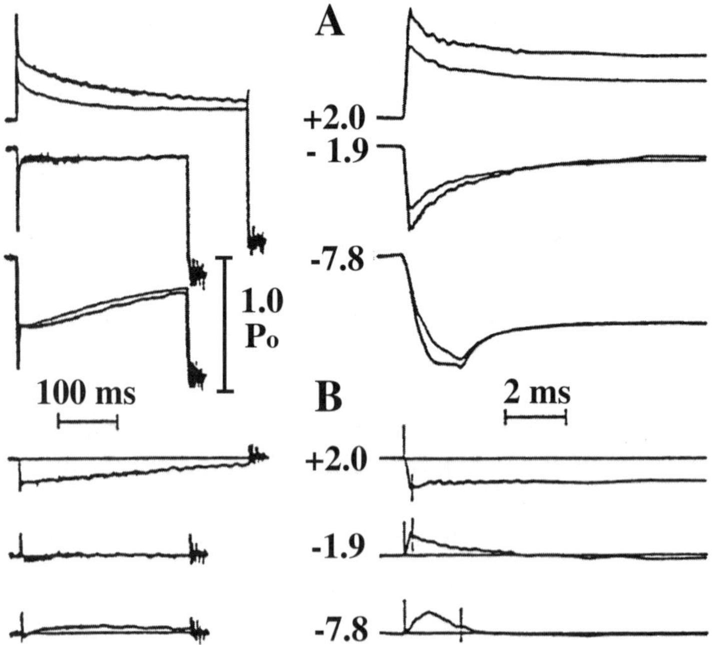

Figure 4.7. *Tension transients at two ionic strengths. From Seow and Ford (1993), with permission.*

of the detained bridges. There are, of course, many potential pitfalls with this analysis, but the experiment shown in Fig 4.7 is one where the results appear to justify the method.

The increased relative stiffness, indicating less average force per attached bridge, is shown by the force difference that develops during the step. The long lasting difference following the stretch further indicates that the increased stiffness is genuine, and not due to some transient artifact. Following releases, however, the force difference is very short-lived, suggesting that bridges attach in an initial state where they resist stretch but are easily detached by release. They have little effect on unloaded shortening velocity because they pose little resistance to shortening. Such an initial attached state might seem improbable had not Huxley (1973) postulated its existence to account for Hill's (1964) later heat data, as explained in Chapter 3.

Dual effect of phosphate

As mentioned above, phosphate release appears to be closely associated with the power stroke in muscle. Not only is it associated with a large free energy drop, but early experiments on the specialized flight muscles of insects showed that it greatly altered the tension transients. Clues to the contractile mechanisms have therefore been sought in studies of the effects of phosphate. Two possible reasons for the close association of phosphate to the power stroke have been advanced. One, suggested in part by Huxley's multi-state scheme in Fig. 4.3, is that phosphate release must occur before the power stroke, and as a result, the release gates the power stroke. The other possibility is that the release occurs after the power stroke and thereby locks the energy release by preventing reversal of the power stroke. The answer is not fully known yet, but some recent experiments suggest that the actual role of phosphate may be more complex, and more interesting.

Increased phosphate concentration inhibits force generation in skinned fibers, but the nature of the effect depends upon pH. At all pH's between 6.3 and 7.4, 40 mM phosphate reduces force by approximately half and the stiffness to a lesser extent, so that the force/stiffness ratio is decreased. These results suggest less average force per attached bridge and detention of bridges in low force states that precedes the power stroke. At low pH (<6.5) phosphate also depresses velocity, suggesting that it inhibits a transition early in the attached phase of the cycle and causes bridges to be detained in a low force state. At high pH (>7.0) it increases velocity.

Phosphate could have two effects on the cycling bridges; it might act by remaining attached to the bridges, or it might act by re-binding to bridges after its initial release. The second effect could be different from the first if an additional transition occurs between the initial release and the re-binding. If this is the case, the two effects might be distinguished by using an ATP analogue that is less readily hydrolyzed. Uridine triphosphate (**UTP**) is such an analogue. When used as the substrate for actomyosin, it is hydrolyzed at about half the rate of ATP. This observation suggests that it might be ideal for testing mechanisms attributed to the slow phosphate release. When UTP is substituted for ATP in skinned fibers at high pH (7.4), their responses are very much like phosphate at low pH, i.e. it slows velocity and decreases both the force and the force/stiffness ratio. This finding suggests that phosphate has a second effect to increase velocity at high pH. This interpretation is supported by the interaction of phosphate with ADP in the responses of skinned fibers.

Interaction of phosphate and ADP

Low concentrations of phosphate (1-2 mM) completely reverse the increases of force and stiffness produced by 5 mM ADP without affecting the decreased velocity. Since ADP increases force and stiffness by detaining bridges in high-force states, this effect of phosphate must act by reducing the detained high-force bridges. The force reduction might be accomplished either by reversal of the power stroke or by detachment of high force bridges.

The concomitant reduction in stiffness and the absence of a further decrease of velocity argue against reversal of the power stroke. Thus, it seems likely that phosphate detaches the bridges detained by ADP. This possibility suggests both a new variation of the crossbridge cycle and an explanation of several older puzzles.

If phosphate binds to and detaches high force bridges with bound ADP, the detached bridges will be similar chemically to new bridges that have not been attached after binding ATP; both will have ADP and phosphate bound. They will be energetically different, however, in that they will have expended some of the energy of ATP hydrolysis on the power stroke. The consideration that there might be several classes of chemically similar but energetically different bridges raises the possibility that the bridges might undergo more than one attachment-detachment cycle per ATP molecule. In this variation of the crossbridge model, phosphate release permits the release of a quantum of energy by gating the power stroke, while re-binding allows the bridge to detach and preserve some of the energy of hydrolysis for further work. Some additional evidence in favor of the postulated multi-stage energy release can be found in a comparison of the quasi-steady state shortening in the presence and absence of added ADP and phosphate. Fig. 4.8 shows the effect of the added reaction products on the time course of force when the skinned fibers are subjected to a ramp shortening. The thick traces were obtained in the presence of the added constituents (test), and the thin traces were obtained in their absence (reference). As indicated, the ramp speed required to obtain the same low steady force is slightly higher under the reference conditions. More importantly, under the reference conditions the fiber is able to maintain a relatively high force for a short distance before force declines to its final steady level. By contrast, force falls immediately to its steady level under the test conditions. This finding suggests that the isometric bridges had a store of reserve energy not available under the test conditions and that this store was exhausted by the rapid shortening.

Similar results are obtained in the reverse type of experiment,

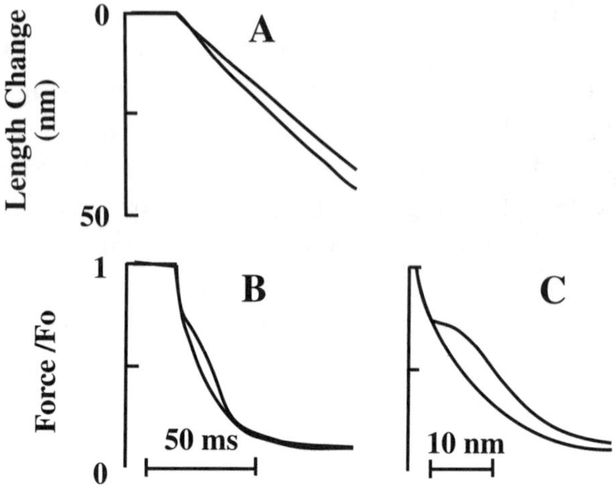

Figure 4.8. *Effect of 5 mM ADP and 1 mM phosphate on the response of skinned fibers to steady ramp shortening.*

where force is suddenly changed from isometric to isotonic and the time course of shortening observed. In the presence of ADP and phosphate the shortening velocity is constant and slow. In the absence of the reaction products, velocity is initially higher and declines gradually to the low level seen under the test conditions. A possible explanation of these results is that under the reference conditions, the isometric muscle has a population of bridges that is heterogeneous with regard to the amount of ATP hydrolysis energy remaining and that rapid shortening consumes this energy. By contrast, under the test conditions the ready detachment of bridges with bound ADP has slowed the rate at which bridges are recharged with ATP and reduced the average store of energy per bridge. Additional support for such a scheme comes from some earlier findings that are otherwise difficult to explain.

Some previously unexplained observations that could be explained by the proposed mechanism are: 1) the finding of Huxley and Simmons (1971) that the area under the T_2 curve, indicating the energy available from a single power stroke of all the attached bridges, is only a small fraction of the energy of ATP hydrolysis; 2) the finding of Lombardi et al., described at the end of Chapter 3, that the bridges appear to regenerate the ability to undergo a power stroke in a time that is much shorter than the complete cycle time for crossbridges; 3) the finding of Kushmerick and Davies, described in this chapter, that the chemomechanical efficiency of intact muscle working over new ranges of filament overlap is much higher than similar work done by previously overlapped filaments; 4) the finding of Ford and Gilbert, described in Chapter 2, that the extra heat of rapid shortening is liberated after the shortening ends but more rapidly than the chemical reactions that replete creatine phosphate; and 5) the controversial finding, described immediately below, that single myosin heads can propel isolated thin filaments for much greater distances than are expected from the possible ranges of crossbridge motion.

In these last experiments, myosin heads are fixed to glass slides at the bottom of suspensions of thin filaments. When the thin filaments come in contact with a head, they are propelled through the solution. By measuring the ATPase rate and the number of such interactions, it is possible to calculate the distance that a thin filament is moved by each ATP hydrolyzed. Such calculations have usually suggested movements that are much larger than the distance a myosin head could move. One obvious explanation is that the heads undergo more than one cycle of attachment, power stroke, and detachment per ATP hydrolyzed, and that the filament travels a large distance during the detachment parts of the cycle.

Another teleological explanation for the effect of low levels of phosphate to detach bridges with bound ADP is that it may help to avert a metabolic disaster during fatigue, as explained in Chapter 8. By promoting detachment of bridges from thin filaments when ADP begins to accumulate, phosphate may prevent their imposing a load on other, normally cycling bridges. In addition, as explained at the end of chapter 5, high force bridges, both rigor bridges without ATP and bridges with ADP bound, increase activation through a cooperative mechanism. If the ADP bridges remained attached during fatigue, muscle activation might become sustaining to the point that it produced exhaustion.

THE TRANSFORMER EFFECT

In his Sherrington Lectures, Huxley (1980) posed the question of how the making and breaking of chemical bonds, which produce movements of a few tenths of a nm, cause movements of a few tenths of a meter at the ends of the muscle. He called this amplification, which could be as much as 10^9, the transformer effect. The final stage, which produces both the greatest amplification and the variability among muscles, is caused by the large number of half-sarcomeres in series. This amplification is equal to the number of half-sarcomeres in series, and can be as many as 10^6. The next stage is produced by the repetitive action of crossbridges in parallel, with each crossbridge producing movements of about 12 nm and the half-sarcomere functioning over a range of about 240 nm, for a further 20-fold amplification. The last stage, which accounts for an amplification of 10 to 40 fold, appears to result from angular movement within the actomyosin complex, although the exact site of this movements remains to be discovered. The final discovery will define the exact chemical bonds that result in the primary movement which generate the angular motion that drives the engine.

SUGGESTED READING

HUXLEY, A.F. (1974) Muscular contraction (Review lecture) *J. Physiol.* **243**: 1–43

HUXLEY, A.F. (1980) *Reflections on Muscle.* (Princeton Univ. Press, Princeton, NJ).

HUXLEY, A.F. and SIMMONS, R.M. (1971) Proposed mechanism of force generation in striated muscle. *Nature.* **223**: 533–538.

HUXLEY, A.F. and SIMMONS, R.M. (1973) Mechanical transients and the origins of mechanical force. *1972 Cold Spring Harbor Symp. Quant. Biol.* **37**: 669–680.

Chapter 5

ACTIVATION

Much of this chapter can be summarized by a single word, and that word is **calcium**. Calcium activates muscle when it is released from internal stores and binds to thin filaments, where it causes a structural change that exposes attachment sites for crossbridges. Relaxation occurs when calcium is re-accumulated and the structural changes on the thin filaments are reversed. The history of the discovery of these mechanisms is a fascinating story that is the subject of this chapter. The final section of the chapter explains the phenomenon of cooperativity in activation, the understanding of which is vital to an understanding of cardiac function.

CALCIUM

Credit for the discovery of the role of calcium in activating muscle belongs properly to Heilbrun (1940) and Heilbrun and Wiercinski (1947). Using micropipettes, they injected various solutions into muscle fibers and found that calcium in low concentrations caused an immediate local contraction. Other salts had no effect. This simple and straightforward experiment and its conclusion were not accepted, however, mainly because they were too far in advance of the field. Neither they nor anyone else could offer any explanation of how calcium had its effect. It was also true that the injections caused a substantial local bulge in the fiber, providing a ready criticism for anyone wanting to discount the obvious conclusion.

The persuasive experiments demonstrating the role of calcium were not published until many years later by Ebashi and Lipmann

(1962), who studied membrane vesicles, and by Weber (1959, 1966), who studied the influence of ionic environment on the ATPase rates of isolated myofibrils. Prior to her experiments the literature on activation was confusing. Part of this confusion arose because ATP binds both calcium and magnesium and because MgATP is the substrate for the ATPase. When ATP without Mg is added to myofibrils, it will not be hydrolyzed, in part because it is not the proper substrate and in part because it binds calcium and prevents the activation of the ATPase. Thus, for example, it is possible to produce convincing experiments suggesting, incorrectly, that magnesium is the agent responsible for activation. When added to a mixture of myofibrils and ATP, it initiates hydrolysis, both because it provides the proper substrate by binding to ATP and because it displaces calcium which can then activate the filaments. By calculating the ion concentrations, Weber showed that all of these effects could be explained by the tight binding of calcium to the filaments. Ionized, "free" calcium in concentrations less than 1 μM were sufficient to activate the filaments when enough magnesium was present to provide the MgATP substrate. Substantially more total calcium is required for activation, but most of this is bound to the myofilaments. Unless precautions are taken, myofibrils absorb sufficient contaminant calcium during their preparation to activate MgATP hydrolysis as soon as it is added.

INTERNAL MEMBRANES

The early problem encountered by muscle biochemists was not how to activate the actomyosin ATPase but how to inhibit it. Unless stringent precautions are taken to remove calcium, isolated myofibrils will hydrolyze ATP until it is gone. This observation initiated the search for the physiological relaxing factor.

Myofibrils are so dense that they are easily removed from solution by centrifugation. Purification of these fibrils is accomplished by washing, resuspending and recentrifuging them several times. In the course of studying such washed myofibrils, Keilley and Meyerhof (1948) found a separate ATPase which could be removed from the supernatant by higher centrifugation. Since it could be separated by centrifugation, it was known to be made of particles, but it was also found to be non-filamentous. The role of this Keilley-Meyerhof ATPase was at first unknown but the finding that the work-producing myofilaments were relaxed by this other ATPase led to some interesting speculations of a cocked spring nature. The true mechanism turned out to be simpler.

Since the isolated myofibrils always hydrolyzed ATP, it was guessed that muscle must contain a relaxing factor that became separated from the myofibrils in the course of their isolation. To identify this factor, various fractions of the muscle were recombined with the myofibrils to relax them. Marsh (1952) found that after myofibrils were centrifuged from a suspension, there remained a particulate fraction that inhibited the myofibrillar ATPase. He named it "Marsh relaxing factor", but before long it was found to be the same as the Keilley-Meyerhof ATPase.

The suggestion that particles might cause relaxation was anathema to investigators of a chemical bent, who continued to search for a soluble relaxing factor. Electron microscopy was in its infancy when these searches were being made, and the structure of the particles was not known until the pioneering work of Ebashi and Lipmann (1962). They made three major discoveries: 1) the particles were made of membranous vesicles that resembled some newly discovered internal membranes in muscle; 2) the vesicles could remove calcium from solution against a gradient of more than 10,000:1; and 3) the vesicles hydrolyzed ATP while sequestering calcium. ATP hydrolysis stopped when calcium had been removed from solution and could be restarted by adding more calcium. An electron micrograph of their particles is shown in Fig. 5.1. These are clearly membranous vesicles. In addition, there were frequent **triads**, consisting of a small central vesicle (arrow) with two larger vesicles on either side. These triads resemble the t-system and terminal cisterns of the sarcoplasmic reticulum described below.

In the discussion of their paper Ebashi and Lipmann wondered how calcium might have affected the myofibrillar ATPase. Un-

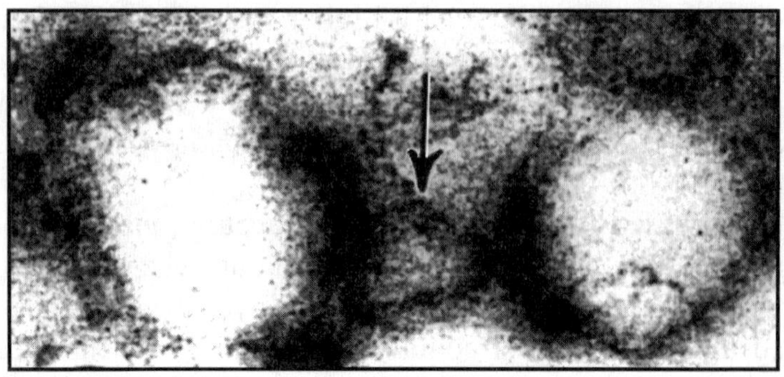

Figure 5.1. *Electron micrographs of the particles studied by Ebashi and Lip-mann (1962). Reproduced with permission.*

known to them, Weber was simultaneously carrying out the experiments described above.

t-system

A.F. Huxley built his interference microscope for the purpose of resolving structural changes in sarcomeres of living and contracting muscle. In addition to discovering the constant A-band length that led to the sliding filament hypothesis, he used it in one other pioneering study done in collaboration with R.E. Taylor. They pressed fine glass pipettes against the side of a single fiber to pass electric current that caused local depolarizations. Depolarization of the center of the I-band, where the Z-line is located, produced a local contraction, with the A-band on either side moving toward the center of the I-band (Fig. 5.2). The extent of inward spread of contraction varied directly with the strength of the depolarization. From this finding they concluded that the "influence of the action potential" spread inward along the Z-line (known eponymously as Krause's membrane) and the title of their first paper was "Function of Krause's membrane" (Huxley and Taylor, 1955) At this time, the internal membranes were unknown, at least to most of the scientific community. It was made known to Huxley abruptly when

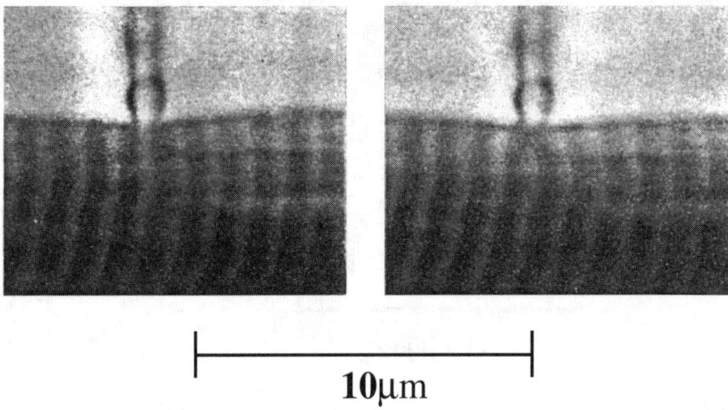

Figure 5.2. *Localized contractions produced by local depolarizations only over the center of the I-band of frog fibers. From Huxley and Taylor (1958), with permission.*

he first presented this work at a meeting. At the end of his presentation, Robertson presented an electron micrograph of muscle showing a set of tubular membranes running across a lizard muscle fiber near the A-I junction, at some distance from the Z-line.

It is now known that muscle has two distinct sets of internal membranes, t-system and sarcoplasmic reticulum. Fig. 1.6 is a three dimensional representation of their relationships made by Peachey as part of his Ph.D. work in Huxley's laboratory. A more recent electron micrograph of a longitudinal section of muscle in Fig. 5.3 clearly shows the two structures. The junctions of the terminal cisterns of the sarcoplasmic reticulum with the t-tubules closely resemble some of the membrane triads seen by Ebashi and Lipmann (Fig. 5.1). But the tissue fixatives used in the early electron microscope studies frequently destroyed membranous structures. The picture shown by Robertson at the end of Huxley's seminar showed only the t-tubules, the sarcoplasmic reticulum having been destroyed in fixation (Robertson, 1955).

Robertson's picture suggested that it might be the tubular membranes, rather than the Z-line, that carried the influence of the

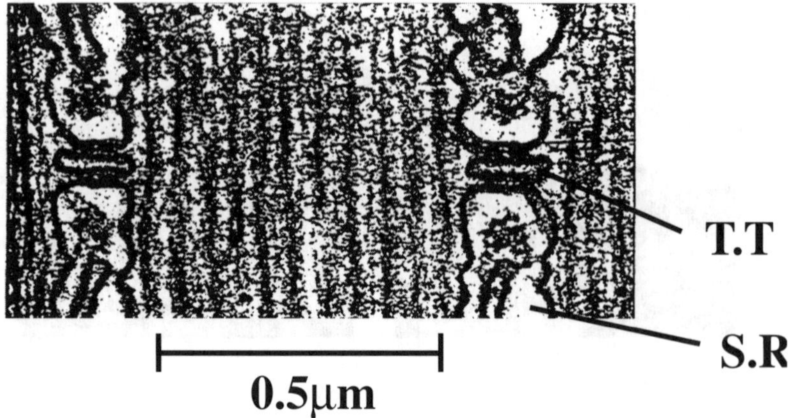

Figure 5.3. *Electronmicrograph showing the t-tubules (T.T.) and sarcoplasmic reticulum (S.R.)*

action potential inward. Proof came from an outgrowth of further structural studies. While a single set of t-tubules run inward at the Z-line in frog fibers, in many muscles, including crab and lizard muscle, there are two sets of t-tubules in each I-band near the A-I junctions. Huxley and his colleagues therefore repeated their experiments in crab and lizard muscle and found that the sensitive area was not at the Z-line, as in frog, but at the A-I junction. Furthermore, the finding that the membranes were tubular suggested that there would be discrete openings at the surface. Huxley and Taylor (1958) repeated their studies making use of this new knowledge. In their original experiments they had found that some Z-lines failed to respond when depolarized. They accepted this failure originally as the outcome of technically demanding experiments. The suggestion that there might be discrete openings to the t-system offered another explanation. They therefore modified their apparatus so that the fibers could be rotated about their long axis and stimulated at various points around the circumference. Using this new apparatus, they showed that only some parts of the circumference were sensitive, as would be expected if there were discrete areas where the t-system met the sarcolemma.

The observation that the t-system consisted of tubules connected to the surface membrane raised the question of whether the tubules were open to the extracellular fluid. The question was answered by studying the diffusion of extracellular substances that would not pass through membranes. H.E. Huxley (1964) used the electronmicroscope to show that ferritin, a large, electron-dense protein, diffused in and out of the fibers as expected of tubules open the extracellular space. Endo (1964, 1966) working as a post-doctoral fellow in A.F. Huxley's laboratory, arrived at a similar conclusion using fluorescence microscopy and a fluorescent dye. The finding that the t-tubules open onto the extracellular space indicate that the membranes are continuous with the surface membrane.

The finding that the t-tubules were continuous with the surface membranes suggested that the influence of the action potential spread inward by an electrical depolarization of the tubular membranes. This suggestion raised the further question of whether the electrical conduction was a regenerative action potential or whether it occurred by simple passive spread. The original cine films of Huxley and Taylor's (1955, 1958) showed graded responses (Fig. 5.2) suggesting a passive, decremental spread. On the other hand, it was possible that the fibers were not fully responsive, perhaps because of the extracellular fluid composition used for the interference microscope. In addition, they may not have attempted to work with fully responsive fibers because an action potential of the surface membrane would induce a twitch of the entire fiber, causing it to tear itself on the pipette. The issue of the inward spread of the impulse was therefore re-examined by Costantin (1974), who depolarized the local membranes to just short of threshold for a contraction, and then applied a small second pulse to produce a brief local contraction. He found that such stimulation caused the local contraction to pass across the fiber in a manner suggesting a regenerative action potential in the t-system.

Sarcoplasmic reticulum

The sarcoplasmic reticulum regulates activation by sequestering calcium during relaxation and releasing it to produce activation.

The experiment of Ebashi and Lipmann (1962), described above, showed that the ion pumps of these membranes could reduce calcium to less than 0.1 uM against concentration gradients exceeding 10,000:1. Analysis of the ATPase of the sarcoplasmic reticulum has subsequently shown that two calcium ions are pumped for each ATP molecule hydrolyzed. It is also now known that calcium within the sarcoplasmic reticulum is partially bound to calsequestrin, a calcium binding protein inside the membrane vesicles.

CONTRACTILE PROTEINS

Weber's demonstration of the activating effect of calcium led the husband and wife team of Ebashi and Ebashi (1964) to investigate the influence of calcium on the contractile proteins. In the course of purifying thin filaments they found a distinct difference between native thin filaments isolated directly from muscle and reconstituted filaments made by first purifying the g-actin monomers and then causing them to link into filamentous f-actin. The reconstituted thin filaments were calcium insensitive and would induce MgATP hydrolysis by myosin irrespective of the calcium ion concentration. On the other hand, native thin filaments were calcium regulated. They found the difference to be due to the presence of a protein complex called troponin-tropomyosin that was present on native thin filaments but removed by the process of dissolving and re-constituting f-actin. Subsequent experiments showed that tropomyosin is long protein that binds to the f-actin, covering 7 actin monomers. Troponin is now known to be a complex of three proteins that binds to the tropomyosin. When only the tropomyosin is bound to the reconstituted thin filaments, ATP hydrolysis proceeds at a rate that is slightly faster than that obtained in the presence of pure f-actin. Perhaps the tropomyosin serves as a guide to prevent the myosin attaching in awkward states that would impede movement through the power stroke. When the troponin complex is bound to the tropomyosin, the ATPase is inhibited unless calcium is present. The relationships between the ATPase rate and calcium ion concentration expressed as the negative logarithm of the ion concentration (**pCa**) is shown for various protein preparations in Fig. 5.4

These observations led to the suggestion that relaxation was caused by a steric blocking of the crossbridge binding site on actin by the tropomyosin and that activation occurred when calcium binding forced tropomyosin to move aside. Evidence of the postulated movement of the tropomyosin was first provided by H.E. Huxley (1973). He stretched muscles to the point where there was no overlap of the thick and thin filaments, so that there was no force developed on activation, and then stimulated them while observing the x-ray diffraction pattern. Changes in the pattern were taken to represent an activation-induced, lateral movement of the tropomyosin to a position nearer the groove in the actin helix. A schematic representation of the thin filament cross-section and the postulated tropomyosin movement is shown in Fig. 5.5.

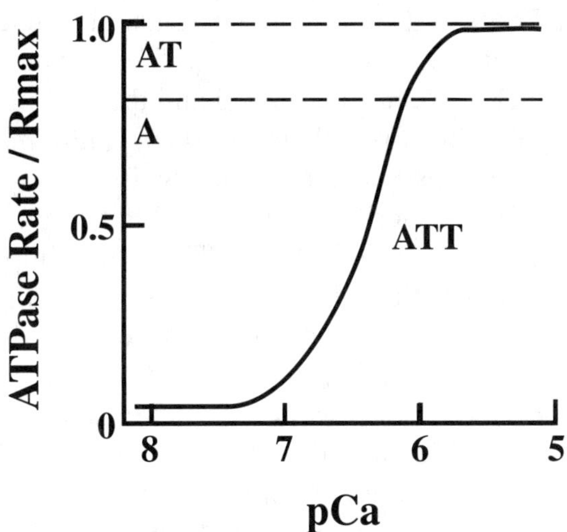

Figure 5.4. *ATPase rates for myosin mixed with purified actin filaments (A), thin filaments with just tropomyosin (AT), and thin filaments plus tropomyosin and troponin (ATT) as a function of pCa.*

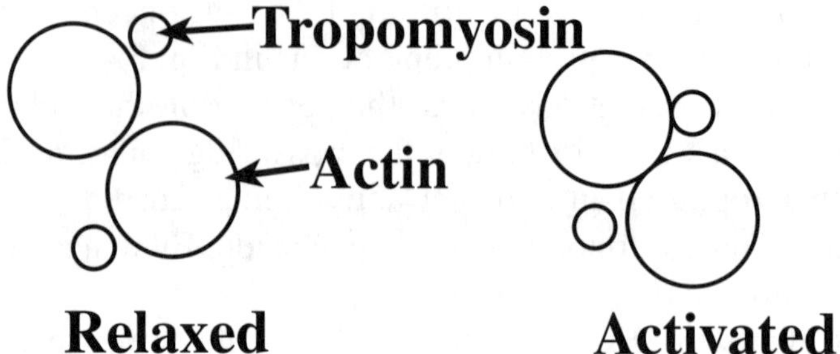

Figure 5.5. *Schematic cross-section of a thin filament showing the tropomyosin shift due to activation.*

CALCIUM TRANSIENTS

The biochemical experiments described above showed that calcium was both necessary and sufficient to produce contraction. Proof that calcium was actually released into the cytoplasm during activation and removed in advance of relaxation required the development of rapid methods of assaying free calcium in the cytoplasm. Most such methods are optical and the first of these experiments was published by Jöbsis and O'Connor in 1966. They injected a calcium sensitive dye, murexide, into the lymph sac of a toad and the next day sacrificed the animal to study the optical absorption changes in its muscle fibers. Their superimposed records of twitch force and optical signal (Fig. 5.6) shows that intracellular ionized calcium rises in advance of force and falls nearly to baseline by the time peak force is achieved.

Jöbsis and O'Connor's work was published in a journal intended for rapid publication of preliminary results, where it is expected that a more complete description will follow later. A more extensive description never appeared. It seems likely that the success of these experiments depended on an element of chance, both in obtaining satisfactory optical absorption records from moving fibers and in the diffusion of the dye from the lymphs sac into the fibers. Further records of this sort therefore had to await the devel-

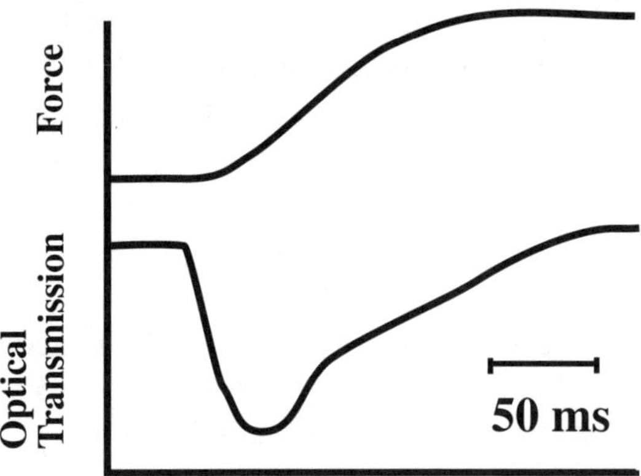

Figure 5.6. *Time course of force and optical signal due to intracellular calcium following a single stimulus. Adapted from Jöbsis and O'Connor (1966).*

opment of more reliable calcium indicators. In spite of the limitations of their methods, the records produced by Jöbsis and O'Connor are now known to be typical of the calcium transients.

In 1962, Shimomura and Johnson described a protein, aequorin, that is responsible for the luminescence of jellyfish. The protein emits light when calcium binds to it, and the flash of the jelly fish is activated in the same way as contraction in muscle, calcium is released into the cytoplasm. Ridgway and Ashley (1967) first used aequorin to observe the calcium transients in muscle fibers. Their records were similar to those in Fig. 5.6.

A difficulty with optical experiments in muscle is that movement can alter with the amount of tissue and dye in the optical path. Thus, measurements of optical absorption are often confounded by movement. An advantage of aequorin is that it luminesces, so that it is only necessary to capture the same fraction of light from the specimen at all times, a technically easier task than the absorption measurements of Jöbsis and O'Connor. It does, however, have several disadvantages; the intensity of its luminescence does not vary linearly with calcium ion concentration, it must be isolated from jelly fish, and it must be pressure-injected into cells. Optical measurements of intracellular calcium have been greatly facilitated by the recent invention of fluorescent dyes that shift their absorption and/or emission spectra when calcium binds. The amount of calcium bound, indicating the concentration of ionized calcium, thus causes a change in the ratio of the absorption and/or emission at two wavelengths. Such **ratiometric** methods provide signals that are proportional to the calcium ion concentration and independent of both the amount of dye in the optical field and the size of the field. As with the aequorin signal, the time course of the calcium transient measured with these is similar to that first found by Jöbsis and O'Connor.

SKINNED FIBERS

The first skinned fiber preparation was described by Natori in 1954. He isolated segments of muscle fibers under oil and then peeled off the surface membrane. His skinned fibers exhibited some puzzling behavior. He used a sharp wooden stick to transfer bits of other fibers to the skinned fibers under oil, and the application of this material frequently produced contraction. He also used wooden sticks to apply droplets of salt solutions to the fibers and found that most salts caused contraction, unlike the experiments of Heilbrun and Wiercinski, described above, where only calcium would cause activation. At the time, the role of calcium in activation was unknown. Podolsky later found that droplets on a wooden stick contain calcium leached from the wood, but in kindness to Natori, he never published this finding.

Podolsky did, however, publish several important studies of the role of calcium in activating skinned fibers. Using skinned fibers in oil, he found that calcium, but rarely other salts, caused immediate activation (Podolsky and Hubert, 1961). He also showed that the spread of activation in response to an applied droplet was less than expected from the amount of calcium in the drop and concluded that the spread was limited by the rapid uptake of calcium by the sarcoplasmic reticulum (Costantin and Podolsky, 1965). Using oxalate to precipitate calcium in dense spots that could be seen in electronmicrographs, he and his colleagues showed that calcium sequestered by the sarcoplasmic reticulum is concentrated in the terminal cisterns near the Z-lines (Costantin et al. 1965). Hellam and Podolsky (1969) developed a method for attaching skinned fibers to a force transducer and changing the aqueous environment of the fibers. They used the apparatus to measure the relationship between the ionized calcium concentration and developed force shown in Fig. 5.7. For these experiments they used a newly invented chemical **EGTA** (ethylene glycol tetra-acetic acid) to buffer the calcium. This agent has the advantage of a much greater selectivity for calcium over magnesium, unlike many other calcium binding agents. When calcium is buffered in this way, its level is usually expressed as **pCa**, the negative logarithm of the ion concentration, in a manner analogous to pH.

In the course of their force measurements Hellam and Podolsky noticed that force rose much more slowly than expected from the simple diffusion of calcium into the fibers. The high affinity of EGTA for calcium derives from the slow rate at which calcium dissociates from it. Hellam and Podolsky speculated that the slow

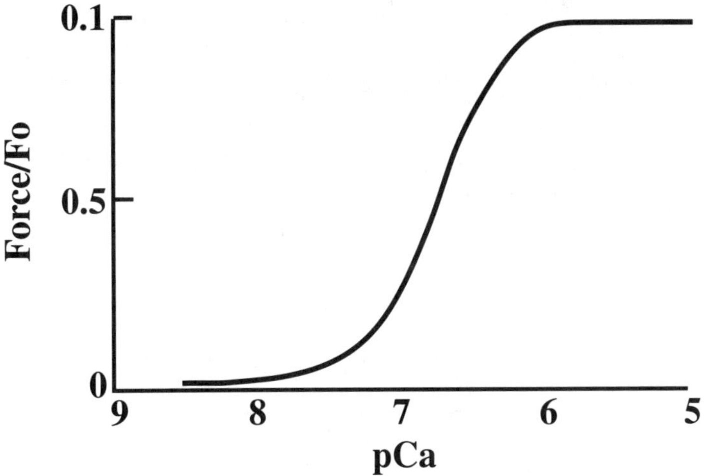

Figure 5.7. *Relationship between developed force and calcium ion concentration expressed as pCa. Adapted from Hellam and Podolsky (1969), with permission.*

rise of force was due to a combination of the slow dissociation of calcium from the buffer and its rapid sequestration by the sarcoplasmic reticulum. To confirm this speculation, Ford and Podolsky (1972a) studied the factors affecting the delay in force development and found that the fibers would accumulate calcium to an average intracellular concentration of about 2 mM before they contracted. More interestingly, they found that the fibers produced transient contractions, presumably from the release of sequestered calcium, when the buffer concentration was low. In spite of the absence of the surface membrane, the contractions appeared to be coordinated along the fiber length. This led to the speculation that the calcium was released by a regenerative process in which calcium in the intracellular space produced the release of additional calcium from internal stores. Proof of this speculation came from two types of experiments. In the first, the fibers were loaded with calcium and then exposed to a low concentration of free calcium, which produced an immediate contraction followed by relaxation

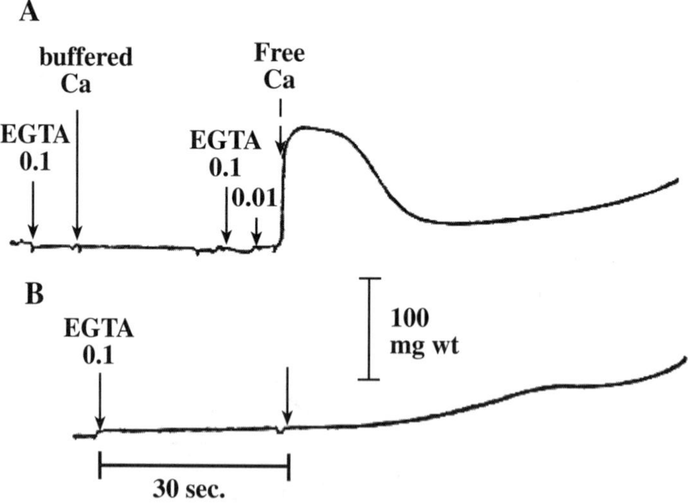

Figure 5.8. *Force records from a skinned fiber upon immersion in 0.1 mM unbuffered calcium solution with preloading in buffered calcium (A) or no preloading (B). From Ford and Podolsky (1970), with permission.*

(Fig. 5.8 A). There was no immediate contraction if the fibers were not pre-loaded (Fig. 5.8 B). These findings suggested that the sarcoplasmic reticulum was capable of sequestering calcium more rapidly than it could diffuse into the fibers and that the initial contraction was due to the release of stored calcium. This suggestion was confirmed by the measurements of radioactive ^{45}Ca movements shown in Fig. 5.9.

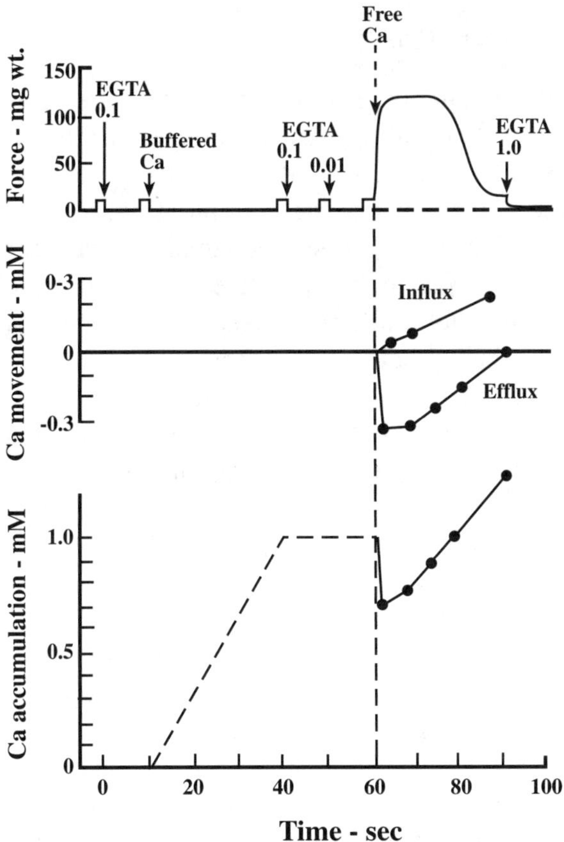

Figure 5.9. *Time course of ^{45}Ca uptake and release by skinned fibers during calcium loading and stimulation with low levels of free calcium. Adapted from Ford and Podolsky (1972b), with permission.*

Almost simultaneously with Ford and Podolsky's work, similar conclusions were derived from skinned skeletal fiber responses to caffeine by Endo et al. (1970), who gave the phenomenon its name, **calcium-induced calcium release**. Both groups found that this release required unphysiologically low magnesium ion concentrations, raising the question of why such a distinctive mechanism was present if it were not physiological. It now appears that it is the physiological mechanism in cardiac but not skeletal muscle.

SIGNAL TRANSMISSION

While the mechanisms described above have been known for a quarter century, the nature of the signal transmission between t-system and sarcoplasmic reticulum was discovered only recently, using the techniques of genetic manipulation, and may not yet be completely understood. Three mechanisms have been postulated. Two of these are likely to be operative in striated muscle; chemical transmission by the calcium-induced calcium release appears to operate in cardiac muscle and mechanical transmission in skeletal muscle. The third mechanism, electrical coupling, seems highly unlikely but will be described here because the evidence against it represents both the types of experiments and logic employed in these studies.

Electrical coupling

It seems highly likely that the calcium flow from the sarcoplasmic reticulum carries a net charge that polarizes the membranes. This likelihood is supported by optical studies suggestive of such membrane voltage changes. These studies include both the use of voltage-sensitive dyes that bind to membranes (Bezanilla and Horowicz, 1975) and early birefringence changes following an action potential of the surface membrane (Baylor and Oertliker, 1975). If the membranes increased their calcium conductance in response to voltage changes, then the polarization caused by the calcium release would produce an all-or-none release. Such a regenerative response does not occur, however. When the muscle is activated by a sustained depolarization of the surface membrane, both the contractile response (Hodgkin and Horowicz, 1960) and the optical signals (Bezanilla and Horowicz, 1975; Baylor et al., 1975) return to rest levels when the surface membrane is returned to its resting potential. Furthermore, since the membrane area and current carried by calcium in the sarcoplasmic reticulum is much greater than the current flow and membrane area of the surface membrane, the continued control of calcium release by the surface membrane suggests that signal transduction must be unidirectional. If this were not the case, if there were bidirectional transmission, the greater current flow in the sarcoplasmic reticulum would maintain the depolarization of the surface membrane. Taken together, these considerations argue strongly against bidirectional, electrical coupling.

Two examples of unidirectional signal transduction are chemical transmission and mechanical coupling.

Mechanical coupling

It is possible to disconnect the t-system of muscle from the surface membrane by the technique of **glycerol shock**. A fiber placed in a saline solution containing high concentrations of glycerol shrinks transiently because the higher osmotic concentrations in the extracellular fluid cause water to flow out of the fiber. After a short time, however, the glycerol diffuses into the fiber and water follows, restoring the normal volume. The reverse happens when the glycerol is removed from the extracellular fluid; the fiber swells transiently and then recovers its normal volume. If the swelling is sufficiently great, the t-tubules are electrically disconnected from the surface membrane (Eisenberg and Gage, 1967). By comparing the membrane properties before and after glycerol shocking it is possible to deduce the electrical properties of the t-system. One important property is the **gating charge movement**.

In the course of their description of the ionic basis of the action potential, Hodgkin and Huxley postulated the existence of small, intramembrane charge movements resulting from movement of structures responsible for gating the membrane channels. They were unable to observe such charge movements because the tiny electrical signals were buried in the noise of their single recordings. The discovery of these gating currents required the development of digital recording techniques that could be used to average many repetitive signals. With this **signal averaging** technique, random noise is reduced to the point where the gating currents can be measured. They are recognized as a fixed charge movement that occurs only over a defined range of voltages. An obvious place to look for charge movements was the t-system of muscle, and Chandler and his colleagues found such movement in abundance (Schneider and Chandler, 1973; Chandler et al., 1976 a&b).

Their finding of these movements raised the question of their function. It was possible, for example, that the movement simply reflected the opening of membrane channels that allowed activating ions to flow into the muscle. For several reasons this seemed unlikely. The charge movement appeared to be associated with a calcium channel, but the calcium flow developed much slower than would be required by the brief action potentials that passed through the muscle. In addition, it was possible to remove calcium from the extracellular space without immediately interrupting mechanical activation after a depolarization. For these reasons, Chandler et. al. offered a different interpretation. They suggested that the charge movement might reflect the movement of a mechanical link between the t-system and sarcoplasmic reticulum. They postulated that one end of the link was attached to a charge buried in the t-system membrane, such that it moved in response to voltage changes in the membrane. The other end was imagined to serve as a gate in the calcium channel of the sarcoplasmic reticulum membrane in such a way that movement of the link opened and closed the channel. This mechanical linkage would cause unidirectional transmission of the signal as long as there were not a charge in the portion of the link within the sarcoplasmic reticulum membrane. Although

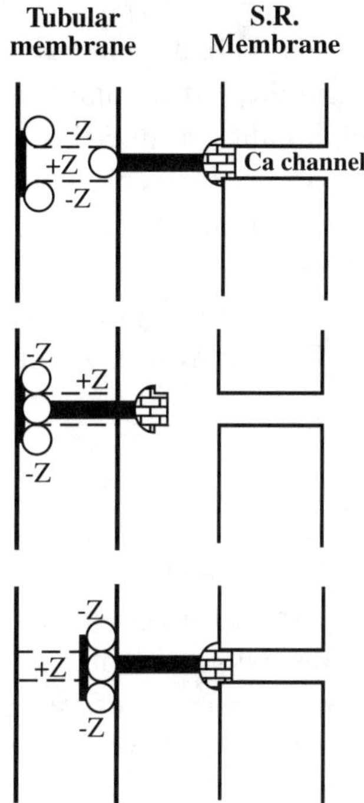

Figure 5.10. *Schematic representation of the mechanical link between t-system and sarcoplasmic reticulum in skeletal muscle. From Chandler et al. (1976b), with permission.*

the postulate was mainly speculative, it appears to be correct. A schematic representation of their linkage is shown in Fig. 5.10.

Chemical transmission

Chemical transmitters are an obvious example of unidirectional signal transmission, and the calcium-induced calcium release has always been an obvious candidate for the mechanism of signal transmission in muscle. The problem with this postulate is that the conditions required to elicit the response in skeletal muscle were unphysiological. To demonstrate the release in skinned skeletal fibers, they must first be loaded with calcium and the magnesium ion concentration must be less than physiological. In spite of these discouraging observations, Fabiato and Fabiato (1978) persisted in miniaturizing the techniques of skinned skeletal fibers to adapt them to the very much smaller skinned cardiac muscle cells. They then showed that small and perhaps physiological calcium releases could be obtained under conditions approaching the normal intra-cellular conditions. At the time it seemed a bit whimsical to pursue a mechanism proven to be unphysiological in the preparation where it was discovered, but it turned out that they were correct.

Two mechanisms

It is now known that calcium is released from the sarcoplasmic reticulum through a channel to which ryanodine binds tightly. It is therefore called the **ryanodine receptor**. It is also known that there is a receptor in the t-system that binds the calcium channel blocker, dihydropyridine, and is therefore called the **dihydropyri-dine receptor**. This receptor is similar in structure to the L-type calcium channel found in many tissues but it is structurally and functionally different in cardiac and skeletal muscle. In cardiac muscle this receptor functions as a true calcium channel and ad-mits calcium to a space very near the ryanodine receptor that re-leases calcium by the calcium-induced calcium release. If calcium is withdrawn from the extracellular fluid, or if cadmium is added to block the calcium channel, contraction is inhibited. Depolariza-tion of skeletal muscle also causes the calcium channel to open, but the opening is slower, and much less calcium flows. In addi-tion, the muscle will continue to contract for a long time after cal-cium is withdrawn from the extracellular fluid.

On the wall of the Marine Biological Library in Woods Hole there is this quote from Pasteur: "In science, chance favors only the prepared mind." This statement aptly applies to Kurt Beam, whose training was in electrophysiology. When he learned that Powell had discovered a recessive mutation in mice that was lethal because their muscles failed to contract, he obtained samples of the fetal muscles. When these muscles were explanted in tissue cultures, they formed **myotubes**, long syncytia formed by the fusion of many cells. Beam et al. (1986) found that these cells lacked dihydropyridine receptors. He then enlisted Tanabe and others to inject DNA encoding the receptor from normal mouse cells. After a few days in culture, these "rescued" myotubes began to contract spontaneously and to show electrophysiological signs of having dihydropyridine receptors (Tanabe et al., 1988). Subsequent experiments showed that if the DNA for cardiac receptors were injected, the cells behaved as cardiac cells. That is, the contraction could be stopped by withdrawing calcium or adding cadmium to the extracellular fluid, and the calcium channels appeared to have a large ion conductance and a rapid opening rate. By contrast, when skeletal receptor DNA was injected, the cells behaved as skeletal muscle, being relatively insensitive to the calcium in the extracellular fluid, and the calcium channels had slow opening kinetics and a low ion conductance (Tanabe et al., 1990). Finally, they made hybrid combinations of the receptor to show that a peptide loop on the cytoplasmic side of the membrane conferred the skeletal properties on the receptor (Tanabe et al., 1990).

Taken together, these experiments demonstrated that signal transmission from t-system to sarcoplasmic reticulum occurs by different mechanisms in cardiac and skeletal muscle, as had been postulated earlier but not proven.

COOPERATIVITY

It has been known for a long time that filament sliding in muscle causes inactivation. An example is shown by the records superimposed in Fig. 5.11. An **afterload** is a load that a muscle will shorten against when the developed force reaches the level of the load and which serves to lengthen the muscle as it relaxes. Fig. 5.11 show the length and force records superimposed for an isometric twitch (solid) and an afterloaded (dashed) contraction. As indicated, the force lasts longer at the end of the isometric contraction than at the end of an afterloaded contraction. Investigations into this **shortening deactivation** have generally shown that

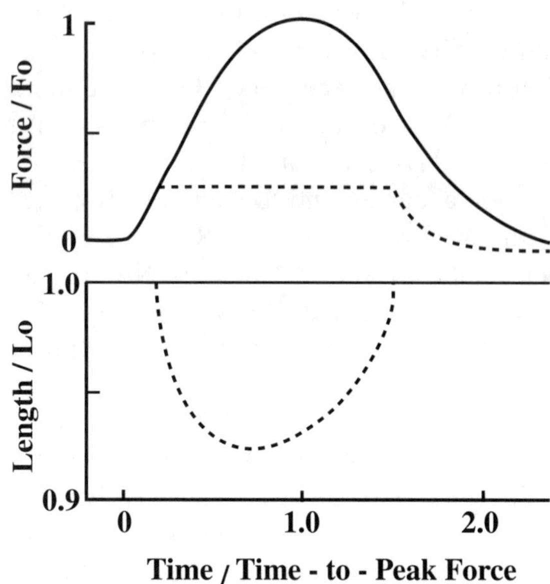

Figure 5.11. *Shortening deactivation. The isometric contraction (interrupted curves) last longer than the contraction that includes shortening (solid curves).*

the deactivation varies directly with the amount of shortening, with the speed of shortening, and with time in the twitch. Studies have also shown that lengthening has similar effects (Brady, 1965). The responsible mechanism was first described by Bremel and Weber (1972). They found that calcium binding to thin filaments was increased in the presence of myosin and low concentrations of ATP. Since the ATP limitation is likely to detain bridges in rigor-like states, they reasoned that these rigor bridges increased the calcium affinity of the thin filaments. They further argued that such responses were expected on "thermodynamic grounds." Because calcium binding to troponin caused a physical shift in the position of the tropomyosin, it seemed likely that a similar physical shift in the tropomyosin, as might be caused by an attached crossbridge, would be transmitted backward to the troponin in such a way that the binding of calcium would be facilitated.

As part of his Ph.D. studies in A.M. Gordon's laboratory, Godt (1974) showed that the force-pCa curve in muscle was displaced to the left in the presence of a low ATP concentration (Fig. 5.12) and interpreted this as being due to the same cooperative mechanism. The more firmly attached "rigor" bridges increase calcium affinity of the thin filaments. Finally, Gordon and Ridgway (1978), as well as other (see Vanderboom et al., 1998) have shown that there is a transient increase in myoplasmic calcium when the fibers shorten rapidly. Such a transient calcium increase suggests a decreased myofilament affinity for calcium during the shortening, when fewer crossbridges are attached.

As expected, there are several types of cooperativity. The first is the normal activation mechanism in which the binding of calcium promotes the binding of crossbridges. The second is the reverse mechanism by which the binding of crossbridges promotes the binding of calcium. The third mechanism is a crossbridge-crossbridge interaction in which the binding of one crossbridge holds the tropomyosin aside so that another crossbridge can bind, even in the presence of very low calcium ion concentrations.

The force-pCa curves in Fig. 5.12 are steeper than would be expected from simple calcium binding, and this provides addi-

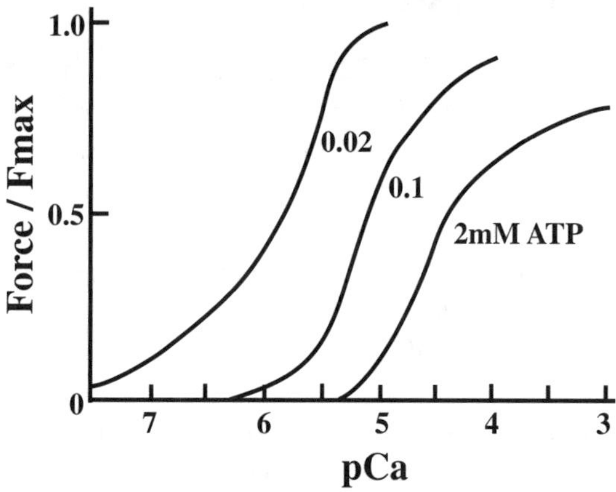

Figure 5.12. *Force-pCa curves in the MgATP concentrations (in mM) indicated by numbers next to curves. Derived from the data of Godt (1974) and Ferenczi et al. (1984).*

tional evidence of cooperativity. The extent of the cooperativity is often quantified using the **Hill coefficients** derived from fitting the curves. As mentioned at the end of chapter 2, these coefficients were first developed by A.V. Hill to describe the cooperative binding of oxygen to hemoglobin.

An interesting feature of the cooperative mechanism is that it extends beyond the domain of a single tropomyosin molecule. As described with Fig. 5.4 above, removal of troponin activates filaments in the absence of calcium. Thus, skinned fibers can be activated continuously by extraction of the troponin. Brandt et al., (1984) and Moss et al. (1986) have shown that removal of only one or two troponins per thin filament greatly increases cooperativity of calcium binding. Moss et al. (1985) have speculated that this is due to a physical contact at the ends of neighboring tropomyosins which causes them to move in concert.

A final point about the cooperative activation is that it explains shortening deactivation in twitch contractions. This is very important in cardiac muscle, which produces only twitches because the process of shortening abbreviates the contraction.

SUGGESTED READING

BREMEL, R.D. and WEBER, A. (1972) A cooperation within actin filament in vertebrate skeletal muscle. *Nature New Biol.* **238**: 97–101.

ENDO, M., TANAKA, M., and OGOWA, Y. (1970) Calcium induced release of calcium from the sarcoplasmic reticulum skinned skeletal muscle fibres. *Nature.* **228**: 34–36.

FORD, L.E. and PODOLSKY, R.J. (1970) Regenerative calcium release within muscle cells. *Science.* **167**: 58–59.

HUXLEY, A.F. and TAYLOR, R.E. (1955) Function of Krause's membrane. *Nature.* **176**: 1068.

TANABE, T., BEAM, K.G., ADAMS, B.A., NIIDOME, T., NUMA, SHOSAKU, S. (1990) Regions of the skeletal muscle dihydropyridine receptor critical for excitation-contraction coupling. *Nature.* **346**: 567–569.

Chapter 6

LENGTH DEPENDENCE OF ACTIVATION

It is important that all muscles function over a length range where the force increases with length. One reason for this need is illustrated in Fig. 6.1, which shows the consequences of sarcomere inhomogeneity superimposed on a length-tension diagram. At long lengths, called the **descending limb** of the curve, where force decreases with increasing length, inhomogeneities become unstable. Regions of muscle that have shorter sarcomere lengths are stronger and will stretch the longer and weaker segments, thereby worsening length differences. By contrast, sarcomere length inhomogeneities are reversed when the length-force relation has a positive slope, over the **ascending limb** of the curve. The longer sarcomeres are stronger and can therefore stretch the shorter and weaker sarcomeres. It should also be emphasized that it is the slope of total force, and not just the developed force, that must be positive.

In addition to the need to maintain sarcomere homogeneity, illustrated in Fig. 6.1, cardiac muscle has a further need for an even steeper dependence of force on length. The requirement arises because the tension in the wall of a hollow vessel is proportional to the vessel radius. This relationship, which was first determined by the French mathematician Laplace and is called the **Laplace relationship**, is explained in greater detail in Chapter 14. It is mentioned here to emphasize the importance of the mechanisms responsible for the steep ascending limb in cardiac muscle. This steep ascending limb is often called the **Frank-Starling** relation-

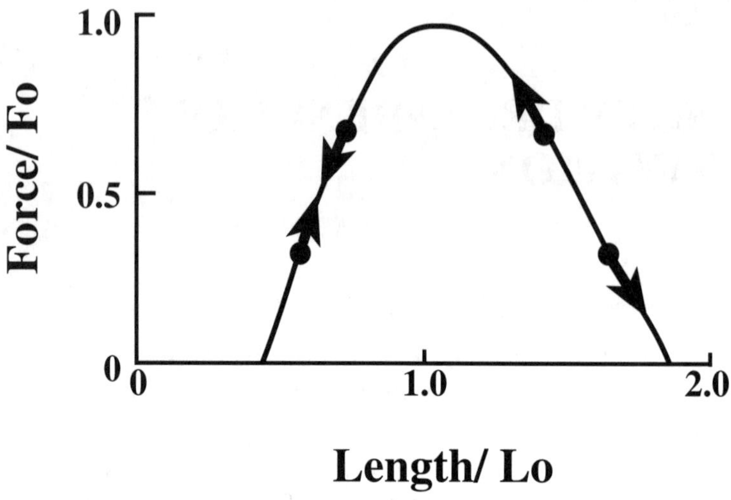

Figure 6.1. *Length-force relation showing that length inhomogeneities are worsened at functionally long lengths and reversed at short lengths.*

ship because Frank, and later Starling, described how the heart is stronger when it begins contracting from a larger volume. To understand why the Frank-Starling relationship is so important, consider what happens when the heart skips a beat. Since blood continues to return to the heart, the first beat after the missed beat begins contracting from a larger volume. The larger radius of the ventricle increases the tension in the muscles of the wall. In the absence of a greater strength of contraction, the greater wall tension would lessen the extent of shortening and produce less emptying, so that the following beat would begin from an even larger volume. In this manner, hearts without a steep ascending limb would soon dilate until they had reached some elastic limit, either rupturing or coming to rest on parallel elastic structures. Of course this does not happen; mechanisms have evolved to increase the strength of contraction at the longer muscle lengths.

There are several mechanisms that contribute to the positive slope of the length-tension relationship. Perhaps the evolution of a multiplicity of such mechanisms has occurred because the phenomenon is so vital to muscle function. The purpose of this chapter is to describe how the ascending limb arises.

FILAMENT INTERACTIONS

Although filament interference may be the least important physiological mechanism related to the Frank-Starling relationship, it is described first because it achieved great prominence as the result the length-force relationship first described with Fig. 3.1 in Chapter 3 and shown again in Fig. 6.2. As indicated, force reduction begins at about the length where the thin filaments meet in the center of the sarcomere and the slope of the relationship becomes steeper at lengths where the thick filaments abut the Z-lines.

The relationship between force and thin filament overlap with crossbridges at lengths longer than 2.0 µm demonstrated very convincingly the role of crossbridges in force development. It was so persuasive that filament overlap at short lengths was accepted uncritically by some as the dominant cause of the positive slope of the relationship at short lengths. Additional support for this acceptance came from the knowledge that a muscle stimulated to contract below its minimum rest length will elongate to at least this minimum length on relaxation. The responsible mechanism is not known, but it seems to be a passive property, sometimes called a **restoring force**. Sarcomeres appear to distort an elastic element that restores minimum length as it returns to its rest state during relaxation. The finding that force declined with increasing overlap and that the decline became steeper when the thick filaments met the Z-line suggested that some filament arrangement produced the restoring force.

The relationship between filament overlap and force at sarcomere lengths below 2.0 µm in Fig. 6.2 suggest several possible mechanical mechanisms for the positive ascending limb of the

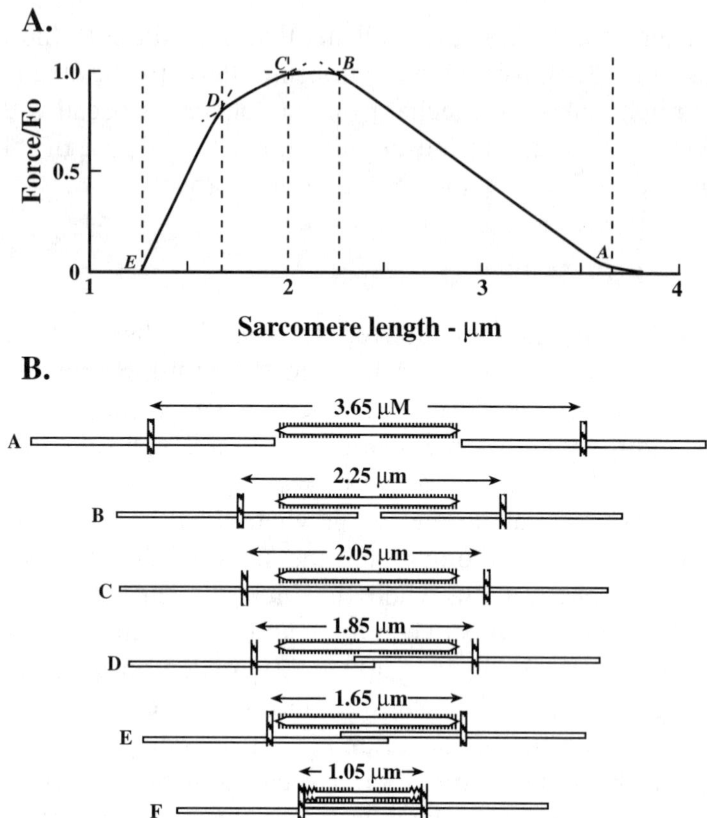

Figure 6.2. *Sarcomere length-force relationship. Filament orientation at each of the corners in the graph (A) are shown in B. From Gordon et al. (1966). Same as Fig. 2.3.*

length-tension curve. These mechanisms include: 1) mechanical resistance to shortening by the thin filaments meeting in the center of the sarcomere; 2) mechanical resistance to shortening when the thick filaments abut the Z-line; 3) interference with the normal force-generating mechanism by thin filaments entering the zone of overlap from the other side of the sarcomere; and 4) crossbridges operating in the lengthening direction by attaching to thin filaments with an opposite polarity after they have crossed the center of the sarcomere. Huxley considered all of these possible mechanisms as well as a few others. In his original proposal of the cross-bridge theory, Huxley (1957, p. 308), mentioned evidence from earlier work of Katz (1939) that fibers became partially inactivated at short lengths. He also considered the possibility (Huxley, 1965) that volume changes observed in fibers contracting to short lengths (Sato, 1954) might impose an osmotic load on the contractile elements. Thus, there appeared to be several possible mechanisms that could cause the positive slope of the length-force relationship, and it now seems likely that each contributes, at least slightly. The mechanisms that rely on the filament relationships in Fig. 6.2 are likely to make only a small contribution under physiological conditions, however, because the positive slope often extends to sarcomere lengths greater than those where the thin filaments meet, as explained in detail below. It now seems most likely that the major mechanism responsible for the positive slope is length-dependent activation.

DECREASED ACTIVATION AT SHORT LENGTHS

The data in Fig. 6.2 suggesting filament interactions as the cause of the ascending limb were obtained by A.F. Huxley in collaboration with two post-doctoral fellows, A.M Gordon and F.J. Julian. This is mentioned here because a major study suggesting an alternative explanation of the positive slope of the length-force relationship also came from two other post-doctoral fellows in Huxley's laboratory, S.R. Taylor and R. Rüdel. As was his policy, Huxley elected not to be an author on their papers because he did not participate in the actual experiments and only gave advice.

To understand Taylor and Rüdel's experiments, it is first necessary to understand the methods developed by H. Gonzalez-Serratos (1971), a Ph.D. student in Huxley's laboratory. The object of the Gonzalez-Serratos study was to measure the rate of inward spread of activation. Two methods were critical to his success. The first was the use of a microscope with a wide numerical aperture to cut a thin optical section through the specimen. Only a portion of a specimen in the optical field of a microscope is in focus, and the wider the numerical aperture, the narrower the depth of focus. With a very wide numerical aperture it is possible to focus on a thin section through a fiber and thus observe structural differences across its diameter. The second method relied on the observation that myofibrils become wavy when shortened to lengths less than those where restoring forces cause sarcomeres to maintain a minimum rest length.

If the ends of an unrestrained relaxed fiber are pushed towards each other, the fiber bends. If this bending is prevented by embedding the fiber in a gel, the internal fibrils form microbends or waves. When the fiber is activated, the myofibrils shorten and the waves straighten. Thus, wave straightening can be used to signal activation. Gonzalez-Serratos used high speed cinemicrography to measure the rate at which wave straightening moved to the center of the fibers.

Taylor and Rüdel (1970) extended the methods developed by Gonzalez-Serratos and used more advanced optics to observe finer waves than had been seen previously. They then examined fibers activated at very short lengths. To everyone's surprise, they found that a fiber shortened below 1.6 μm became wavy in its center. This waviness is shown in the right-hand panel of Fig. 6.3. Their

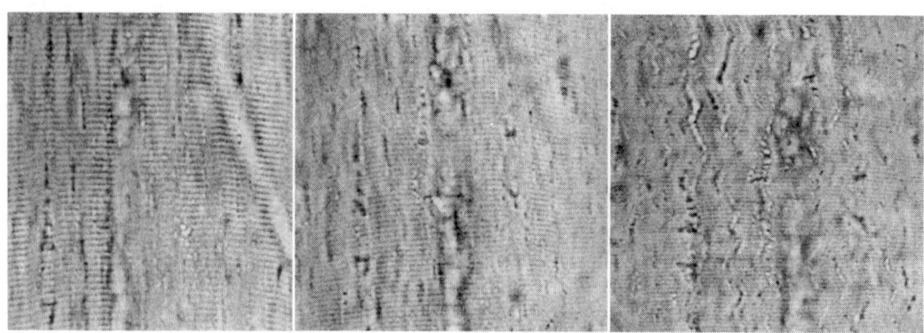

Figure 6.3. *Micrographs of a fiber stimulated to contract at three sarcomere lengths, 2.2, 1.7, and 1.4 μm (from left to right). Adapted from Taylor and Rüdel (1970) using digital images kindly supplied by Dr. S.R. Taylor.*

findings indicated that the center of the fibers became inactivated at short lengths and suggested that length-dependent activation was a major mechanism contributing to the ascending limb of the curve.

An additional experiment by Rüdel and Taylor (1971) provided further convincing evidence of the role of length-dependent activation in generating the ascending limb of the force-length curves. Caffeine is known to enhance activation. They therefore investigated the influence of caffeine on the force-length curves. Their results (Fig. 6.4) showed that caffeine partially reversed the force decrease at sarcomere lengths below 1.6 µm. This was an especially surprising result because the part of the ascending limb of the length-force relationship that is most convincingly attributed

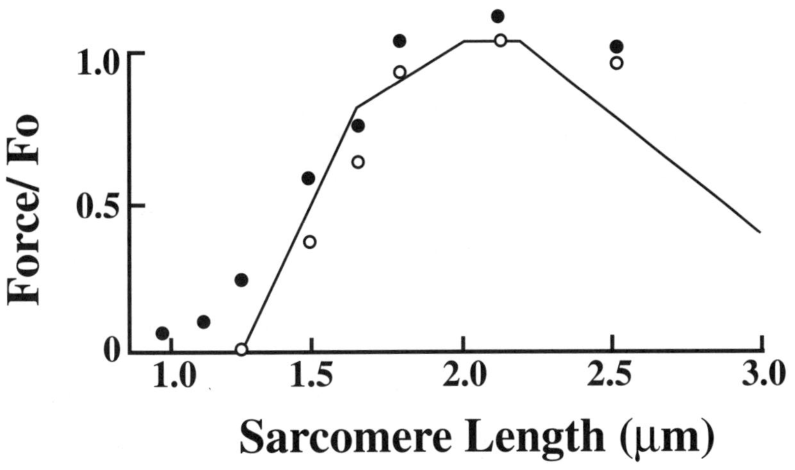

Figure 6.4. *Force generated in the presence (closed circles) and absence (open circles) of caffeine superimposed on the length-force relation of Gordon et al. (1966). Adapted from Rüdel and Taylor (1971) using data kindly supplied by Dr. S.R. Taylor.*

to mechanical effects is the steep slope below 1.6 μm where the thick filaments touch the Z-lines.

The earliest descriptions of single-fiber-length tension relationships are attributed to Ramsey and Street (1940). They showed that fibers develop irreversible and damaging contraction bands when stimulated to contract to very short lengths. More recent electron micrographs show thick filaments projecting through the Z-lines in these contraction bands, suggesting that the irreversible contraction bands result from the thick filaments getting stuck in the Z-line. Rüdel and Taylor's experiments suggest that muscles have evolved strategies to prevent them from shortening to an extent that would cause damage.

A point emphasized in Chapter 10 is that skeletal muscle is protected from overshortening by the skeleton, and that the relevance of avoiding very short lengths applies largely to cardiac muscle. A point made in the rest of this chapter is that, under physiological conditions, both cardiac and skeletal muscle use the mechanisms of length-dependent activation over all of their functional length range.

The responsible mechanisms for the inactivation at short length are not yet fully understood, but at least two classes of mechanism were initially suspected. Both were based on the observation that intact muscle maintains a constant filament lattice volume and a constant overall volume, so that fiber diameter and distance between filaments vary inversely with sarcomere length. The diameter increase at short lengths is likely to stretch the t-system, and this stretch might impede the inward spread of electrical depolarization. The increased lateral filaments spacing could also have a mechanical effect to decrease activation. The best current evidence is that the first of these mechanisms does not occur and that the second is likely to be the dominant mechanism.

Length-independence of calcium release

Taylor and Rüdel's findings led many muscle physiologists to assume that the length-dependence of activation is caused by greater calcium release at longer lengths. In fact, this does not appear to occur, although there are some residual uncertainties about the quantitative aspects of this conclusion, as will be apparent shortly. Most of the studies have been carried out in cardiac muscle, in part because of the greater importance of the ascending limb of the length-force relationship in that preparation. The main reason for the greater interest in cardiac muscle, however, is that the heart has intrinsic mechanisms which regulate the strength of contraction. These changes in **inotropic** state are extremely important to animal physiology and therefore the object of great attention. Since the strength of contraction is also altered by the muscle length, a complete understanding of cardiac function requires that the various mechanisms be distinguished. Much of what we know about these different mechanisms grew out of the work of Allen and Blinks and their colleagues.

John Blinks was one of the pioneers in the use of the luminescent protein aequorin to study calcium transients in muscle. The earliest studies were done in skeletal muscle because the individual fibers are large and more easily injected with aequorin. They also release much more calcium than the tiny cardiac muscle cells and therefore produce a larger light signal. But Blinks was primarily interested in cardiac muscle. For a time he was unable to obtain signals from cardiac cells because their small size precluded large signals. Success came as the result of a collaboration with David Allen, who became a post-doctoral fellow in Blinks' laboratory at the Mayo Clinic after completing his M.D./Ph.D work at University College London. Allen's strategy was to microinject very many individual cells with aequorin until the combined signal from all the cells was sufficiently large to be recorded. Frequently the multiple microinjections took more than 8 hours, so that Allen began recording data at a time in the day when many people were going home. His perseverance paid handsomely, however, and more of his data will be described in Chapters 8 and 17, on fatigue and inotropy.

Surprisingly, Allen and Blinks (1978) found that the amplitude of the calcium transient in cardiac muscle increased directly with interventions that increased inotropic state but decreased at long lengths. These initial studies were done in frog heart muscle, but similar results were obtained later from mammalian cardiac muscles by Allen and Kurihara (1982) and Blinks and Endoh (1986). The later results showed some minor differences, both from the earlier work and from each other, but confirmed the more important conclusion that most of the length-dependence of force generation was not associated with a large change in the calcium transients.

Allen and Kurihara found that the length-dependent force changes occurred in two stages, as had been described earlier (Parmley and Chuck, 1973). The largest change occurred immediately when length was changed with a smaller change in the same direction progressing over several minutes. In these experiments the muscles were stimulated regularly every 5 sec. and the isometric force and light responses are plotted in Fig. 6.5. There was very little change in the light response immediately after the step, when the greatest force change occurred, and a slow change in the light signal in association with the slow force change.

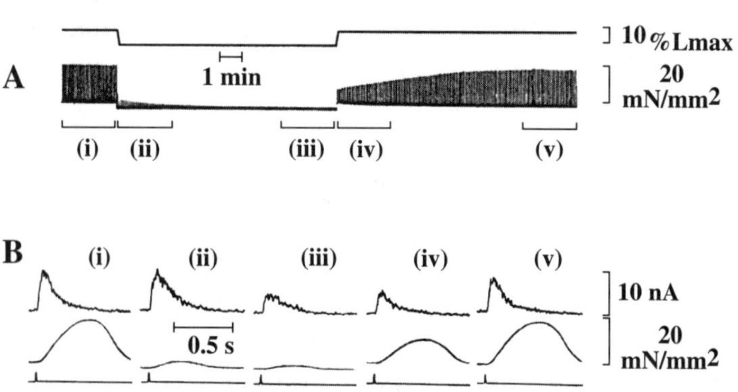

Figure 6.5. *A) Slow records of isometric force (lower) and light (upper), indicating calcium, from cardiac muscle injected with aequorin and stimulated regularly every 5 sec. B) Fast records of signal averaged over the periods indicated by the brackets under the records in (A). Adapted from Allen and Kurihara (1982), with permission.*

The qualifying phrase "very little" is used here because their best interpretation of the data was that the immediate change in light intensity was opposite to the force change, i.e. the amplitude of the calcium transient diminished initially following a stretch. The reason for the uncertainty arose because the records were so noisy that reliable measurements could not be made from a single contraction. The responses from 32 consecutive contractions, indicated by the brackets in Fig. 6.5 A, were therefore signal averaged to obtain the records shown in Fig. 6.5 B. While these signal averaged records suggest that the amplitude of the initial calcium transient is unchanged Allen and Kurihara concluded that the averaging process was likely to obscure changes that occurred immediately following the steps. Furthermore, the finding of a progressive slow change following a step and an unchanged average amplitude for a period immediately following the step suggests a biphasic response. For example, the unchanged average response superimposed on the slow rise following a stretch implied that the immediate change after the step might have been a decline in amplitude, with the unchanged average value then being due to the slow and progressive rise over the averaging period. This interpretation would have been in keeping with the earlier findings of Allen and Blinks that the calcium transients varied inversely with length and force, at least in the short term.

Allen and Kurihara also found that while the amplitudes of the calcium transients were little changed immediately after a length step, the time course of these transients was abbreviated (Fig. 6.6), suggesting that calcium released from the sarcoplasmic reticulum was quickly bound to the thin filaments.

Allen and Kurihara's (1982) experiments, as well as the later experiments of Blinks and Endoh (1986), suggest strongly that the myofilament calcium sensitivity is increased at long lengths. There is an uncertainty over the relationship between the size of the calcium transient and the amount of calcium release because the greater sensitivity of the myofilaments is likely to result from a tighter calcium binding to the thin filaments. The amplitude and

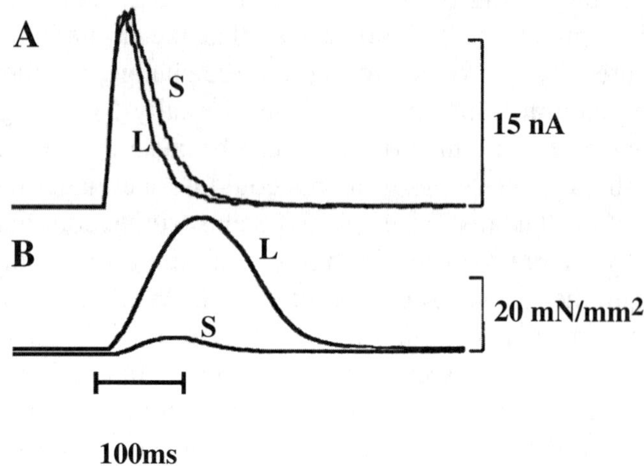

Figure 6.6. *A) Superimposed aequorin light signals at long, L, and short, S, lengths. B) Force records. Redrawn from Allen and Kurihara (1982), with permission.*

time course of the calcium transient results from a balance between calcium release and calcium binding to the thin filaments. Tighter calcium binding to thin filaments is likely to abbreviate the duration of the transient, if not reduce its amplitude. Thus, Allen and Kurihara's finding of an abbreviated calcium transient was expected, and it suggested that the amplitude of the transient would not be a reliable indicator of the amount of calcium released. A greater thin filament calcium affinity at longer lengths might reduce the free calcium transient at these lengths and thus mask a small effect of increasing length to increase calcium release.

The finding that slow force changes are associated with changes of the calcium transient in the same direction suggests that there is a small and direct effect of length on the amount of calcium released from the sarcoplasmic reticulum in the steady state. Such slower changes are likely to result from length-dependent changes in the amount of calcium sequestered in the sarcoplasmic reticulum.

Although the altered calcium binding to thin filaments might influence the size of the calcium transients, one conclusion is very clear from these experiments; the sensitivity of the contractile elements to calcium is strongly length-dependent.

Length-dependence of calcium sensitivity

Using skinned skeletal fibers from *Xenopus laevis*, Endo (1973) showed that the calcium sensitivity of the myofilaments at low calcium ion concentrations reaches a maximum at sarcomere lengths of about 3.0 μm (Fig. 6.7). At maximally activating concentrations of calcium the force varies with filament overlap. At partial levels of activation, however, the steepness of the negative relationship declines and then becomes positive, reaching a peak at sarcomere lengths of about 3.0 μm. This length correlates ap-

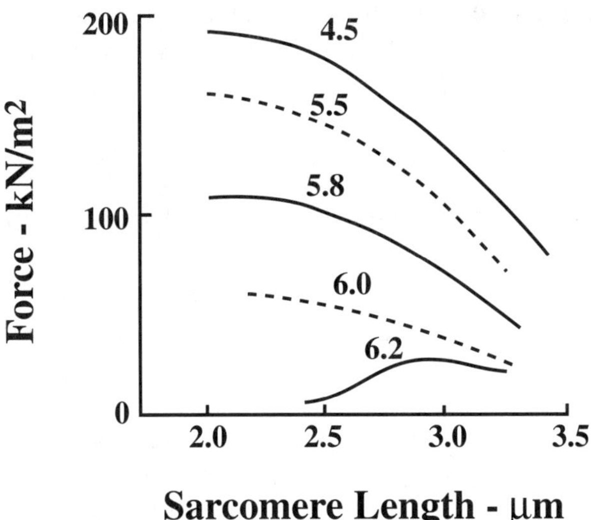

Figure 6.7. *Isometric force vs. sarcomere length at different calcium concentrations. Numbers next to curves indicate Calcium concentration in μm. Adapted from Endo (1973).*

proximately with the position of the peak in the physiological length-force relationship in mammals, described below.

The mechanisms responsible for this length-dependence of calcium sensitivity are not known. The constant volume behavior of intact fibers is not seen in skinned fibers. Diameter varies inversely with sarcomere length, but to a much lesser extent than in intact fibers (Matsubara and Elliott, 1972). There are two contrary points that might be made from this observation. It might be taken as evidence that diameter changes are not responsible for the length-dependence of activation in intact muscle. On the other hand, the variations in force shown in Fig. 6.6 are not very large. There might therefore be a stronger length dependence of calcium sensitivity in intact fibers where the diameter changes are larger.

A final point to be made about the data in Fig. 6.7 is that it is obtained in a steady state. Since equilibrium binding is determined as the ratio of the rate of binding to the rate of unbinding, it is possible that the relatively small steady-state changes reflect larger changes in the dynamic rate constants. Thus, it is possible that much larger changes in calcium sensitivity would occur during twitches, when rates rather than steady-state ratios determine the force level.

Possible causes of length-dependent calcium sensitivity

Several mechanisms for the phenomena described here have been proposed. One possibility, alluded to above, is that activation or crossbridge attachment is increased when the filaments are closer together. Another possibility is that the passive structures which maintain sarcomere structure act through an unspecified mechanical linkage to increase activation at long lengths. It has also been recognized that the cooperative interactions between crossbridges and calcium binding, described at the end of the last chapter, will amplify the effects of length-dependent activation and steepen the length-force relationship at partial levels of activation.

LENGTH-FORCE RELATIONS IN INTACT MUSCLE

The ascending limbs of the length-force relationships of both cardiac and skeletal muscle extend beyond lengths expected from filament overlap alone, but these relationships are different for the two muscle types. The relationships are explained here and the possible functional implications are discussed in Chapter 10.

Skeletal muscle

Rack and Westbury (1969) were interested in the basic mechanisms of tremor, a neurological abnormality. In a circuitous route to their objective, they carried out one of the most comprehensive studies ever done on the physiological length-force relation in skeletal muscle. They studied cat **soleus**, a muscle that extends the ankle, leaving the muscle connected at its proximal origin with the nerve and blood supply intact. Only the distal insertion was cut, to measure force at the achilles tendon. Before disconnecting the tendon, however, they correlated the joint angle with muscle and sarcomere length and found that the sarcomere length ranged from 3.0 to 2.0 µm, as the joint angle rotated from 30 to 150°.

They next developed a method for stimulating the muscle in a more physiological manner. The advantage of studying the soleus muscle is that the its nerve derives from five separate spinal roots (Fig. 6.8 A). They could either stimulate all the roots simultaneously (Fig. 6.8 B) or in rotation (Fig. 6.8 C). Fig. 6.9 shows that simultaneous stimulation of all the roots at a slow rate produced twitches while rotatory stimulation at the same rate produced smooth tetani.

These experiments showed that smooth, graded tetanic forces could be produced with different frequencies of rotatory stimulation (Fig. 6.10). This was a new type of recruitment in a muscle made of individual fibers that generate brief, all-or-none twitches. Instead of a few motor units producing maximum tetanic force while the others rested, the load was spread among all the fibers,

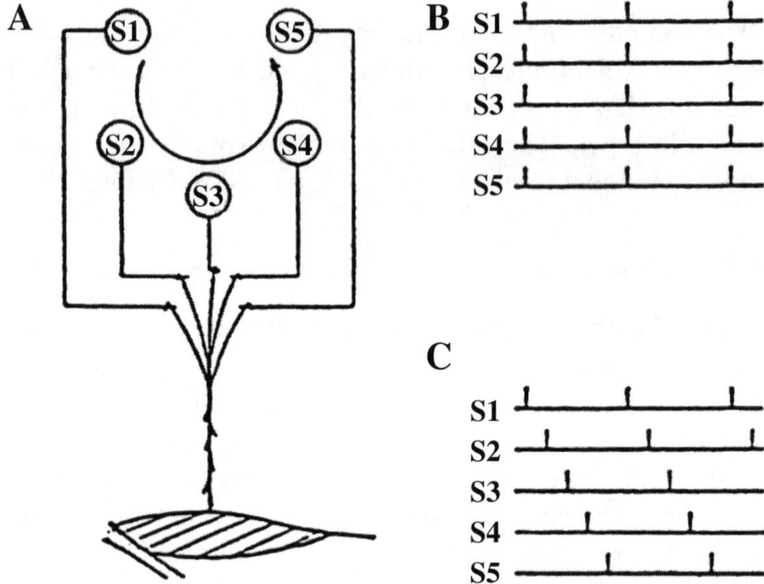

Figure 6.8. *Simultaneous vs. rotatory nerve stimulation of cat soleus muscle. From Rack and Westbury (1969), with permission.*

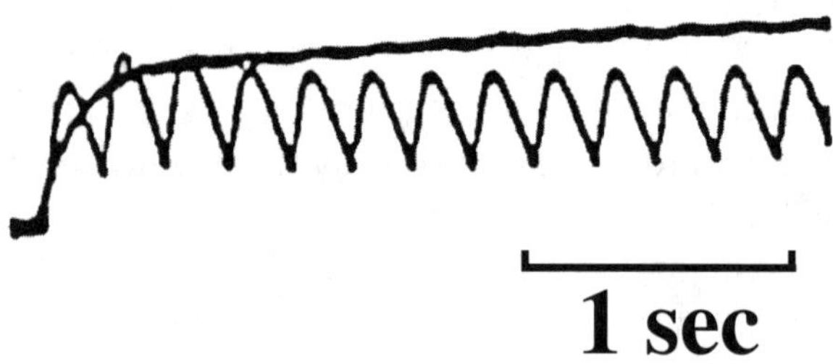

Figure 6.9. *Superimposed force records of a smooth tetanus produced by rotatory stimulation and unfused twitches produced by synchronous stimulation at the same frequency. Redrawn from Rack and Westbury (1969), with permission.*

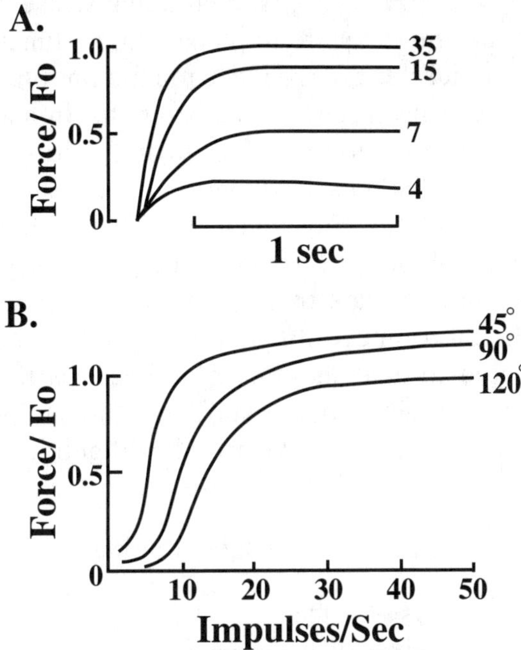

Figure 6.10. *A) Individual tetani generated by the stimuli at frequency indicated at the end of the records. B) Force vs. frequency at indicated joint angles. Redrawn from Rack and Westbury (1969), with permission.*

with the total force dependent on the duty cycle of the individual fibers. Since the central nervous system does not require separate nerve roots to distribute the impulses, much more complex stimulus patterns can be achieved by the body.

Rack and Westbury next used rotatory nerve stimulation to obtain the length-force curves in Fig. 6.11. At all frequencies, the curves reached their peak at a sarcomere length of about 3.0 μm, corresponding to the maximum physiological length. Only at maximum stimulation rates did the curves begin to approach the shape expected from the filament overlap relationships. A final point about their data is that the passive force in the muscle did not become appreciable until the muscle was stretched to sarcomere lengths approaching the longest encountered in the body.

Rack and Westbury's work is important because it reveals how multiple mechanisms conspire in physiological function. It is unlikely that most muscles ever produce maximum force. Natural selection is likely to favor safety over strength. In addition, maximum strength is not required for most function. Nature has therefore arranged that muscles function on the ascending limb of the length-force relationship, even though less than maximum force is generated. Finally, the curves in Fig. 6.11 suggest a mechanism for the muscle tears that sometimes occur with unaccustomed and therefore uncoordinated activity. If a muscle is suddenly stimulated to near maximum activity, particularly while being stretched, it might produce sufficient force to place it on the descending limb of the curve, where it could pull itself apart.

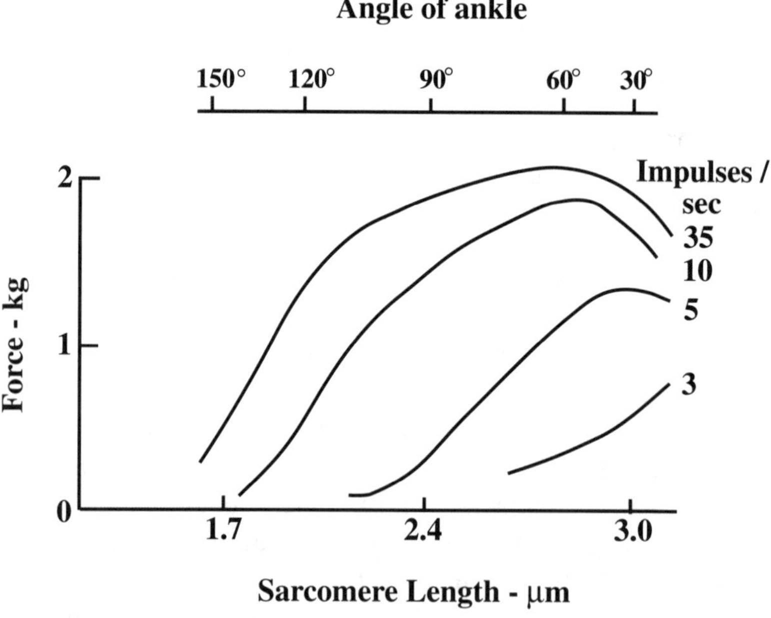

Figure 6.11. *Tetanic length-force relationships at different frequencies of rotatory stimulation of cat gastrocnemius muscle. Numbers indicate stimulation frequency. Redrawn from Rack and Westbury (1969), with permission.*

Cardiac muscle

The sarcomere length-force relationship muscle in Fig. 6.12 shows that cardiac muscle functions over a smaller percentage of its length than skeletal muscle, so that the ascending limb of the active (solid) curves is much steeper. The entire functional range is about 25% of its maximum length. An issue discussed in Chapter 15 is that the architecture of the heart provides that this relatively small operating range produces nearly complete ventricular emptying. Note also that the peak of the curve occurs at sarcomere lengths longer than those where the thin filaments overlap.

The passive length-force (interrupted) curve in Fig. 6.12 shows that substantial resistance to stretch occurs as the muscle approaches Lmax, the length at which peak force is developed. A point to made in Chapter 10 is that the heart does not have a skele-

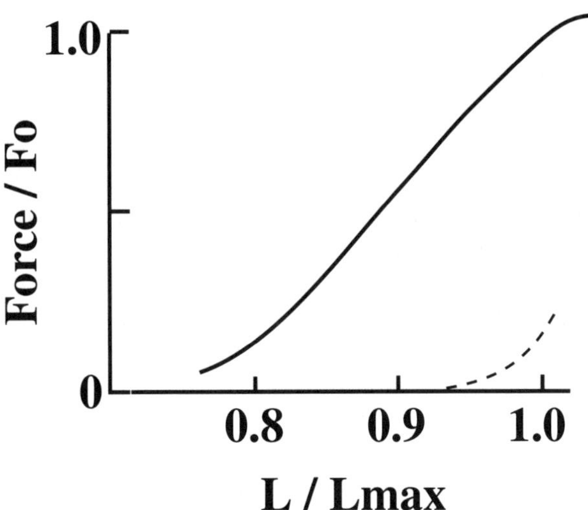

Figure 6.12. *Sarcomere length-force relationships in cardiac muscle. Active (upper) and passive (lower) force.*

ton to protect the muscle from overstretch and thus requires intrinsic passive structures to prevent movement of the muscle to the descending limb of the Frank-Starling relationship. The absence of a skeleton also requires that the heart have some mechanism to prevent it from shortening too far and developing irreversible contractures.

At least four mechanisms operate to insure that damaging contraction bands do not occur: 1) The muscle produces only twitches, and the absence of sustained activation provides that the muscle does not shorten too far during a single activation; 2) the muscle has strong restoring forces which insure that it is returned to a minimum starting length before the onset of each twitch; 3) the myofilament sensitivity to calcium is diminished at short lengths; and 4) active shortening inactivates the muscle, probably through the cooperative mechanisms described at the end of Chapter 5, and thus prevents sustained shortening. Together, these mechanisms usually prevent heart muscle from developing contraction bands. The importance of this protection is illustrated by the findings of Walley and Cooper (1991), described in Chapter 10, that under extremely severe conditions damaging contraction bands can be produced.

SUGGESTED READING

ALLEN, D.R and BLINKS, J.R. (1978) Calcium transients in aequorin-injected frog cardiac muscle. *Nature*. **273**: 509–513.

TAYLOR, S.R. and RÜDEL, R. (1970) Striated muscle fibers: inactivation of contraction induced by shortening. *Science*. **167**: 882–884.

Chapter 7

MECHANICAL MANIFESTATIONS OF ACTIVATION

An entire chapter is devoted to the effect of activation on contractile performance because it is vital to an understanding of cardiac function. Since cardiac muscle only twitches, its active state is never constant, so that an understanding of cardiac function requires a knowledge of the effects of variations in activation.

When a muscle is stimulated, its ability to contract rises and falls. It is easy to imagine, therefore, that something called **active state** rises and falls with a well-defined time course. Although the concept is simple, a major problem arose because the earliest attempt to define active state gave an incorrect result. Subsequent attempts to define its time course produced such conflicting results that the concept fell first into disrepute and then into disregard, as attention turned to mechanisms rather than definitions. More recent studies have provided sufficient knowledge of the mechanisms to define both the events that occur as the muscle is activated and the changes in contractile parameters that result from changes in activation. This knowledge is used here first to provide a mechanistic definition of active state and then to propose an operational definition. The first is a conceptual definition, independent of any parameter that might be measured. The second identifies a parameter that is correlated empirically with the mechanistic changes. A final section of the chapter is a brief history of the concept of the active state. This history would be allowed to rest unremembered were it not for a wealth of literature in cardiac function that is based on unsound principles.

A MECHANISTIC DEFINITION OF ACTIVE STATE

The preceding chapters explain that contractile activity is generated by myosin crossbridges that cyclically attach to actin, undergo a force-producing power stroke, and detach. A definition of activation must therefore account for the number of bridges available to cycle. This is determined by the number of thin filament sites available for crossbridge attachment. Thus, the definition proposed here is that active state is defined as the *number of thin filament attachment sites available for crossbridge attachment.*

In considering this definition, it is as important to recognize what is not included. Since filament sliding reduces the number of attached crossbridges, the active state is not defined by the number of attached bridges, or any parameter, such as developed force or stiffness, determined by the number of attached bridges. Because of the cooperativity between calcium and crossbridge binding, there is not a unique relationship between the amount of calcium bound to thin filaments and the number of activated sites. Thus, the amount of calcium bound to thin filaments would not define the number attachment sites available, if it could be measured. Furthermore, since filament sliding decreases the thin filament affinity for calcium and thereby promotes more rapid calcium sequestration, there is not a fixed relationship between the amount of calcium released and the level of activation achieved. Although the level of activation is likely to depend strongly on the calcium made available, the absence of a fixed relationship between this calcium and the level of activation precludes using the size of the calcium transient as a reliable indicator of the extent of activation unless other conditions, such as the extent of shortening, are fixed.

AN OPERATIONAL DEFINITION OF ACTIVE STATE

The only difficulty with the proposed mechanistic definition is that the number of activated thin filament sites cannot be measured directly. Some other parameter is therefore required to signal activation. Since the function of muscle is to contract, it would be useful if this signal were a mechanical parameter that reflected the instantaneous contractile ability. Furthermore, since the range of activity encompassed by contraction is shortening and force production, the parameter must be either force, velocity, or some combination of the two. Finally, since force and velocity are related inversely, the search for a parameter to define activation leads to a consideration of force-velocity curves. What is needed is some single-valued parameter that reflects changes in these curves. For purely practical reasons, it is proposed that the *maximum power capability of the muscle be used to signal changes in activation*. It is important to emphasize that while this is a single parameter, it cannot be determined by a single measurement. At each time that a measurement is required, sufficient force-velocity data must be obtained to determine peak power.

FORCE-VELOCITY PROPERTIES

Since force and velocity are inversely related, an isolated measurement of either force or velocity alone provides little useful information. Although active force is often measured with the assumption that the contractile elements are isometric, the series elastic elements can allow substantial movement of the contractile elements when force is changing. Consequently, it is necessary to define the complete curves when accurate measurements are required. The presumed advantage of making these measurements with isotonic loads is that the constant load does not alter the series elastic element length, so that overall muscle velocity reflects contractile element velocity, but this presumption can be fallacious when length changes in the series elements are damped by viscosity.

The theoretical effect of halving activation is shown in Fig. 7.1. Two ways of reducing the number of active crossbridges are illustrated. In the first (Fig. 7.1 B), the number of crossbridges in parallel is halved by reducing muscle cross-sectional area. The velocity achieved at any finite load is exactly half that observed under the reference condition (Fig. 7.1 A), but the two curves converge as the load approaches zero, so that the maximum velocities are the same (Fig. 7.1 D). The two curves superimpose when force is normalized to the isometric value (Fig 7.1 E.).

The effect of halving activation is illustrated in Fig. 7.1 C. In this case, half the thin-filament sites for crossbridges attachment are blocked. If the bridges act independently, the effect will be the same as reducing cross-sectional area; all forces at finite loads are halved, but the curves converge to the same maximum velocity at zero load and superimpose when normalized to isometric force.

Fig. 7.2 shows curves fitted to experimental force-velocity data. Skinned fibers were fully and partially activated by calcium buffered with EGTA in alternate contractions. Partial activation

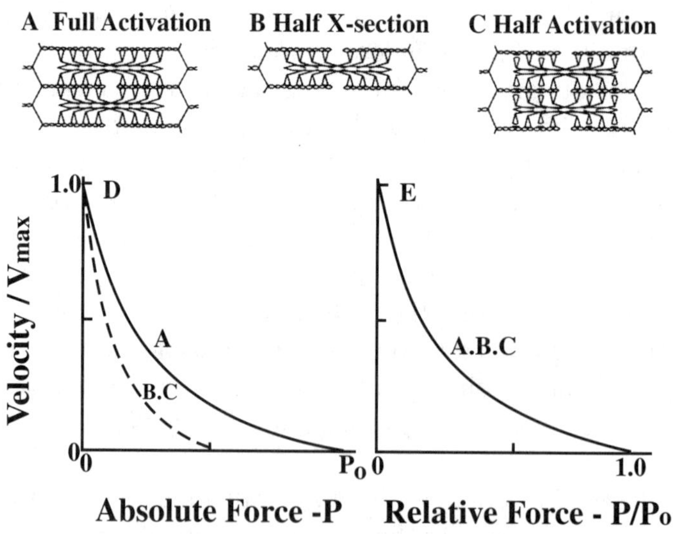

Figure 7.1. *Theoretical effect of halving the number of independent cross-bridges in parallel. From Ford (1991), with permission.*

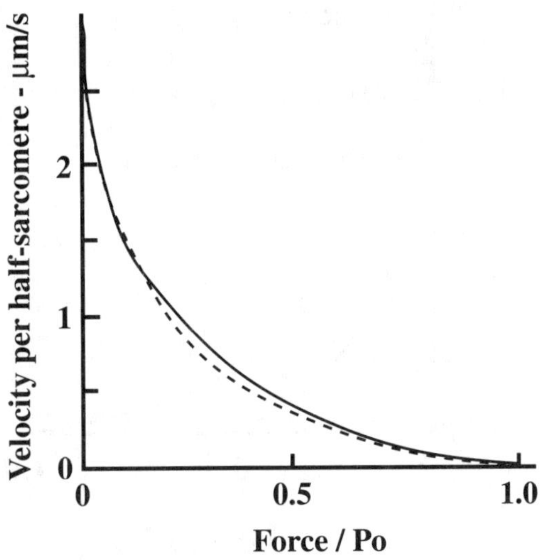

Figure 7.2. *Force velocity relations at full (solid curve) and partial (interrupted curve) activation. From Podolin and Ford (1986), with permission.*

produced isometric forces that averaged 43% of the value at full activation. When the isotonic forces were normalized to the isometric value, the curves superimposed. These results suggest strongly that maximum velocity is independent of the level of activation, at least over the activation levels studied.

Although skinned skeletal fibers appear to have the same maximum velocities at full and partial activation, the same is not true of intact cardiac muscle. Fig. 7.3 shows force-velocity curves with different maximum velocities typical of cardiac muscle, together with a graphical explanation of the reason for the difference. According to this interpretation, the contractile elements cannot be fully unloaded because of an internal load. Since a fixed internal load appears as a relatively larger load on the partially activated filaments, it depresses maximum velocity to a greater extent at lower levels of activation. These curves can be analyzed to determine whether the differences in velocity can be accounted for by

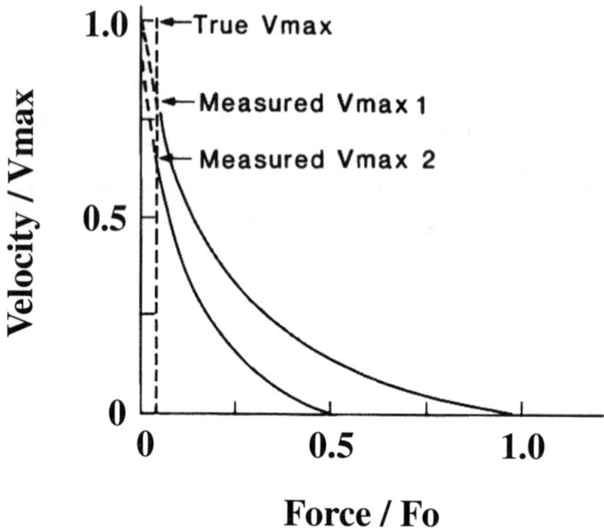

Figure 7.3. *Force-velocity curves from cardiac muscle at two levels of activa-tion analyzed in terms of a small internal load. From Ford (1991) with permis-sion.*

an internal load and the value of the load. The curves in Fig. 7.3 are least squares fits of experimental data from cardiac muscle. Analysis showed that all the velocity changes can be explained by an internal load equivalent to a 4–6% of isometric force in the upper curve (Chiu et al., 1987). Because the curves approach the ordinate at a sharp angle, small changes in the internal load have a substantial effect on the apparent maximum velocity.

PARAMETERS OF THE FORCE-VELOCITY CURVES

The following discussion is a consideration of the variables that might be used to signal changes in the force-velocity curves. The three parameters considered, isometric force, maximum ve-locity, and maximum power are superimposed on an isometric twitch of cardiac muscle in Fig. 7.4.

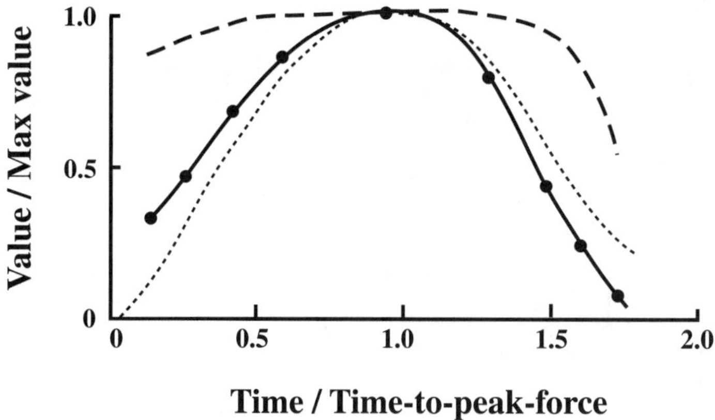

Figure 7.4. *Isometric twitch force (dotted curve), maximum velocity (dashed curve), extrapolated isometric force (solid curve), and maximum power (dots). From Chiu et al. (1987), with permission.*

Isometric force

The curves shown in Figs 7.1 and 7.2 suggest that isometric force might be the ideal variable for signalling changes in the force-velocity curves, since the curves can be made to superimpose when normalized to this force. The reason for not using this variable is that the isometric force measured during the rise and fall of activation does not lie on the curve.

Fig. 7.5 shows how this occurs. The records for five contractions, including an isometric twitch and shortening under four different loads are superimposed. For the highest load, an afterload, shortening began when the force had risen to the level of the load and the velocity accelerated almost immediately to its maximum. For the lower loads, the muscle was released quickly to the load at the same time in the contraction and the muscle again accelerated almost immediately to its maximum. Velocity was measured as the slope of the length record shortly after the onset of shortening, and the four force-velocity points are plotted on the lower curve in Fig. 7.6.

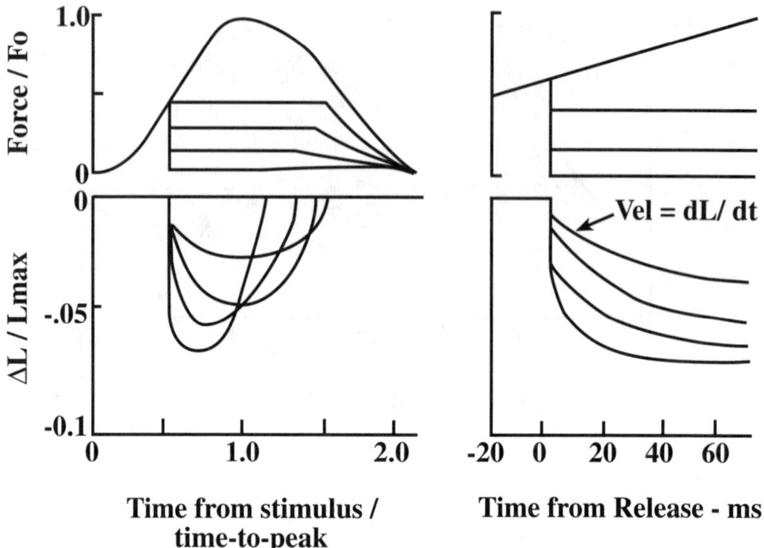

Figure 7.5. *Superimposed force and shortening records for three separate con-tractions and three different loads. Slow speed records on the left show the en-tire twitch. Higher speed records on the right resolve the rapid events following the step.*

The interrupted portion of the curve in Fig. 7.6 represents an extrapolation to higher forces and lower velocities than those achieved with the afterload. As indicated, the isometric force achieved at the time of the measurement, equivalent to the after-load, was substantially less than the zero velocity intercept of the curve and does not lie on the curve. When the measurements are made at the peak of a twitch or the plateau of a tetanus, the iso-metric force coincides almost exactly with the zero force inter-cept, as does the upper solid curve in Fig. 7.6. The qualification "almost" is used, however, because most force-velocity curves overshoot the isometric force by a small amount.

Fig 7.6 shows that there are two very different values for iso-metric force. One is the force achieved at each instant, and the other is the extrapolated value determined as the zero intercept of the force-velocity curve. Since the latter value lies on the force-ve-

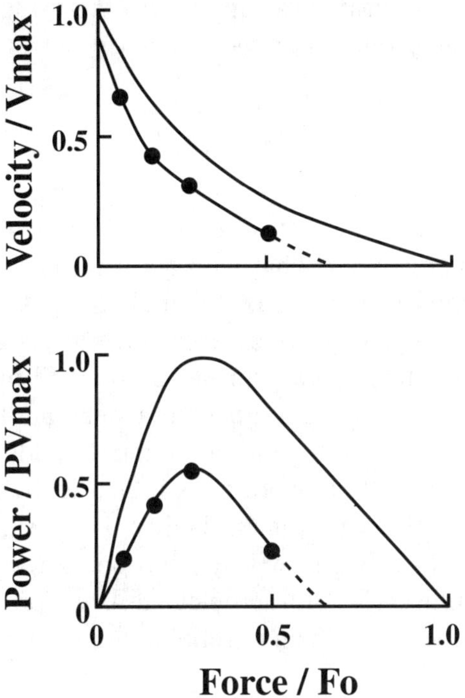

Figure 7.6. *Force-velocity and force-power relations obtained at the peak of a twitch (upper curve) and during the rise of force (lower curve). The four points on the lower curve represent that data derived from the records in Fig. 7.5 and the interrupted portion of lower curve represents extrapolation to zero velocity.*

locity curve, it will indicate a shift to a new curve. The problem with this value is that it requires a long extrapolation of the curve. The former value would be of little use in defining the curves because its zero velocity value does not lie on the curve.

It was originally thought that the disparity between the measured and extrapolated isometric force was due to internal shortening, but more recent measurements have disproved this. In addition, high afterloaded velocities are predicted by Huxley's original crossbridge theory. The delay between activation and the rise of force is due to the finite time required for crossbridge attachment to achieve its steady value. When the muscle is allowed to shorten, the more rapid crossbridge cycling drives the bridges to approach their steady cycling rate much more quickly.

Maximum velocity

The results in skeletal muscle illustrated by Figs. 7.1 and 7.2 indicate that maximum velocity does not change with activation. Since it does change somewhat with activation in cardiac muscle, it has been proposed that it might signal changes of activation in that preparation. There are, however, several problems with the use of this parameter. Since it must be obtained by extrapolation of curves that approach the ordinate at a steep angle, there can be substantial uncertainty in the measurement. In addition, it does not change very much with activation, as indicated by the interrupted curve in Fig. 7.4. Finally, if its variability is determined by a passive internal load, its value will depend on factors that are independent of the muscle's contractile capability. For all of these reasons, maximum velocity is an unreliable index of activation.

Maximum power

The instantaneous power output of a muscle is the product of force times velocity. Maximum power is determined by interpolation of force-power curves, such as those in Fig. 7.6. The determination by interpolation rather than by extrapolation, as is required for maximum velocity or extrapolated isometric force, greatly lowers the uncertainty of the experimental measurement. In an analysis of force-velocity data in skinned fibers, Podolin and Ford (1986) found the standard errors of maximum velocity and extrapolated isometric force to be 2 and 4 times higher, respectively, than the standard error of maximum power.

If the force-velocity curves can be superimposed by normalizing force to the isometric value, as in Figs. 7.1 E and 7.2, maximum power should be exactly proportional to the extrapolated isometric force. The circles (maximum power) superimposed on the solid line (extrapolated isometric force) in Fig. 7.4 shows that even when such superposition of the curves does not occur, maximum power correlates empirically with the extrapolated isometric force. Although the major reason for preferring maximum power as the variable to signal activation is its lesser error, there are two additional reasons for this choice. The first is that the need for estimating activation arises mainly in cardiac muscle, which does not normally function isometrically or at maximum velocity. Its principle function is to do work, and since maximum power indicates its instantaneous work capability, it expresses activation in terms of the muscle's main function.

The second reason for preferring maximum power is that it is normalized to the muscle mass. This normalization occurs because force is normalized to cross-sectional area and velocity to muscle length, so that the product of the two denominators is length times cross-sectional area, equal to volume. This normalization to muscle mass is especially useful in the heart for two reasons. Several other parameters such as oxygen consumption and energy liberation are also normalized to muscle mass, so that these measurements can be compared directly with the level of activation. Even more importantly, the irregular shape of the heart makes it difficult or impossible to normalize force and velocity accurately, whereas muscle mass can be estimated accurately. Furthermore, muscle force and velocity differ across the walls, so that only average parameters can be assessed. By contrast, the power of the heart muscle, measured as force times velocity, must equal the power of the ventricle, measured as pressure times blood flow, which can be measured with relative accuracy.

Time course of active state

The values of extrapolated isometric force and maximum power in Fig. 7.4 suggest that activation rises and falls in advance of isometric force. The intracellular calcium transient superimposed on twitch force in Fig. 7.7 further suggests a mechanistic reason why

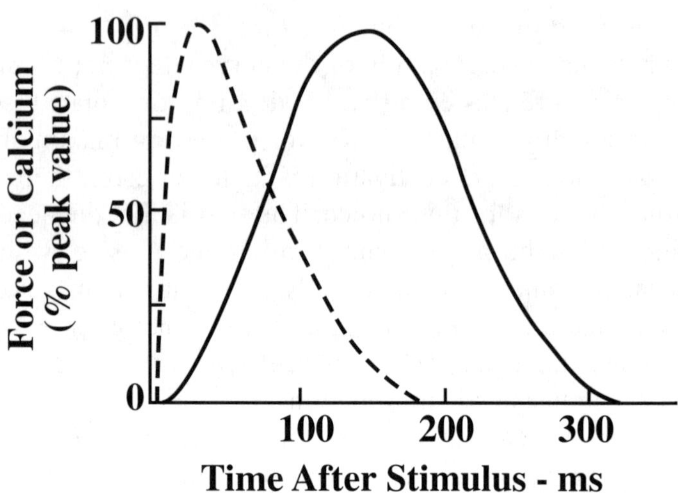

Figure 7.7. *Time course of a calcium transient in cardiac muscle (interrupted curve) superimposed on twitch force (solid curve). Adapted from Blinks and Endoh (1986).*

this time course is expected. Calcium rises and falls well in advance of force and the figure makes the point that two parameters that are intimately associated with activation have very different time courses. Calcium rise is the first intracellular event in the excitation-contraction process and twitch force is the final outcome. The time course of something called activation is likely to fall between these two extremes. This point should be remembered in following history of the study of active state.

A BRIEF HISTORY OF THE CONCEPT OF ACTIVE STATE

The earliest definition of active state is usually attributed to Hill (1949a). He postulated that the rise of activation was rapid but the rise of force was delayed by internal shortening. The rising force elongated the series elastic elements and allowed shortening of the contractile elements. To test this hypothesis, he stretched the muscle early in the twitch and found that its ability to bear a load developed quickly.

During a stretch applied early in the twitch, force rose from a low value to somewhat more than its full value and then remained nearly constant (Fig. 7.8). From this observation he concluded that the active state was equivalent to the "force the contractile elements could bear if they were neither lengthening or shortening", i.e. if they were isometric. The finding that this measure of active state rose quickly led him to calculate that the presumed rise of activation was more rapid than could be accounted for by inward diffusion. On the basis of these calculations he concluded that the inward spread of activation was due to a "process and not a substance", and extended this conclusion to suggest that Heilbrun and Wiercinski (1947) were incorrect in their conclusion that calcium caused activation.

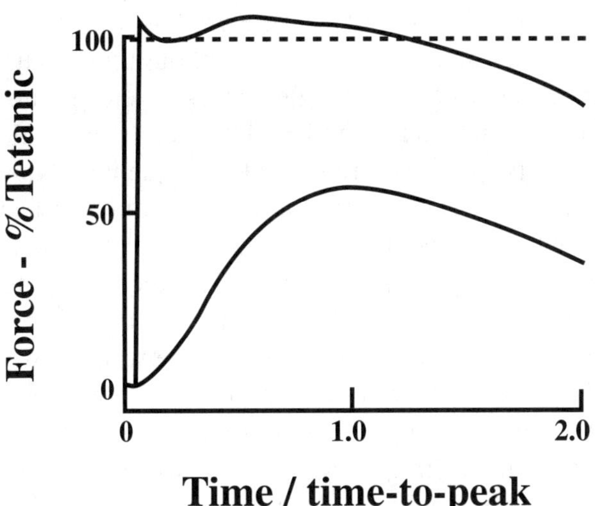

Figure 7.8. *A rapid stretch applied to a muscle early in a twitch causes force to rise to a high and nearly constant value. Adapted from Hill (1949a).*

Afterloaded contractions

Releasing a muscle from its isometric state to an isotonic load, as in Fig. 7.5, is technically difficult because the levers have inertia which interacts with the compliance of the muscle to cause significant oscillations. The afterloaded method of measuring force-velocity properties shown in Fig. 7.9 was therefore devised by Magnus Blix (1895) to obtain smoother records. In a typical experiment, the muscle is hung vertically, with the upper end attached to a lever pivoted about a fulcrum. A **preload** hung from the other end of the lever serves to stretch the muscle to establish its starting length. An adjustable stop is then lowered to just touch the lever above the muscle, so that no further stretch will occur when the **afterload** is added to the preload at the other end of the lever. When the muscle is stimulated, force rises isometrically until it equals the value determined by the afterload. Once this occurs, the muscle first shortens and then lengthens isotonically under the afterload.

The inset of the Fig. 7.9 shows records of force, length, and velocity superimposed for four contractions with shortening at four afterloads. The velocity is usually measured as the peak value observed shortly after shortening begins, although an earlier, initial value is sometimes used. The single force-velocity points derived for each contraction are graphed and extrapolated to determine maximum velocity. As illustrated, this technique largely avoids the oscillations that are usually stimulated by a quick release to an isotonic load when levers are used. But it is also necessary to give Blix credit for some cunning in devising two techniques for reducing the effective inertia of the lever. One was to shorten the distance from the fulcrum to the point where the weights were hung, so as to reduce their effective inertia. The other was to hang the weights from a long elastic band that greatly delayed acceleration of the weights. This technique was used in some laboratories well into the 1970's, and quick release experiments, such as those shown in Fig. 7.5, only became feasible with the development of electromechanical servo systems that could be critically damped.

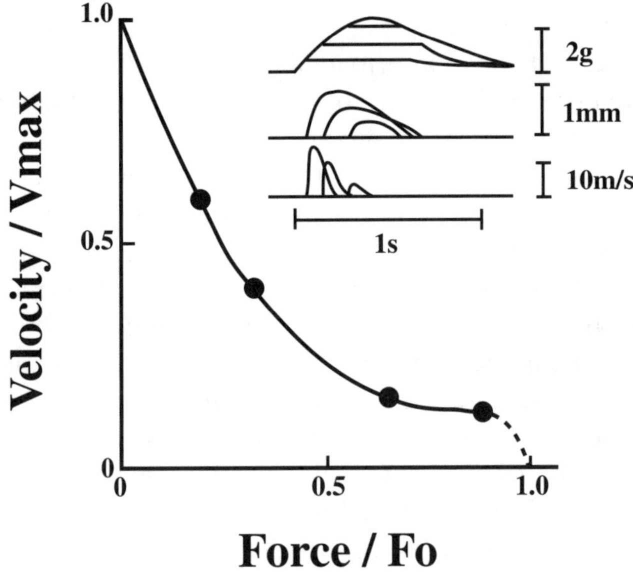

Figure 7.9. *Force-velocity curves derived from afterloaded contractions shown in the inset. Negative velocity records not shown.*

The problem with these afterloaded experiments is that each point is obtained at a different time in the twitch and therefore at a different level of activation. In addition, if there is a substantial series elastic compliance in the preparation, as there usually is in heart muscle, each force-velocity point is obtained at a substantially different sarcomere length.

Hill was correct in his conclusion that the inward spread of activation was due to a process, and we now know that it is due to electrical spread along the t-system. The remainder of his conclusions were wrong, however, and would not be remembered had they had not generated so many later controversial studies. For example, the assumption that active state rose almost immediately allowed the technically simple but incorrect method of measuring force-velocity properties using a series of afterloaded contractions. The method was adopted uncritically in many practical applications, where the users were not in a position to evaluate the method from first principles.

The use of these afterloaded force-velocity curves became widespread in cardiac physiology. Furthermore, the simplifications of the measurements did not end with the use of afterloaded contractions. The finding that the velocity rose almost immediately to its peak at the onset of shortening was taken as evidence that the contractile elements were shortening at the same velocity immediately before the contraction began. MacPhearson (1953), working in Hill's laboratory, extended this conclusion to measure the force-velocity relations using single isometric contractions. He reasoned that the internal shortening could be calculated from the rate of rise of force (dF/dt) and the known stress-strain characteristics (dL/dF) of the series elastic elements that were being stretched. Multiplying one by the other yielded the velocity (dL/dt). It would have been a wonderful method if activation had remained constant throughout the measurement period.

Hill's suggestion that the slow rise of force was due to internal shortening is easily tested experimentally, and the first such test was provided by Jewell and Wilkie (1958), Hill's colleagues at University College London. First they compared the rise of force at the onset of a tetanus with the rise of force after a quick release that reduced tension to zero. It was substantially slower, suggesting that the muscles were more completely activated at the time of the release than at the outset of the tetanus. Next they calculated the time course over which force would rise if the series-elastic and force-velocity properties at the onset of the tetanus were the same as they measured at the plateau of the tetanus. This calculated rise was faster than either of the observed time courses, suggesting that the slow rise was not due just to internal shortening. Taken together, these experiments indicated that the rise of activation was substantially slower than Hill had postulated.

A very different explanation of the slow rise of tetanic force was suggested by Huxley in his original (1957) description of the crossbridge theory. He postulated that, even if activation were instantaneous, the rise of force would reflect the relatively slow approach of crossbridge attachment to its steady isometric level. Using "reasonable rate constants" for attachment, he showed that the calculated rise of force was very similar to that observed experimentally. Finally, in much later experiments, he showed that when the sarcomeres are held absolutely isometric with a mechanical servo, the rise of force is not much faster than is usually observed (Ford et al., 1986).

A possible explanation of Hill's finding of the early development of the muscle's ability to bear a load is shown in Fig. 7.10. It is now known that the decay of isometric force following a stretch can be extremely slow and that there may even be a very long lasting component of residual tension (L. Hill, 1977). Both the slow decay and the residual tension can be explained in terms of crossbridges that are stretched to the point that they do not complete their power stroke and therefore detach at a slow rate. It seems likely that the rise of force associated with the stretch is due to a few bridges resisting the stretch and that the force in these bridges decays slowly (dotted curve in Fig. 7.10). The addition of new,

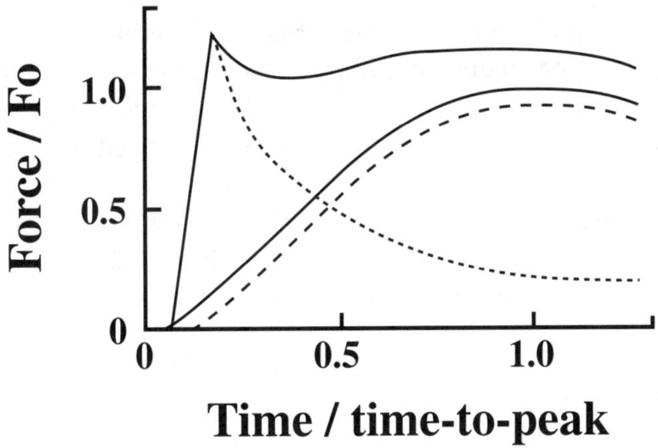

Figure 7.10. *Theoretical effect of a stretch applied early in a twitch. Adapted from Ford (1991), with permission.*

normally cycling bridges would be expected to increase force (interrupted curve). It seems likely that under optimum conditions the rise of force in these new bridges could match the fall of tension in the stretched bridges, so that a nearly flat force response (solid curve) is achieved. It should also be observed that the qualification "nearly" is used here because both Hill's original responses (re-drawn in Fig. 7.8) and the calculated responses in Fig. 7.10 fell from an early peak immediately following the stretch and later rose slightly.

The assessment of the time course of activation in skeletal muscle is relatively unimportant because the muscles can be stimulated to produce a constant level of activation, and the complications of varying activation are avoided. This is not true in cardiac muscle. Except under very non-physiological conditions, the preparations produce only twitch contractions. If the level of activation varies continuously during the twitch, as we now believe it does, an understanding of cardiac function demands that the time course and effects of activation be understood. There is now good evidence that the time course of activation follows the time course indicated by maximum power in Fig. 7.4, but this was not always the case. Brady was the first to attempt to assess the time course of activation in cardiac muscle. He began by attempting to repeat Hill's stretch experiment in cardiac muscle, but was unable to do so. His results are summarized in Fig. 7.11. He found that a length change of either direction, stretch or release, caused the force following the length change to be less than it would have been if the isometric contraction began from the final length. We now interpret such results in terms of the filament sliding detaching cross-bridges and thereby interrupting the cooperativity of activation.

CONCLUSION

The main points to be learned from this chapter are that activation results in the availability of thin filament sites for cross-bridge attachment, and the level of activation is determined by the number of such available sites. Changes in the number of these

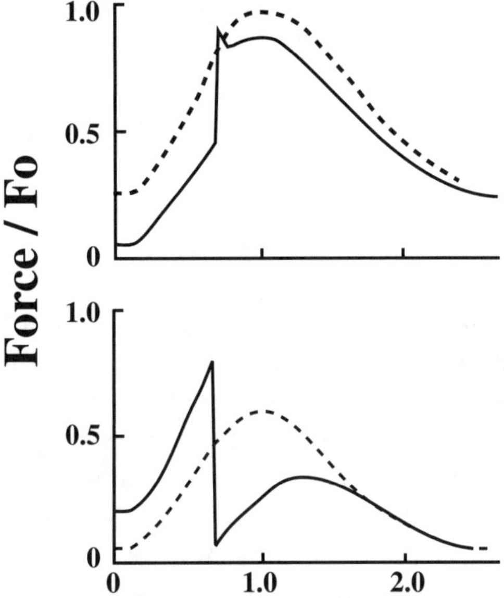

Figure 7.11. *Effect of stretch (A) and release (B) on the time course of twitch force. Adapted from Brady (1965) and Ford (1991).*

sites cause changes in the force-velocity relations, and these changes in mechanical ability can be used to quantify changes in activation. It is proposed that the most robust parameter for signalling these changes is maximum power. Three advantages of this parameter are: 1) it is determined by interpolation rather than by extrapolation; 2) it measures the muscle's ability to do work, which is cardiac muscle's principal function; and 3) it is normalized to muscle mass, so that it can be compared with the results of many other types of experiments, including those on whole heart.

SUGGESTED READING

FORD, L.E. (1991) Mechanical manifestations of activation in cardiac muscle. *Circulation Res.* **68**: 621–637.

Chapter 8

FATIGUE

There are many situations where a muscle's performance is limited by its metabolic stores. Natural selection has therefore provided muscles with mechanisms for accommodating to these stresses without becoming irreparably damaged. Chapter 5 explains that ATP depletion leads to the formation of rigor bridges, which increase activation through cooperative interactions with the thin filaments. Depletion of this substrate could therefore lead to a disastrous spiral in which cooperativity maintains activation, so that ATP depletion becomes self-sustaining. When muscle activity goes on unchecked it is often fatal because it produces severe hyperthermia. The massive muscle destruction that occurs in this circumstance also leads to kidney failure from myoglobin impaction of the renal tubules. Such a clinical disaster is rare and is usually initiated by severe hyperthermia, caused either by extremely vigorous activity in a hot environment or from the syndrome of **malignant hyperthermia**. This latter condition is caused by a dominantly inherited sensitivity of the muscle activating system to some anesthetic agents. When given these agents, affected patients develop spontaneous contractions which increase body temperature. The increased temperature in turn produces more muscle activation, more heat, and eventually death. Interestingly, when recognized early, the uncontrolled muscle activity can be blocked with **dantrolene**, a drug that blocks calcium release from the sarcoplasmic reticulum. This observation suggests that even under the most extreme conditions, contraction does not escape the control of its activating system.

These considerations suggest that fatigue, the muscle weakness that develops with sustained, vigorous activity, is a physiological process that has evolved for the preservation of the animal. This chapter addresses the mechanisms that might be responsible for this adaptation. An understanding of the processes involved could be useful for a rational planning of athletic endeavors and the treatment of heart disease associated with myocardial ischemia. Since fatigue results from muscular activity, it appears that there is negative feedback of the contractile process on itself. It is therefore fruitful to examine the effects both of accumulation of reaction products and depletion of substrate when seeking the causative agents. The products of contraction include heat as well as molecules. Of the many substrates that support contraction, only a few of these, such as oxygen and glycogen, are likely to be limiting.

HEAT

Temperature changes alter muscle contraction in many ways that will not be reviewed here. The point to be made here is that muscular activity increases body temperature and that reflexes limit physical activity to avoid severe hyperthermia. In studying running in cheetahs, Taylor found that the animals stop running when body temperature reaches about 41°C. It is also well recognized that heat-loss mechanisms are important for endurance, and one group of evolutionary biologists have postulated that man has evolved to be a long-distance runner in a desert environment. Apparently, man is able to outdistance almost all other animals in the desert.

METABOLITES OF CONTRACTION

ATP hydrolysis yields ADP and inorganic phosphate, but the ADP is immediately rephosphorylated from creatine phosphate, so that the products of contraction are creatine and phosphate. ADP accumulation or ATP depletion occur only under extreme duress. Furthermore, ADP accumulates as ATP is depleted. Since the ADP concentration is normally near zero, the relative increase of ADP is very much greater than the relative decrease of ATP. As explained below, ADP accumulation in the presence of phosphate is likely to inhibit the contractile machinery. Thus, it appears that all these metabolic changes prevent significant ATP depletion and the development of rigor crossbridges, which might sustain activation through a cooperative mechanism.

Although the phosphate liberated by ATP hydrolysis has a hydrogen ion that would be liberated to lower the pH at the normal intracellular pH, this hydrogen is absorbed by the creatine so that the activation of the contractile machinery produces almost no change in pH. Contraction does, however, cause an intracellular pH decrease that has some role in fatigue. The responsible hydrogen ions are not produced directly by crossbridge activity but by the metabolic reactions that restore creatine phosphate. The hydrogen ions derive from CO_2 under aerobic conditions and from lactic acid under anaerobic conditions.

If the contractile machinery is physiologically inhibited by its own metabolism, the inhibition would have to be produced by phosphate, creatine, or by creatine phosphate depletion. Creatine and creatine phosphate appear to have no direct effects on crossbridge function, so that the only immediate product of contraction that would inhibit further activity is phosphate. Its effects are large and probably operate through two reactions.

Effects of phosphate

As described in Chapter 4, very high concentrations of phosphate inhibit force generation and power output of skinned fibers when the ATP levels are near normal and there is little or no ADP present. These effects appear to result from detention of bridges in low-force states and inhibition of the power stroke. The detained bridges impede work of the normal bridges and thus reduce muscle efficiency. Such bridges are seen only at very high phosphate concentrations, tens of millimolar, as would only be found near the terminal stages of exhaustion. At very much lower concentrations, a few mM, phosphate detaches bridges detained in high force states by ADP. These levels are likely to occur whenever ADP accumulates. In addition, since phosphate at low concentrations detaches bridges with ADP bound, it removes the detained bridges from states that might impede shortening and diminish muscle efficiency. Thus, phosphate appears to be the ideal metabolite for preventing the untoward effects of ATP depletion and ADP accumulation. The question remaining, therefore, is whether there are other mechanisms that limit contraction before ADP accumulation begins.

GLYCOLYSIS AND RESPIRATION

The energy for skeletal muscle contraction comes from the oxidation of glucose. Under aerobic conditions, one mole of glucose yields 32 moles of ATP as well as 6 moles of CO_2. Of the 32 moles, 2 are derived from the initial breakdown of glucose to pyruvate in twelve reactions known collectively as **glycolysis**. The remaining 30 molecules come from the breakdown of pyruvate in the **tricarboxylic acid cycle**, sometimes called the **citric acid cycle**, or the **Krebs cycle**, in the mitochondria. Most of the glucose used by muscle comes from internal stores of **glycogen**, branched chains of glucose, but lesser quantities may be come directly from the blood by diffusion across the cell membrane.

A schematic diagram of the glycolysis pathway with numbered reactions is shown in Fig. 8.1. Note that the 6-carbon fructose is cleaved to two 3-carbon moieties in reaction 6, so that the yield per molecule of glucose is doubled from that point onward. The cell first invests two molecules of ATP, in reactions 5 and 1 or 3, to phosphorylate the glucose to fructose-1,6-bisphosphate before it can be metabolized further. In addition, **nicotinamide-adenine dinucleotide** (**NAD**) is reduced in reaction 8, and this energy transfer molecule must be re-oxidized if more glucose metabolism is to occur. In the absence of hypoxia, this oxidation occurs in the mito-

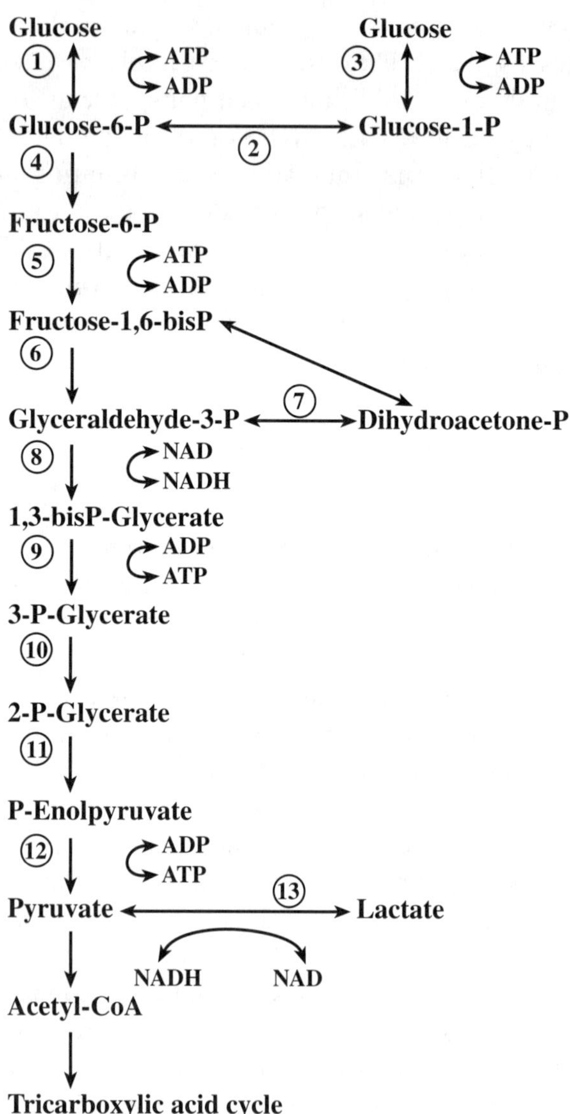

Figure 8.1. *Glycolysis.*

chondria, but it can occur anaerobically in the cytoplasm, deriving energy from the conversion of pyruvate to lactate in reaction 13.

Under anaerobic conditions, pyruvate is inhibited from entering the tricarboxylic acid cycle and pyruvate is converted to lactate. This reaction produces NAD and allows rephosphorylation of ATP from glycolysis to occur outside the mitochondria and does not require oxygen. The energy yield from anaerobic glycolysis is therefore only about 6% (2/32) of the total available from the complete oxidation of glucose, but it is sufficient to sustain muscle contraction in sprints. It results in lactic acid production, whereas the oxidation of glucose results in CO_2, which combines with water to form carbonic acid.

Most of the energy source for cardiac muscle contraction comes from the metabolism of free fatty acids and lactic acid, again utilizing the tricarboxylic acid cycle and oxidative phosphorylation by the mitochondria, but requiring a continual supply of oxygen. Temporary depletion of oxygen during periods of increased activity is minimized by high intramuscular concentrations of myoglobin, which binds and stores oxygen, but since these stores are finite, the heart cannot tolerate anything more than very brief periods without oxygen.

The metabolic pathways that provide energy for contraction are described here to address the issue of how muscular activity is reduced by limitations of energy supplies. Since the end-products of these pathways are acids, either CO_2 or lactic acid, the effects of acidosis should be considered. There are several conditions, both physiological and pathological, that result in metabolic insufficiency, including **hypoxia** (the lack of adequate oxygen), **ischemia** (the lack of adequate blood flow), and depletion of glycogen, which only occurs under conditions of extreme exhaustion. These conditions will be considered separately.

ACIDOSIS

Acidosis, both locally within the muscle and globally throughout the body, occurs as the result of muscular activity. It develops mainly as a consequence of the catabolism of glycogen, rather than from the hydrolysis of ATP, and the acids are CO_2 and lactic acid. Acidosis produces at least two reflexes that promote its reversal. One is dilation of pre-capillary sphincters in the blood supply to muscle thereby increasing flow blood to accelerate delivery of oxygen and hasten the removal of CO_2 and lactate. The other is increased breathing that rids the body of the excess CO_2 and increases oxygen uptake.

There is some uncertainty over the extent to which acidosis limits itself by limiting muscular contraction. Not long after the

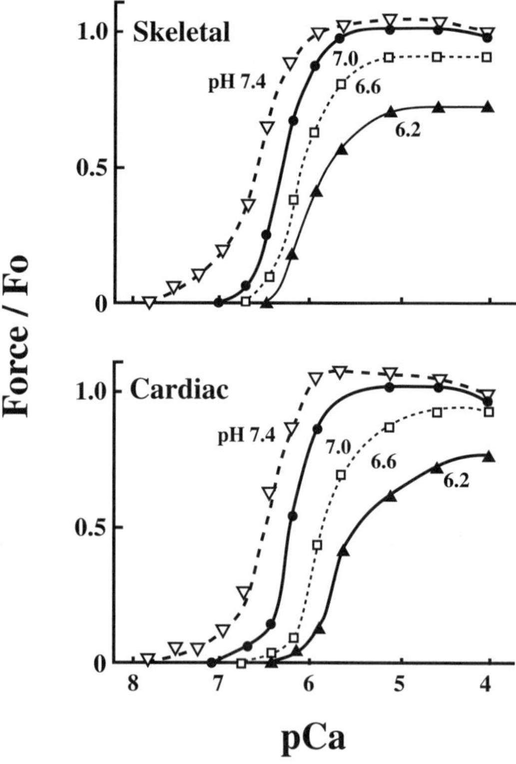

Figure 8.2. *Force vs. pCa in skinned muscle cells at the four pH values indicated. Adapted from Fabiato and Fabiato (1978), with permission.*

role of calcium in activation was discovered, Katz and Hecht (1969) speculated that hydrogen ions might compete with calcium for binding to thin filaments in cardiac muscle. Fabiato and Fabiato (1978) later showed that decreasing pH from 7.4 to 6.2 decreased force at full activation and increased the concentration of calcium needed to produce half-maximal activation of skinned muscle, both cardiac and skeletal (Fig. 8.2). In addition, they showed that the sarcoplasmic reticulum accumulated and released less calcium at lower pH. Thus, it is certain that acidosis inhibits contraction, both by reducing myofilament calcium affinity and by reducing calcium released into the myoplasm. There is however, a question remaining as to whether other mechanisms supervene before a sufficiently low pH is reached.

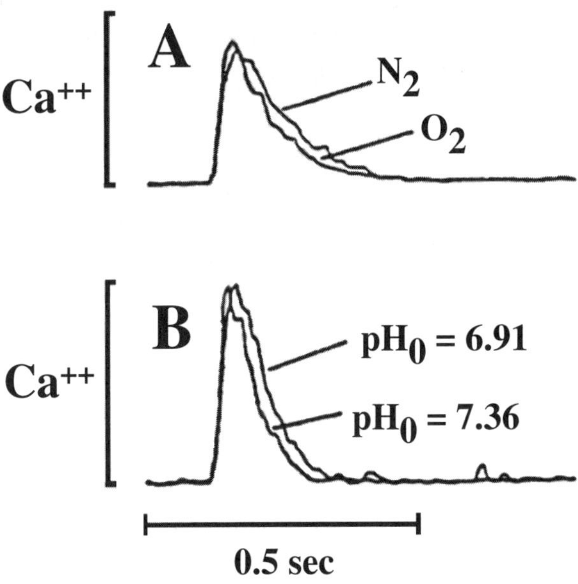

Figure 8.3. *Calcium transients recorded from aequorin injected muscles. A) Oxygen (O_2) vs nitrogen (N_2) perfusion, and B) at the two extracellular pH's indicated. From Allen and Orchard (1983), with permission.*

Allen and Orchard (1983) measured the calcium transients in isolated cardiac muscle during hypoxia caused by substituting nitrogen for oxygen. Under the conditions studied, force fell to about 1/3 of its control value with no decrease in the calcium transient. From these results they concluded that the decreased force resulted from a diminished sensitivity of the myofilaments to calcium, as expected from the results of Fabiato and Fabiato (Fig. 8.2). In support of their conclusion, they found that the calcium transients were slightly prolonged (Fig. 8.3 A), as expected if calcium binding by the myofilaments were weaker. A similar prolongation was seen when the extracelluar pH was lowered (Fig 8.3 B), suggesting that the effects of hypoxia were due to acidosis. In a later NMR study, however, Allen et al. (1985) found that pH actually increased slightly during the early part of the force reduction due to hypoxia and then fell by only a small amount, from a control level of 6.98 to a stable level of 6.85, as force fell to a steady 35% of its control level. During this same time, phosphate rose by several millimolar and ADP rose from undetectable levels to being just detectable. These studies suggest that acidosis is not the major cause of the force decline associated with hypoxic conditions that do not totally abolish force. A possible reason is that some other condition supervenes to inhibit contraction, thereby preventing severe and possibly damaging intracellular acidosis. While the identity of this other condition is not known with certainty, a good candidate is the phosphate, perhaps in concert with the accumulation of ADP.

HYPOXIA

This is the least severe of the metabolic limitations and occurs frequently during vigorous activity. Arterial oxygen desaturation probably never occurs in the healthy athlete, and muscle hypoxia occurs because the blood flow is insufficient to deliver the oxygen required by the tissue metabolism. Lactic acid is produced by the hypoxic muscles and accumulates progressively during activity that exceeds the anaerobic threshold of the muscles and their blood supply. The endurance of the animal during such activity is determined largely by its ability to tolerate the resulting acidosis. Unlike CO_2, which is readily cleared by breathing, lactic acid must be cleared by metabolic activity, which takes much longer. Most of it enters the blood and is cleared elsewhere in the body; some is metabolized by the heart and other organs, and some is re-formed into glycogen by the liver.

There is a general belief among athletic trainers that lactic acidosis is to be avoided as much as possible during practice. This belief derives from the finding that the total exercise in a work-out can be greatly extended when anaerobic conditions are avoided. To the extent that muscle growth and development depends upon their repetitive use, and to the extent that such growth and development is desirable, this would seem to be a wise strategy. It should also be recognized, however, that there is a great deal more to be learned about athletics training, so that the optimum work-out for endurance training is probably still unknown.

ISCHEMIA

The major difference between ischemia and hypoxia is that limitation of the blood flow restricts the buffering of acid and oxygen to the local organ or muscle and thus deprives this tissue of access to the normal total body buffers. It seems likely that ischemia produces much more local acidosis than hypoxia alone. The diffusion of CO_2 is much more rapid than that of oxygen, and until blood flow is limited, carbon dioxide is rapidly carried off from tissue beds. When the blood flow to a muscle is limited, however, the CO_2 does accumulate. In addition, the resulting hypoxia produces high local concentrations of lactic acid. Thus, ischemia is likely to cause an acidosis that will inhibit contraction. This local acidosis also produces pain that reflexly limits muscular activity. Ischemia is frequently a problem in atherosclerotic conditions, where blood flow is limited by fixed obstructions to arterial flow.

Angina pectoris is the chest pain that occurs when the heart's metabolic needs exceed the ability of coronary circulation. It can be correlated with lactic acid in the blood drained from the coronary circulation. The term **intermittent claudication** is used to denote pain in the legs that develops with walking when there is arterial insufficiency to a leg. It usually resolves after a short rest, hence the adjective "intermittent." The term itself is derived from the French verb meaning *to limp*, because it was first described in horses by a French veterinary student. When horses develop leg pain while being driven, they cannot rest, and therefore limp.

Ischemia also occurs when arteries are occluded by pressure. It seems likely that sustained isometric contraction will generate sufficient intramuscular pressure to limit blood flow and cause ischemia. This may be the reason that sustained contraction causes local pain to develop more rapidly than repetitive activity.

GLYCOGEN DEPLETION

It is well known that even well trained marathon runners may encounter a strong resistance to running near the end of the race. This is called "hitting the wall," and it appears to be due to glycogen depletion in the active muscles, so that there is no longer a ready substrate for replenishing creatine phosphate. A similar state occurs with lesser activity in rare inherited absences of the enzymes responsible for converting glycogen to glucose. Muscular activity can only be carried out by the metabolism of glucose delivered directly from the blood or from the metabolism of other substances, such as fats, that are not as well metabolized in skeletal muscle. When this occurs, the runner usually stops. A strategy called **carbohydrate loading** has been developed to avoid this disappointing outcome in an endurance contest. It requires eating an excess of starch for 1 to 2 days prior to the race. In its extreme form, a diet nearly deficient of carbohydrate is eaten for several days, followed by two days of almost exclusive carbohydrate. It is probably effective in very long distance endurance events, such as marathons, but it is not clear that such loading would be of any benefit in shorter events. In addition, since glycogen is utilized in the muscle where it is stored, carbohydrate loading would be of lesser benefit in endurance events, such as triathlons, that use different muscle groups at different times.

FATIGUE ASSOCIATED WITH DIFFERENT ACTIVITIES

For very brief athletic events, such as weight lifting, maximum muscle output can be achieved with very little metabolic support from the remainder of the body. For endurance events the muscle activity must be matched to the ability of the cardiovascular system to supply the necessary oxygen and carry off the waste products of contraction. If the initial muscular activity is too vigorous, the subsequent effort will be reduced by accumulation of metabolic deficits, and endurance athletes must learn to "pace" themselves to avoid fatigue early in the event. Short sprints are run almost without breathing, so that acidosis develops quickly and the runners are very breathless by the end of the race. By contrast, distance running usually occurs under nearly aerobic conditions that produce much less intense breathing. While it is often said that such a race should be run to exhaustion, this is usually not possible, perhaps because the metabolic derangements produced by the sustained activity prevent the necessary violent activity in the active muscles at the end of the race. The lore of the locker room is that 400 meters is the most difficult foot race because it involves sprinting for the maximum possible distance.

Local muscle pain develops quickly during activity that requires maximum contraction for sustained periods, such as arm wrestling. It seems likely that the pain derives at least partially from ischemia caused by occlusion of the blood vessels as well as from the high rate of energy expenditure. Whatever its origin, it can be the main factor that determines the outcome of the event when the contestants are evenly matched.

Chapter 9

SMOOTH MUSCLE

Smooth muscle differs from striated muscle in many respects, not the least of which is that it is less well understood. In contrast to striated muscle, there is also more controversy over basic mechanisms in smooth muscle. The controversy is mentioned here, at the outset, because this chapter will present a view that is decidedly heretical in some circles. This view is presented in an effort to provide a unifying theory of smooth muscle function. Some or all of this theory may later be proven wrong, but an effort has been made to provide sufficient description that new evidence and theories can be incorporated into the framework presented.

In many ways, smooth muscle is more similar to non-muscle motile cells than it is to striated muscle. These motile cells include white blood cells that migrate through body tissues well as some unicellular organisms. The functional similarity between the two cell types is that they both undergo large plastic changes. Motile cells undergo large shape changes while smooth muscles undergo large length changes. The principle underlying the proposed theory is that the need for plastic changes requires both cell types to have filament architectures that can change quickly, and that the necessary mechanisms produce functional differences from striated muscle. It is further suggested that these architectural changes are the reason that smooth muscle lacks sarcomeres.

LENGTH RANGE OF SMOOTH MUSCLE

Anyone who has watched x-rays of hollow viscera enclosed by smooth muscle must be impressed by the large length changes in the muscle. A human urinary bladder can be reduced from a volume of nearly one liter to 1-2 ml in 1-2 minutes. The cavity of a uterus that holds an entire fetus, placenta, and fluid amnion is reduced to less than the size of a human fist very shortly after delivery of the infant. Similarly, the gut undergoes enormous circumferential changes. The large length changes in muscle of urinary bladder were confirmed by Uvelius (1976). He correlated the pressure-volume relationship of bladder with the length-tension relationship of bladder muscle strips and the length changes in the individual muscle cells dissected from bladders fixed at different volumes. Some of his results are summarized in Fig. 9.1. The length tension relationship of strips of muscle (Fig. 9.1 A) showed a more than 7-fold range of functional length. If the bladder contracts uniformly in all dimensions, the volume change is proportional to the third power of the circumference change. The 7-fold length range observed by Uvelius would therefore accommodate a 343-fold change in volume. There is a small and constant volume of tissue enclosed by the muscle, so that the volume change of the bladder contents undergoes a larger relative change than the muscle envelope. Thus, the 343-fold range of volume enclosed by muscle is sufficient to produce a 500- to 1000-fold change in the urine volume within the bladder.

Uvelius' photographs of the single cells dissected from the walls of bladders fixed at four different volumes (Fig. 9.1 B) exhibited a 7-fold range of length, similar to that of the muscle strips. This observation makes the important point that the large volume changes are caused directly by length changes of the muscle cells and are not due to architectural changes within the bladder wall or to change in extracellular connections.

The length-force relationships described in Chapters 2 and 6 show that skeletal muscle has a total length range of about 3-fold. At a sarcomere length of 3.65 um there is no force developed because there is no overlap of the actin and myosin filaments. At sarcomere length of 1.2 um there is no force developed because the filaments are so packed together that they develop contraction

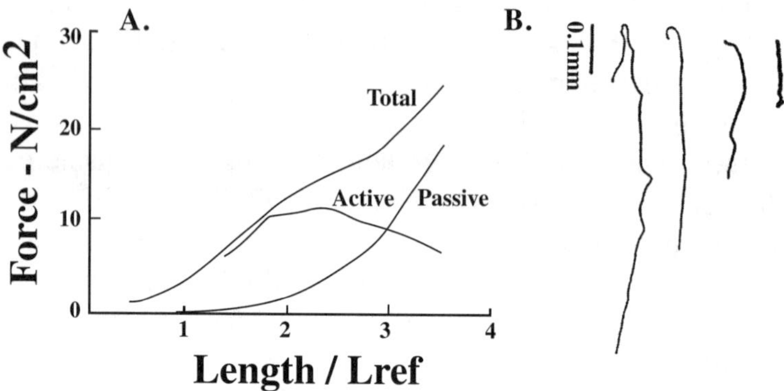

Figure 9.1. *A) Length-force relationship in smooth muscle of the urinary blad-*
der. B) Photographs of individual cells dissected from bladders fixed at differ-
ent volumes. Redrawn from Uvelius (1976).

bands and become irreversibly damaged. By contrast, the length-
force relationship for smooth muscle in Fig. 9.1 A does not extend
to the longest lengths where developed force would reach zero,
and it has more than twice the range of skeletal muscle. Furthermore, smooth muscle is not damaged when it contracts to very
short lengths. These comparisons suggest strongly that the large
length changes of smooth muscle cannot be accommodated by a
fixed array of filaments, as is found in skeletal muscle. Some other
explanation must be sought. The explanation offered here is that
the muscle undergoes plastic changes that places variable numbers
of contractile units in series. By "contractile units" is meant a half-
sarcomere-like structure with half-thick filaments and associated
thin filaments arrayed in parallel across the muscle cell.

FUNCTIONAL EVIDENCE OF STRUCTURAL PLASTICITY

As will be described later, the filament structure of smooth
muscle has been difficult to interpret from micrographs. As a con-

sequence, Pratusevich et al. (1995a) attempted a functional test of whether there are more contractile units in series at different muscle lengths. They reasoned that since each contractile unit was in series with all the others, they would all experience the same force. On the other hand, length changes in each unit would add incrementally to the overall muscle length change. Thus, the length change required to produce a given dynamic force would vary directly with the number of contractile units in series. A good test of this hypothesis was to measure isotonic force-velocity curves. At each isotonic load, the length of the series elastic elements is constant because their force is constant, and all the shortening is produced by the contractile units in series. In addition, if the only effect of adaptations to different lengths is a change in the number of units in series, the velocity at each load should scale in proportion to the number of units in series. This consideration provided an additional test of the theory. If the length change altered only the number of units in series, without changing their performance, the curves obtained at different lengths could be superimposed by multiplying all the velocities by a constant factor. If, on the other hand, the length change altered the contractile function, such scaling would not occur.

The results from one experiment on a dog tracheal muscle are shown in Fig. 9.2. The muscle was adapted at three lengths, 100%, 67%, 33% of the length where rest force began to rise steeply. The force-velocity data for the middle length (solid curve) were fitted with a curve and this curve was scaled up or down by multiplying velocity by a constant scale factor to match the data obtained at each other length. As shown, there was excellent agreement between the predicted result and the data.

The results of multiple experiments are summarized in Fig. 9.3. Over a 3-fold range of length there was only a slight dependence of developed force on adapted muscle length but a strong dependence of shortening velocity. It would be difficult to account for this data with any of the known mechanisms in muscle without a change in the number of contractile units in series.

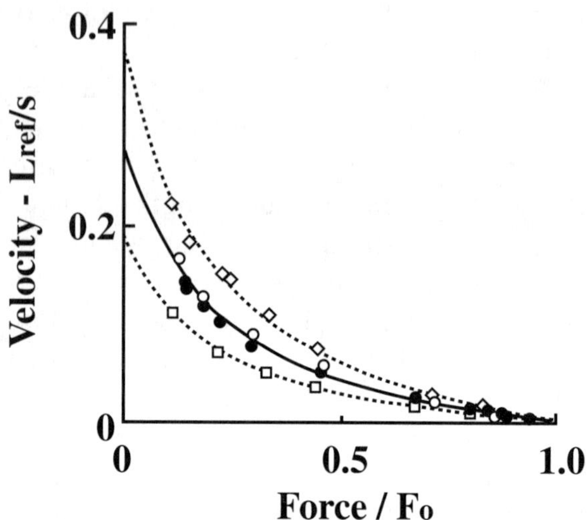

Figure 9.2. *Force-velocity data obtained at three lengths. Solid curve fitted to the middle set of data is scaled by multiplying velocity by a single factor to superimpose the data obtained at each other length. From Pratusevich et al. (1995), with permission.*

THICK FILAMENT EVANESCENCE

The proposed plasticity would not be controversial if it were explained by a simple rearrangement of formed filaments. It seems likely, however, that this plasticity would be greatly facilitated by thick filament evanescence, and therein lies a controversy.

When the significance of the myofilaments in striated muscle was first discovered, similar structures were sought in electron micrographs of smooth muscle. Thin filaments were always seen, but thick filaments were much rarer. Some workers saw thick filaments only infrequently, some saw them sometimes, while others saw them always. The reported shapes of the filaments also varied. Some saw round thick filaments similar to those in striated muscle, others saw ribbon-shaped thick filaments, and some saw giant filaments. Since the filaments are not arranged into sarcomeres, it is necessary to measure each filament carefully to determine

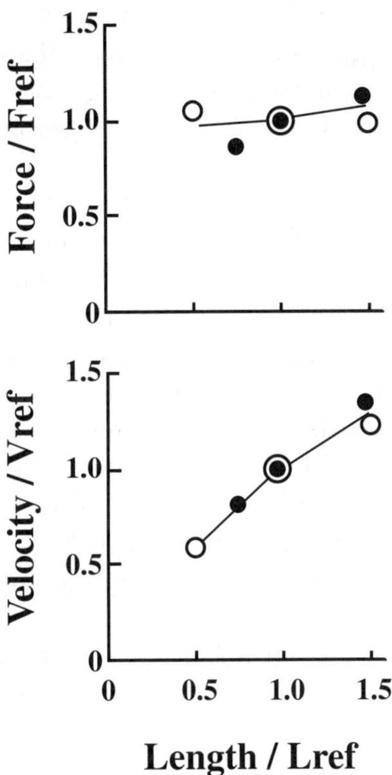

Figure 9.3. *Effect of adapted muscle length on force and velocity. From Pratu-sevich et al. (1995), with permission.*

whether it is thick, thin, or intermediate, so that it seemed likely that some thick filaments might be overlooked if careful measurements were not made. In addition, it became recognized that the thick filaments of smooth muscle are more labile than those of skeletal muscle. Special precautions were required to preserve them during preparation for microscopy, but the question also arose as to whether these precautionary measures precipitated rather than preserved the filaments.

It was soon noted that the thick filaments were more abundant when the tissue was prepared for electron microscopy at a slightly acid pH (Kelly and Rice, 1968) and in high calcium (Schoenberg,

1969). Since these conditions mimic activation, Schoenberg (1969) and Rice el al. (1970) proposed that the thick filaments might be evanescent, reforming during activation and dissolving at least partially during relaxation. This proposal was vigorously resisted by those who had shown that thick filaments were always seen when their methods were used (see review by Somlyo and Somlyo, 1982). Over time, several other laboratories confirmed that these methods reproducibly yielded thick filaments, and it became accepted that thick filaments were a regular presence. In spite of this acceptance, however, there have been sporadic reports of activation increasing thick filament density. Since the electron microscope methods have been controversial, these reports often use other techniques, such as x-ray diffraction (Watanabe et al., 1993) or optical birefringence (Godfrain-DeBecker and Gillis, 1988).

This controversy is described in detail because thick filament evanescence is not widely accepted, even though the proposed structural plasticity would be greatly facilitated if the thick filaments reformed as they are needed. The issue, therefore, is not to determine which of the preparative methods is correct but to seek an explanation for the disparate findings. If thick filaments are evanescent, it might be expected that the preparative techniques which yield more abundant filaments could be extended to produce the giant thick filaments that were seen by some early workers. Even more cogent is the finding that thick filaments are made more sparse by the conditions that mimic relaxation, higher pH and low calcium. This observation indicates that thick filaments are more labile in smooth muscle than in striated muscle and supports the suggestion that the lability contributes smooth muscle function.

Another reason for describing the controversy is that thick filament evanescence may explain several other phenomena in smooth muscle, and these possible explanations must be judged in light of the imperfect understanding of smooth muscle structure. Before describing these other phenomena, the current knowledge of smooth muscle filament structure will be reviewed.

SMOOTH MUSCLE FILAMENT STRUCTURE

There are now known to be three types of filament in smooth muscle. Thin filaments are always seen, and these are attached to structures called **dense bodies**, which are of two types, those attached to membranes and those that are free in the cytoplasm. The thin filaments appear to project from the **cytoplasmic dense bodies** in two directions, with all the filaments pointing from one side having the same polarity with regard to myosin. The **membrane associated dense bodies** have a single polarity, with thin filaments projecting into the cytoplasm. Thus, membrane-associated dense bodies appear to connect the contractile apparatus to the world outside, while cytoplasmic dense bodies appear to serve the same function as Z-lines in skeletal muscle, anchoring the thin filaments and connecting thin filaments of opposite polarity.

Another class of filament, called **intermediate filaments** because they are intermediate in diameter between thick and thin filaments, appears to run diagonally across the cell and connect the dense bodies. Fay et al. (1983) proposed that the dense bodies are arranged something like a fish-net in three dimensions. A schematic arrangement of thin filaments, dense bodies, and intermediate filaments is shown if Fig. 9.4. The intermediate filaments appear to hold the lattice structure together by supporting the dense bodies. It is also possible that they are the parallel elastic structures which prevent the cells from overstretch. This possibility is suggested by some ancillary observations in the functional studies of Pratusevich et al. (1995a) described above. The experiments were made in tracheal smooth muscle, in part because there is little connective tissue in this preparation. The muscles showed very little resistance to stretch until a critical length was reached. A stretch of only a 2–3% above this critical length caused rest tension to rise higher than the developed force, and shortening to 2-3% below this length would lower rest force to less than 15–20% of developed force. Thus, the muscles appeared to have a very stiff parallel structure that was normally slack but that prevented stretch above the critical length. The intermediate filaments might be a good candidate for this parallel structure.

The thick filaments of smooth muscle are the least well understood and the most controversial. They are recognized as round structures somewhat larger than intermediate filaments and similar in size to the thick filaments of striated muscle, although ribbon shapes and very large diameters have been reported. Most microscopists would concede that they are more labile than the thick filaments of striated muscle. It also seems likely that the filaments are **side polar** as opposed to the **end polar** filaments of striated muscle; all the crossbridge heads project from two sides of the filaments without central bare zone (Cooke et al., 1989). This side polar structure is illustrated in Fig. 9.4.

The effect of thick filament evanescence is also illustrated in

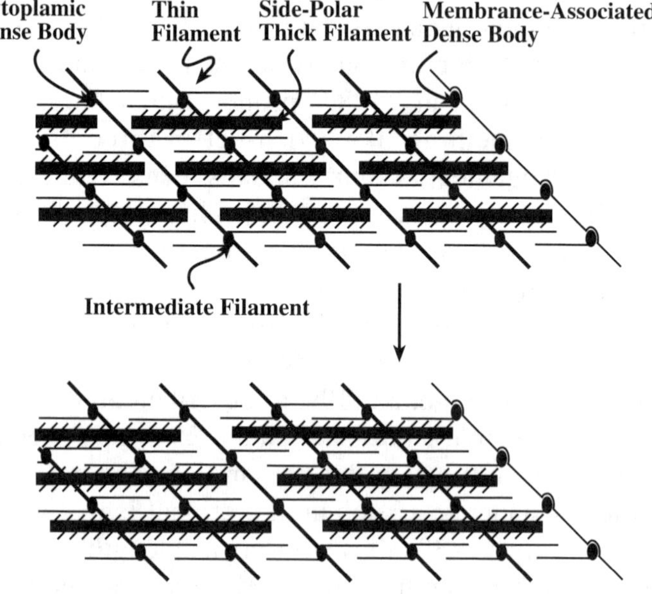

Figure 9.4. *Schematic representation of the filament lattice in smooth muscle. Two thick filament lengths are shown.*

Fig. 9.4. If the filaments do not grow wider or denser as they re-form, they must grow longer. This lengthening would be very difficult to detect in the thin sections cut for electron microscopy because the filaments are frequently truncated by the sectioning, so that serial sections must be examined carefully to find the ends of the filaments. The lengthening will have substantial functional implications, to be described after describing some additional studies supporting filament evanescence.

MYOSIN LIGHT CHAIN PHOSPHORYLATION

Contractile activation of muscles and motile cells also activates enzymes, called **kinases**, that catalyze the transfer of phosphate from ATP to proteins. It is becoming known that a great many proteins become phosphorylated on activation. Other enzymes, called **phosphatases**, act continuously to remove the added phosphate. The functional reasons for these multiple phosphorylations are, as yet, poorly understood, but it appears that these reactions serve to signal activation within the cells. One of the proteins in muscle that becomes phosphorylated is the myosin regulatory light chain. Several functional reasons for this phosphorylation have been proposed in striated muscle, but none are generally accepted. In smooth muscle, however, it was found to have a major role in the activation process (Sobieszeck 1977; Adelstein and Eisenberg, 1980; Hartshorn and Siemenkowski, 1980). Solutions of smooth muscle myosin mixed with actin will not hydrolyze ATP unless the light chains are phosphorylated. Thus, it was proposed and is widely accepted that light chain phosphorylation is the mechanism responsible for activation in smooth muscle.

Myosin light chain phosphorylation has also been found to have another definite effect in smooth muscle. Smooth muscle myosin and the myosin-like proteins of non-muscle motile cells are similar to striated muscle myosin in that they have an S-1 head region that interacts with actin to hydrolyze ATP. They differ from striated muscle and from each other in the tail region and in their affinity for forming thick filaments. Three separate groups of workers nearly simultaneously discovered that thick filaments were formed only when the myosin was phosphorylated (Trybus et al., 1982; Onishi and Wakabayashi, 1982; Craig et al., 1983). In addition, electron micrographs of individual molecules showed the tails to be folded so as to prevent thick filament formation when light chains were unphosphorylated. Phosphorylation causes the tails to unfold and allows filament formation. Thus, it appears that phosphorylation produces two distinct effects, activation of ATPase and structural changes that promote thick filament formation. The issue to be resolved is how these two functions are related.

It should be emphasized that while myosin light chain phosphorylation appears to be necessary to initiate contraction, it is not needed to sustain contraction. Most experiments suggest that the level of phosphorylation normally reaches a peak early in a tetanus and then declines to near baseline levels while force is maintained. It is also possible, however, to prevent the dephosphorylation. Under some circumstances it is possible to dissociate relaxation from dephosphorylation (Gerthoeffer, 1986; McDaniel et al., 1992, Bárány et al., 1994). These observations suggest strongly that light chain phosphorylation is not needed to sustain activation once it is initiated.

SMOOTH MUSCLE ACTIVATION

Although it is generally believed that smooth muscle is activated by light chain phosphorylation, it is also well known that the thin filaments contain several regulatory proteins, and one of these, tropomyosin, is responsible for activation in skeletal muscle. Thus, there appear to be two activating systems in smooth muscle. The proposal is made here that the primary function of light chain phosphorylation is to regulate thick filament formation and that its role in activation is to prevent the non-filamentous myosin from interacting with actin to hydrolyze ATP. The need for a second activating system arises because there will be a period of time when myosin is phosphorylated but not yet incorporated into thick filaments. Absent another mechanism, these individual myosin molecules would be capable of interacting with thin filaments to hydrolyze ATP without doing useful work. The second activating system allows thin filaments to distinguish myosin incorporated into thick filaments. How this distinction is made is unknown, but skeletal muscle tropomyosin confers cooperativity on the activation mechanism. Cooperativity is the type of mechanism that would allow thin filaments to recognize high concentrations of myosin, as are found in thick filaments. Two other regulatory proteins have been found to be associated with thin filaments, **calponin** and **caldesmin**. The functions of these proteins are not understood; they might help in the thick filament recognition process.

The events associated with activation of both smooth muscle and non-muscle motile cells are less well understood, and possibly more complex, than those of striated muscle. Calcium is released into the cytoplasm from the sarcoplasmic reticulum. The signal for this release is transmitted from the surface membrane to the sarcoplasmic reticulum by a chemical transmitter, **inositol triphosphate**, abbreviated **IP$_3$**, The calcium activates calmodulin which facilitates phosphorylation of many proteins, including myosin light chain. Whether calcium has any direct role in activating the thin filaments is uncertain, but it has been suggested that it binds to the calponin on the thin filaments and that calponin has a role similar to troponin in striated muscle.

VELOCITY SLOWING

It is well known that shortening velocity slows during the rise of activation. Dillon et al. (1981) found that the velocity decline correlated with dephosphorylation of myosin light chain. They interpreted these observations as suggesting that the dephosphorylation created a slowly cycling crossbridge that impeded shortening. Mollusks possess muscles that can be stimulated to contract and then go into a state called **catch** in which force and stiffness are maintained with very little energy expenditure, presumably because the crossbridges remain attached but cycle very slowly. These catch muscles allow the mollusk to cling to a rock or maintain its shell closed with very little energy expenditure. Dillon et al. proposed that vertebrate smooth muscle might do the same, and they named their slowly cycling bridge a **latch** bridge. It was a clever suggestion, and the theory became popular. There have, however, been more recent reports suggesting that the velocity slowing and dephosphorylation are not always correlated, casting doubt on the specific latch bridge hypothesis. It should also be mentioned, however, that the hypothesis is but one of a class of metabolic mechanisms that could cause slowing. Anything that decreases crossbridge cycling rate would cause slowing. The depletion of substrate or accumulation of a reaction product, as well as a change in the crossbridge itself, could also be responsible for the slowing. The reason for classifying these theories together as metabolic mechanisms is that thick filament evanescence suggests an entirely different, anatomic mechanism for the slowing.

Series-to-parallel transition

As described above, thick filament formation during activation is very likely to cause the filaments to lengthen. This lengthening, illustrated schematically in Fig. 9.4, would be difficult to recognize in electron micrographs because the only difference in the filament lattice is the length of the thick filaments. This lengthening will, however, have large functional consequences. Since all the crossbridges along one side of a filament operate in parallel, thick filament lengthening will increase the force generated by the individual filaments. If the number or filaments in the cross-section of the muscle is unchanged, the force generated by the muscle will be increased. If the function of the individual bridges is not altered, the velocity of the thick filament relative to the thin filaments will not be altered, but since fewer filaments are required to span the length of the muscle, the overall muscle velocity will be diminished. In summary, thick filament reformation is likely to cause a series-to-parallel transition of the crossbridges arrangement, and this transition will increase muscle force while decreasing muscle velocity. Preliminary analysis of the force-velocity data from smooth muscle suggests that it can be explained quantitatively by this anatomic mechanism (Pratusevich et al., 1995b). Such a coincidence does not, however, constitute proof. The possibility of this mechanism is described here mainly because it is an expected outcome of thick filament evanescence.

A UNIFYING HYPOTHESIS

The suggestion that thick filaments are labile leads to a theory that this evanescence facilitates plastic changes that allow the smooth muscles to accommodate to very large length changes. This theory has two corollaries. The first is that there is more than one activating mechanism. One of these, myosin light chain phosphorylation, primarily regulates the incorporation of myosin into thick filaments but also controls the myosin ATPase to prevent free myosin from interacting with thin filaments. The other activating mechanism is thin filament regulation that both controls activation directly and enables the thin filaments to recognize myosin in thick filaments. The second corollary is that the well-known slowing of shortening velocity is due to a series-to-parallel transition of the crossbridge array caused by thick filament lengthening. Of the three proposals in the hypothesis, the primary postulate of plasticity and two corollaries, only the second corollary, slowing due to thick filament lengthening, requires absolutely that thick filaments form on activation.

Smooth muscle and non-muscle motile cells have many features in common. They both lack sarcomeres, they both possess abundant thin filaments that project from dense bodies. They both have myosin-like motor proteins that must be phosphorylated to interact with thin filament and hydrolyze ATP. These motor proteins have tails that are folded when not phosphorylated, and the tails unfold to allow the formation of thick filaments when the light chains are phosphorylated. Both cell types undergo similar events during activation, including the IP_3-mediated release of calcium from internal stores, the activation of calmodulin, and the phosphorylation of myosin light chains that is required to initiate activation. Both cell types have a requirement for plastic changes of structure. It is proposed here that the many similarities between the two cell types result from their common need for plasticity.

Chapter 10

COMPARISON OF CARDIAC AND SKELETAL MUSCLE

Much of what we know about muscle function has been learned from studies of skeletal muscle. A major issue raised in Chapter 1 and again at the end of this chapter is that skeletal muscle is much easier to study than cardiac muscle, so that the knowledge derived from skeletal preparations is more certain than that obtained from cardiac muscle. On the other hand, cardiac muscle has more intrinsic control, so that the consequences of its derangements are more varied and more frequently seen in a clinical setting. The central role of the heart in sustaining life, and the high incidence of disease and mortality resulting from cardiac malfunction, also make the heart an important object of study. Since some of the knowledge of cardiac function derives from studies of skeletal muscle, it is important to recognize both the similarities and differences between the two muscle types before progressing to the chapters describing their separate functions. These similarities and differences are listed in Table 10.1 and the remainder of the chapter describes the comparisons in the list.

It is generally found that analogous proteins in different muscle types are different, and the reasons for these variations are not completely known at present. The genome often carries more than one copy of the gene for a protein, so that the absence of one gene will not result in absence of the protein. For proteins that have a vital function, such as those of muscle, duplication may confer a decided survival benefit on the species. Multiple copies of a gene also allows divergence of proteins that once subserved the same

Table 10.1. *Heart vs. Skeletal Muscle*

Heart	Skeletal
Actin-myosin filaments	*Actin-myosin filaments*
Slow myosin isoforms	*Fast and slow myosin isoforms*
Ca regulation of activation	*Ca regulation of activation*
Long action potentials	*Brief action potentials*
Intrinsic activation	*Extrinsic (nervous) control*
Nervous modification of intrinsic activation	*Complete nervous control*
Large variation in contractile strength of individual cells	*Small variation in contractile strength of individual cells*
Hormones and drugs have regulatory effects	*Hormones and drugs have small effects*
Small cells	*Large cells*
Electrical syncytium	*Individual syncytia*
Rest tension large and important at functional lengths	*Rest tension unimportant at functional lengths*
Functions over lengths where length tension relationship has a positive slope	*Functions over lengths where length tension relationship has a positive slope.*
Functions at shorter sarcomere lengths	*Functions at longer sarcomere lengths*
No plateau of length-tension relationship	*Distinct plateau of length-tension relationship*
Produces twitches only	*Produces twitches but usually functions with tetani*
Studied with twitches and only over steep part of length-tension relationship	*Studied at constant activation over flat part of length-tension relationship*

function in different tissues, so that tissue specialization through natural selection becomes possible. The functional differences between analogous proteins in different muscles are still in the process of being elucidated, and only those which confer the most obvious functional differences will be described.

These differences between similar proteins are mentioned because it has been observed several times that while cardiac and skeletal muscle use different mechanisms to perform the similar functions, each muscle type possesses the mechanism of the other and uses it in a more limited manner. One obvious example is the different mechanisms for signal transduction between t-system and sarcoplasmic reticulum. There are others.

SIMILAR CONTRACTILE MECHANISMS

The myofilaments of cardiac and skeletal muscle are very similar. Although the proteins are not usually identical, in some animals one of the myosin isoforms in the two muscles derive from the same gene and is therefore identical. The variations in other isoforms produce the differences in speed and power of the various muscle types. There are minor and incompletely understood differences in the actins of the two muscle types. By contrast, the regulatory proteins, and particularly the troponins, are distinctly different, and these differences undoubtedly reflect the functional difference in the two muscle types.

The main point of the observation that the contractile elements are very similar in the two muscle types is that it justifies the extrapolation of mechanical data from skeletal to cardiac muscle.

MUSCLE SPEED

The only function of cardiac muscle is to produce a relatively slow, rhythmic contraction, with the power output of the heart matched to the impedance of the circulation. If energy were expended in developing pressure more rapidly than the circulation could absorb as flow, it would be wasted, and perhaps harmful. As a result, cardiac muscle contracts slowly and has two isoforms of myosin that produce two shortening velocities. On the other hand, skeletal muscle has more varied functional requirements and very different speed and power requirements. At one extreme are the postural muscles that may be required to produce low levels of force for hours. At the other extreme are the **oculomotor** muscles responsible for eye movement, which contract extremely rapidly for very brief periods. In general, the different functional requirements are subserved by specialized muscle cells that operate at different speeds. Slow muscles have slow myosins, fast muscles have fast myosin, and some very fast muscles, such as the oculomotor muscles, have super-fast myosins. Muscles of intermediate speed have mixtures of fast and slow myosins, and often the individual fibers of a single whole muscle will be of different speeds, as might be required for varied function required of the whole muscle. In fact, it is unusual for a muscle to have a large predominance of one fiber type, and this important point must be kept in mind when selecting a muscle for experimental studies. Furthermore, it is possible to change the type of myosin expressed by a muscle by changing its stimulation pattern and loading. Such changes in myosin isoforms occur during normal development of the heart and have served as a model for studying genetic expression.

Of equal importance with the speed of the myosin is the expression of enzymes and other proteins responsible for supporting the metabolism of the muscle. Myoglobin is the most obvious of these because it imparts a red color to muscles, but other enzymes can be recognized by histochemical staining and are used to classify the different fiber types. In general, fast fibers have enzymes that support rapid, anaerobic metabolism while slow muscles have enzymes that support sustained oxidative phosphorylation. Myoglobin is a marker for slow muscles. It binds oxygen more tightly than hemoglobin and therefore accepts hemoglobin carried to the muscle by the blood. The stored oxygen is then used as a buffer to avoid hypoxia in cells that normally contract for extended periods but may be required to have brief periods of increased activity. In a study of animals having a 25,000-fold range of body size, Seow and Ford (1991) found that the muscles of the larger animals, such as cow, were slower and much redder than those of smaller animals, such as mice, but within a given animal, the fast muscles were recognizably paler than the slow muscles.

CALCIUM ACTIVATION

Both cardiac and skeletal muscle are activated by calcium that is released from the sarcoplasmic reticulum and binds to troponin. Thus, the very basic mechanisms of activation are the same, but the differences between the two muscle types begins to become apparent when the details of the two activating systems are examined. As mentioned in Chapter 5, the mechanism of signal transduction from t-system or sarcolemma to sarcoplasmic reticulum is different in the two muscle types. The voltage sensor in the t-system and sarcolemma of cardiac muscle functions as a calcium channel that conducts calcium during the plateau of an action potential. This calcium in turn activates the calcium-induced calcium release from the sarcoplasmic reticulum. The voltage sensor in the t-system of skeletal muscle is capable of conducting calcium currents but mainly functions as a mechanical link to the sarcoplasmic reticulum. Mechanical movements of this link open and close the calcium channels in the sarcoplasmic reticulum.

Cardiac and skeletal muscle have differences in the troponin-tropomyosin complexes which result in differences between calcium and force that at first seem minor but are of substantial significance. It appears that cardiac muscle is inactivated more readily by filament sliding than skeletal muscle. This is probably explained by cardiac muscle having stronger cooperativity in its activating mechanisms. Cardiac muscle also appears to have a steeper isometric length-force relationship under physiological conditions, as explained in Chapter 6.

ACTION POTENTIAL DURATION

The electrical responses of the surface membrane, called **action potentials**, are compared for cardiac and skeletal muscle in Fig. 10.1. Most of the action potential of skeletal muscle is a brief spike that returns to near baseline before contraction begins, whereas the action potential of cardiac muscle has a long plateau that outlasts the peak of the contraction. For both muscle types, the rapid upstroke of the action potential is caused by an inward sodium current that is quickly inactivated. In skeletal muscle, the initial depolarization is reversed almost immediately, both because the sodium current is inactivated and because an outward potassium current increases. The plateau of cardiac muscle action potential is maintained by an inward calcium current that is inactivated much more slowly than the sodium current. As the calcium current inactivates, the outward potassium current increases to hasten repolarization.

The sustained calcium current in cardiac muscle has two effects. One is to prevent rapidly repetitive action potentials that would cause sustained contractions, as explained under the heading of "pump vs. posture," below. The other is to make the strength of contraction sensitive to changes in extracellular calcium.

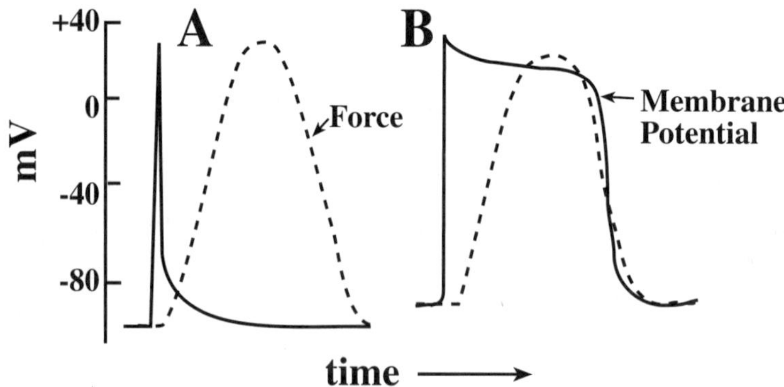

Figure 10.1. *Action potentials and superimposed twitch force records of A) skeletal muscle, adapted from Bucthal and Sten-Knudsen (1959), and B cardiac muscle, adapted from Nathan and Beeler (1975).*

INTRINSIC vs EXTRINSIC ACTIVATION

Many cardiac muscle cells have intrinsic rhythmic electrical activity. Repolarization from the action potential is hastened by an increased outward potassium current that causes a slight hyperpolarization after the resting potential is restored. The restoration of the resting potential causes the potassium current to decrease to its baseline level, with a concomitant rise in the membrane potential. This rise results in another action potential. The farther a cell is from the **sino-atrial node**, that serves as the normal pacemaker, the slower the intrinsic rate. This node is located high in the upper chambers of the heart, called the **atria**. If the sino-atrial node fails to fire on time, a pacemaker lower in the heart will generate an action potential that initiates a heartbeat. The next cells to fire are sometimes other cells in the atrium but more often cells in the **atrio-ventricular node** that forms the junction between the atria and the ventricles. If the atrio-ventricular cells fail to fire after a more extended time, cells in the ventricle will produce a spontaneous but very slow rhythm.

The reason the heart has evolved to have so many reserve pacemakers is that it functions automatically and is capable of being completely independent of the nervous system. This independence of the nervous system is demonstrated clinically in patients with transplanted hearts that have no nerve supply. They demonstrate that adequate the perfusion of the body with blood will occur in the absence of any input from the central nervous system. Skeletal muscle has no automaticity and is controlled completely by the central nervous system.

NERVOUS INNERVATION

The diversity of mechanical activity in the body requires an enormously complex executive control system. Something as commonplace as the upright posture in humans requires continual monitoring and readjustment of muscle tone. It seems likely that intrinsic activity of the muscles performing these functions would greatly complicate the task of the central nervous system. Thus, the membranes of skeletal muscle have evolved to avoid spontaneous activity, even though the mechanisms for such automaticity are present. By contrast, the automaticity in the heart confers an enormous survival benefit on an animal at times when its central nervous system is temporarily inactivated. To meet the varying demands of the body, the heart can adjust its output several-fold by regulating the intrinsic automaticity. While this regulation is accomplished by alterations in nervous tone, the tone modulates rather than creates the intrinsic activity.

VARIATION IN CONTRACTILE STRENGTH OF THE CELLS

It is often said that variations in the strength of skeletal muscle contraction are caused by **recruitment** of varying numbers of cells, i.e. that more or fewer cells are stimulated to contract. This statement arises from the observation that an individual cell produces an all-or-none contraction with each stimulation and shows little variation in the strength of contraction. The work of Rack and Westbury (1969) described in Chapter 6 suggests a different but similar mechanism of recruitment. Groups of cells supplied by a single neuron, called **motor units**, are stimulated synchronously at varying rates. As one group of cells begins to relax another begins to contract, maintaining force constant in the whole muscle even though the force in the individual cells may be varying widely. This might be called a "recruitment in time" to distinguish it from the older view of a "recruitment in number." It is also possible that both types of recruitment occur simultaneously, with the strength of contraction regulated both by the numbers of cells stimulated and by the frequency of stimulation.

Recruitment of skeletal muscle cells is described here because such recruitment almost certainly does not occur in cardiac muscle. Under normal circumstances, every cardiac cell is activated in every heartbeat. It is also true, however, that there is a great need for the heart to vary the strength of its contraction. These variations are accomplished by varying the contractile strength of the individual cells. The responsible mechanisms are described in Chapter 17 on inotropic mechanisms. A point to be made in this chapter is that some of the differences between heart and skeletal muscle, described in the next two sections, may result from the need for changes in contractile strength.

THE ROLE OF HORMONES AND DRUGS

The variation in contractile strength of individual cells is caused by extracellular substances that either interact with receptors on the cell membranes or pass through channels or transporters in the membranes to have a direct effect on intracellular activity. An example of the latter type of substance is calcium. Examples of the former are neurotransmitters such as adrenaline, nor-adrenaline, and acetylcholine. The presence of these membrane receptors, channels, and transporters make the cells sensitive both to the physiological substances that influence contraction and to drugs that mimic or block the actions of these substances. Cardiac muscle appears to be much more sensitive to the interventions that vary contractile strength, but it is frequently observed that similar alterations in contractile responses can be found in skeletal muscle, albeit at a lower magnitude. This is yet another example of both muscle types having the same mechanisms but using them in different proportions. The next section makes the point that the greater sensitivity of cardiac muscle to the external influences is facilitated by the small size of the cells.

CELL SIZE

Cardiac muscle cells have diameters one-tenth or less than those of skeletal muscle cells and lengths that are very much shorter. The difference in cell size confers substantial functional differences on the two cell types. An action potential is conducted much more rapidly in a cell with a large diameter, mainly because the intracellular electrical resistance is reduced. A rapid action potential is required in a skeletal muscle cell to provide longitudinally uniform activation. The absence of such uniform activation would cause the muscle to pull itself apart. The faster the muscle, the greater the need for speed. In general, faster skeletal muscle fibers have larger diameters, although there are many exceptions to this generalization, as might be expected from the observation that use causes hypertrophy.

Although large cells conduct action potentials more rapidly, they have the disadvantage of a smaller surface-to-volume ratio; the diffusion distances to the core of the fiber are greater and there is less membrane area for the transport of material. Slow muscle cells, which have evolved to operate continuously for extended periods, have both smaller diameters and higher myoglobin concentrations, which provide a more steady supply of oxygen. Cardiac muscle, which is nearly the slowest type of striated muscles, has by far the smallest diameter and the most myoglobin.

Although a rapid spread of activation is required for some aspects of whole heart function, the speeds required are much less than in a long skeletal muscle. For other aspects of cardiac function, such as in the atrioventricular node, there is a definite need for slow conduction to allow the contracting atria to fill the relaxed ventricles. The main point to be made from these observations is that the heart does not require the benefits provided by a large diameter. In addition to the benefits of lesser diffusion distances and greater membrane area for the passage of metabolites, the small diameter of cardiac cells make them more susceptible to the effects of externally applied agents. The larger surface-to-volume ratio allows more area for membrane receptors per volume of muscle. Finally, the diffusion of regulatory substances, such as calcium, into the cell are enhanced by the small cell diameters.

CELL-TO-CELL CONNECTIONS

Since skeletal muscle cells are stimulated by their motor neurons, there is no need for the individual cells to be connected. In fact, such connections would only complicate the task of the central nervous system in regulating activity. On the other hand, there is a need for very large cells that can be activated rapidly along very long lengths. It is not clear how many alternative solutions might exist for this problem, but Nature has provided these long and large diameter cells by the fusion of many cells into syncytia. One benefit of this solution is the multiple nuclei in each fiber. It would appear that many of these individual cells persist for the life of the animal, so that a mechanism for ensuring survival is provided. The multiple nuclei in these syncytia may be one such mechanism. If one nucleus fails, its function will be replaced by the others. The distribution of nuclei along the length of the cell also reduce the diffusion distances from a nucleus to the extreme ends of the fiber.

Unlike skeletal muscle, cardiac muscle is activated by its own intrinsic mechanisms. To coordinate the activity of the different cells, they are all connected by **gap junctions**. These are regions where the cell membranes of the two cells are tightly apposed and are spanned by channels that conduct small molecules including the ions that carry electrical currents. Fig. 10.2 is an electron micrograph of a gap junction showing the membranes coming together.

The multiple discrete cells of the heart makes it much less susceptible to injury. When a cell is damaged, the connections through gap junctions to neighboring cells are closed. If the heart were a complete syncytium, damage to any region could completely disrupt cardiac function.

REST TENSION

All muscles contains passive elements that hold them together. The force in these elements at physiological lengths is not

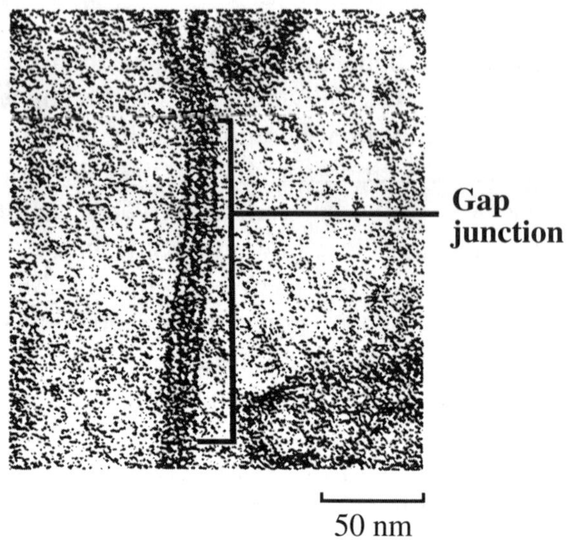

Figure 10.2. *Electron micrograph of a gap junction between two cells.*

great in skeletal muscle because there is little danger of the muscles being overstretched. Unless a bone is broken, the skeleton protects muscles from damaging overstretch. Furthermore, if a skeletal muscle resisted stretch with much passive force in some postures, its antagonist muscle would have to produce sustained force to maintain the posture. In general, sustained muscle activity is only required to maintain extreme body positions.

Cardiac muscle does not possess such a protective skeleton, and as a consequence, the protection must be provided by structures within the muscle itself. In addition, there must be a relatively steep positive slope to the resting length-force relationship to allow the heart to resist increased volumes with increased pressure, in spite of the Laplace phenomenon (described in Chapter 14). These are some of the reasons that cardiac muscle possesses relatively stiff parallel elastic elements that cause steep rises in

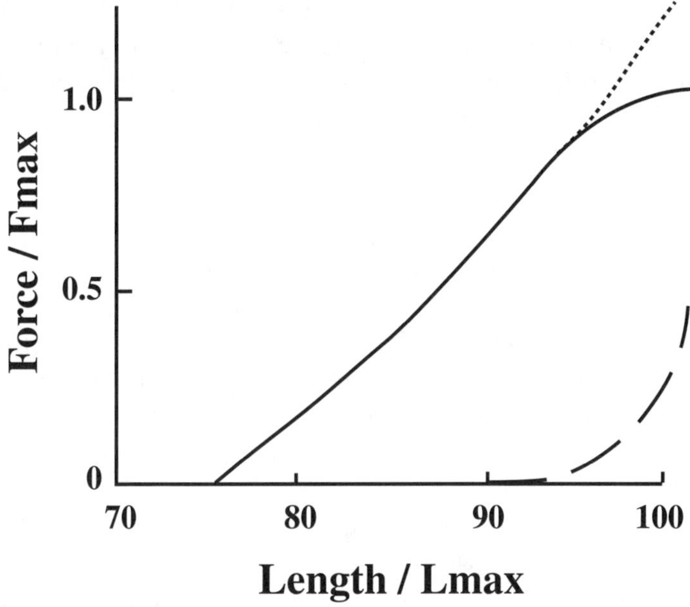

Figure 10.3. *Active (solid), passive (dashed), and total (dotted) force curves for cardiac muscle. L_{max} is the length at which maximum force is developed.*

pressure when muscle length is stretched to near the peak of its active length-force relationship (Fig. 10.3). As described in Chapter 1, not all the responsible structures have been identified, but it is known that cardiac muscles are irreversibly damaged when stimulated to contract at lengths longer than the optimum for force development.

FUNCTIONAL LENGTH RANGE

As described in Chapter 6, both types of muscle function over length ranges where the length-force relation has a positive slope, and natural selection has provided several mechanisms to ensure that this positive slope always occurs under physiological conditions. This positive slope occurs even over sarcomere lengths where the overlap of thick and thin filaments is diminished, particularly in skeletal muscle, which appears to function at longer sarcomere lengths than cardiac muscle. The reason for the different ranges of sarcomere length in the two muscle types may be related to their different functional requirements.

When a muscle is stimulated to contract below a certain length, approximately equivalent to the length where the thin filaments overlap in the center of the sarcomere, the muscle re-lengthens as it relaxes. This appears to be a passive phenomenon caused by an elastic structure that is deformed at short sarcomere lengths, but the nature of the structure has not yet been elucidated. In skeletal muscle such elongation might be nuisance. A muscle that contracted to bring a limb into a desired posture and then relaxed would have to shorten again before it could resist a stretch that reversed the initial change. At least some of the available data (e.g. Rack and Westbury, 1969) suggests that the shortest functional length of skeletal muscle is approximately equivalent to the length at which the muscle begins to passively resist shortening, although this issue has not been studied extensively.

In cardiac muscle, the elongation of the muscle may have several benefits. It helps to reduce the intraventricular pressures during early diastolic filling and may provide a negative pressure to suck blood into the ventricle. In addition, it insures that the muscles begin their contractions from some minimum length. Since the muscles of the heart always shorten to eject blood during systole, it is important that they not begin from very short lengths. In extreme circumstances cardiac muscle can be damaged by prolonged and vigorous stimulation at short lengths and low loads. An example of such damage has been described by Walley and Cooper (1991), who studied the effects of prolonged and severe reductions in blood volume. These volume reductions reduce both the venous pressure that distends the ventricle and the arterial pressure that resists ventricular emptying. In addition, the animals maintain very high levels of adrenaline that increase the force of muscle contraction. Walley and Cooper found that after several hours in this state, the hearts became damaged in a manner suggesting that their muscle had become irreversibly shortened. The reason for describing these experiments here is to point out that such damage does not occur except in extreme circumstances and that one probable reason is that the muscle normally operates over lengths where it relengthens itself.

PUMP vs POSTURE

The heart's only function is to provide rhythmic pumping of blood. The rhythmicity requires that the muscle relax fully immediately after the heart empties. A sustained contraction would be incompatible with life, and so heart muscle produces only twitches. As explained above, relaxation begins before the electrical action potential ends, thereby ensuring that a new action potential cannot be interjected before the muscle relaxes.

By contrast, skeletal muscles must produce contractions that are longer than a single twitch. A muscle that produced only twitches would be useless in maintaining posture or carrying out most activities. Tetani are possible because the cells generate action potentials at a high rate, with the period between them being substantially less than the time required to a achieve peak force.

ACCESSIBILITY TO EXPERIMENTAL STUDY

Several properties described above conspire to make the study of cardiac muscle much more difficult. All measurements of force must be corrected for the force borne by the parallel elastic elements. While this would not seem an impossible task, it is complicated by the transfer of load from the parallel elastic elements to the contractile elements as the muscle shortens. Such shortening occurs both when the muscle is not isometric and when the contractile elements shorten to stretch the series elastic elements during the rise of isometric force. In addition, the internal structure that resist shortening and elongate the muscle during relaxation provide an unmeasurable load on the contractile elements. Although this internal load may be small, the hyperbolic nature of the force-velocity relationship causes the curves to approach the velocity axis at a sharp angle, so that a small load has a proportionately larger effect on maximum velocity. As suggested in Chapter 7, the large effects of these small loads make estimates of maximum velocity unreliable in cardiac muscle.

The inability of cardiac muscle to sustain constant activation levels is also a major complication for the study of its function. Although it can be tetanized under very non-physiological conditions (Forman et al., 1972), these conditions appear to produce maximal levels of activation. Many of the interesting features of cardiac function relate to its ability to vary the strength of its contraction, and studies done at maximum strength shed little light on the relevant mechanisms.

Finally, there is not a satisfactory preparation for the mechanical study of cardiac muscle. Adequate methods have not yet been developed for the study of the very tiny cells in the same way that single skeletal muscle cells are studied. Bundles of cells as are found in trabeculae and papillary muscles have low length/width ratios and frequently are too thick to be studied as skinned preparations. For all of these reasons, a great deal of what we believe we know about cardiac muscle is extrapolated from skeletal muscle studies.

PART II

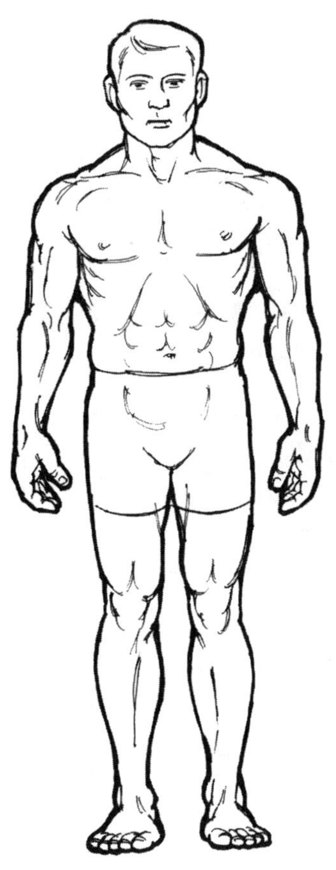

WHOLE BODY FUNCTION

Chapter 11

SOME CONSEQUENCES OF BODY SIZE

It has been known for centuries that small animals have higher metabolic rates and power-to-weight ratios than large animals. Galileo observed that a small dog could carry two or three other small dogs on its back, a man could carry one other man, but a horse was incapable of carrying even one horse. Others observed that food consumption per unit of body weight was higher in small animals. Scientists in the 19*th* century concluded, incorrectly, that metabolic rate was proportional to body surface area, presumed to vary with the 2/3 power of body weight. They made the argument that metabolic rate depended on several factors that varied with surface area, including heat loss through the skin, oxygen diffusion through lung surfaces, and food transport across gut surfaces. In a detailed review, Kleiber (1947) concluded that metabolic rate, measured as heat production per day, varied with the 3/4 power of body weight. His "mouse-to-elephant" series is graphed in Fig. 11.1, where the solid line is a fit to the 3/4 power of body weight. The other lines represent other powers of body weight, which do not fit the data as well.

More recently, McMahon (1973) proposed a unifying theory of size relationships in animals, which he called "elastic scaling." It explains, among other things, the 3/4 power relationship found by Kleiber on the basis of a higher muscle power-to-weight ratio in the smaller animals. A part of the theory is based on earlier calculations by A.V. Hill (1949b) of the relationship between muscle power and body size in different species. One purpose of this chapter is to explain these theories.

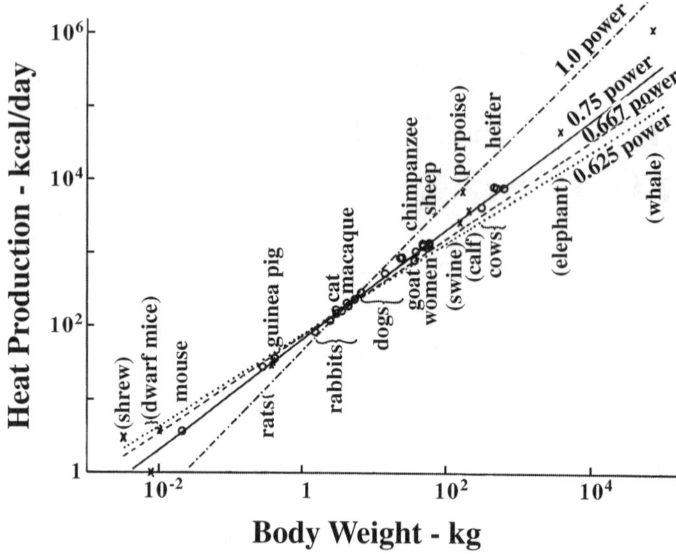

Figure 11.1. *Metabolic rate vs. body size. Solid line is Kleiber's fit to data represented by circles. The data represented by Xs were not fitted because he regarded them as unreliable. Although this is often called his "mouse-to-elephant" series, the fitted data range from mouse to cow, and the overall range is from shrew to whale.*

A second purpose of the chapter is to address the question of whether the relationship between metabolic rate and body size in different species of animals applies to different sizes of animals within the same species, particularly humans. It is common to extrapolate the body surface laws to humans when determining their metabolic rates. These estimates of metabolism are needed as denominators for variety of variables, such as the cardiac output, oxygen consumption, and even drug dosing. If metabolic differences among species are due to inherited differences in tissues determined by natural selection, it seems highly unlikely that the laws of body size which determine these inter-species differences would apply to a single species.

This question has been addressed in two ways. The first, discussed at the end of this chapter, is an examination of the available human data. The second is to ask whether the higher power-to-weight ratios found in small animals is also found in smaller humans. This question was addressed by an examination of the athletic record books, and the findings, discussed briefly at the end of this chapter, were the inspiration for the next chapter.

Allometric relationships

The data in Fig. 11.1 are plotted on logarithmic scales, and an understanding of these plots is needed to appreciate much of the discussion below. Logarithmic plots allow a fair comparison of values that differ over a wide range. Numeric values that differ by a constant factor are separated on the graph by the same distance, irrespective of their absolute values. For example, rabbits are about 10 times larger than rats and elephants are about 10 times larger than cows. In Fig. 11.1, rabbits and rats are separated by the same distance on the body size axis as cows and elephants, in spite of very large mass differences.

More importantly, double logarithmic plots, such as those in Fig. 11.1, linearize the relationship

$$y = A \cdot x^z, \tag{11.1}$$

i.e. taking the logarithm of both sides of the equation yields

$$\log(y) = \log(A) + z \cdot \log(x), \tag{11.2}$$

a linear expression with the exponent in eq. 11.1 becoming the slope in eq. 11.2. Thus, exponents in the relationships can be determined from least-squares, linear fits to logarithms of the data.

Several other logarithmic relationships pertinent to the present discussion are plotted in Fig. 11.1. One is the case in which body dimensions scale **geometrically**, i.e. the linear dimensions of the body scale in the same proportion. In such scaling, body mass is proportional to the cube of body length, L (mass $\propto L^3$)[1]. Linear dimensions are then proportional to the 1/3 power of body weight (L $\propto m^{1/3}$), and areas, such as surface area and cross-sectional area, are proportional to length squared. Areas are also proportional to body mass to the 2/3 power (A $\propto m^{2/3}$). The 2/3 relationship is represented by the dashed line in Fig. 11.1.

Two of the lines represent the McMahon's hypothesis that larger animals are stouter, with heights and lengths proportional to the 1/4 power of weight, cross-sectional areas proportional to the 3/4 power, and surface areas proportional to the 5/8 power. The 5/8 power relationship for surface area and 3/4 power for cross-sectional area are plotted as the dotted line and solid lines, respectively, in the figure. As shown, the data are not well fitted by the 5/8 power representing surface area. Finally, the steepest line in Fig. 11.1 shows the first power relationship that would result if there were no size dependence of metabolic rate on size. This line also does not describe the data.

MUSCLE SPEED AND POWER

Men have argued for centuries over the proper relationship between animal size and speed. A.V. Hill (1949b) brought both physical and physiological reasoning to the debate in an evening lecture to a lay audience at the Royal Institution. He argued that running speed and jumping ability should be independent of body size. He reasoned that acceleration of a limb, or of an animal on the end of its limbs, should be inversely proportional to the limb length, L. This conclusion derives from the following considerations: 1) that a muscle's force, F, is proportional to its cross-sectional area (F $\propto$ A); 2) that muscle cross-sectional area is propor-

---

[1]The proportionality symbol, $\propto$, used here is equivalent to the expression "=k" where k is the proportionality constant. The use of this symbol obviates the need for specifying a separate k for each expression.

tional to body or limb cross-sectional area; 3) that body or limb mass is proportional to the product cross-sectional area times length (m $\propto$ A·L); 4) that the lever ratio of the muscles acting on the bones is the same; and 5) that acceleration, a, is proportional to force divided by mass, i.e.

$$a \propto F/m \propto A/(A·L) = L^{-1}. \tag{11.3}$$

Note that muscle cross-sectional area cancels in this proportionality because thicker muscles have a greater mass to be accelerated. This conclusion is crucial, both here and in the next chapter on athletics, and one of the five considerations enumerated immediately above should be emphasized, namely that muscle cross-sectional area is proportional to limb or body cross-sectional areas area. This is likely to be so in similarly conditioned animals. One of the objects of athletic training is to cause selective hypertropy of the relevant muscles, so that comparisons can only be made among animals that are similarly conditioned for the task at hand.

If proportionality 11.3 holds for different species, their muscles must generate the same force per cross-sectional area of limb or body. This relationship is found empirically, and Hill argued that it avoids dangerously high bone stresses that would result from higher stresses in some species.

Although longer legged animals have lower accelerations, these accelerations are applied over a proportionately longer limb, so that the animals or limbs will be moving at the same speed at the end of the same angular excursion of the joint. If two animals have the same velocity at the moment of toe-off, when beginning a long-jump, they will go approximately the same distance before landing. Most animals run by galloping, i.e. progressing by a series of leaps. If they have the same velocity at the moment they leave the ground, they will have approximately the same stride length and stride frequency. These considerations suggest that all animals will have approximately the same maximum running speed and long-jumping ability. Hill's data for jumping, shown in

Table 13.1, indicate very little size-dependence of jumping ability. As he also pointed out, the relevant accomplishment in the high jump is not the level of the obstruction cleared, but the extent that the animal raises its center of gravity. A horse clearing a 6.5 foot fence may raise its center of gravity no more than a small animal leaping over a 3-foot obstruction.

Hill's data for running speeds showed a 4-fold variation, without a clear dependence of speed on size, but he also included animals that were not well adapted for running. More recent data for fast animals is shown in Fig. 11.2. As indicated, there is a 2-fold variation in speed over a 200-fold range of body weight, with no size-dependence of speed.

Table 11.1. *Hill's jumping distances, in ft.*

| Animal | Long jump | | High jump |
	Running	Standing	
Man	26.7	12	6.9
Rat kangaroo	26		8
Hare wallaby			6
Large kangaroo	26		9
Jumping mouse		12	
Kangaroo rat		12	
Jerboa	8		
Jumping hare	20		
Prairie hare	21		
Hare			4
Peccary			4
Mule deer	15-25		
White tail deer			8
Wapiti			8
Pronghorn antelope			5
Impala	30		8
Horse	25		6.5

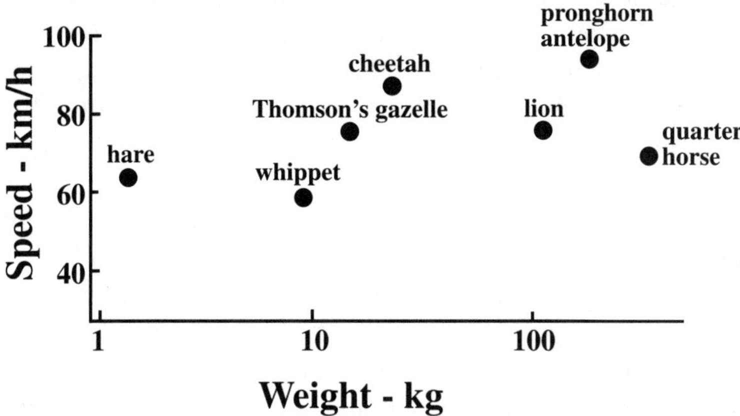

Figure 11.2. *Running speed vs. body weight in fast animals. Adapted from Ford (1984).*

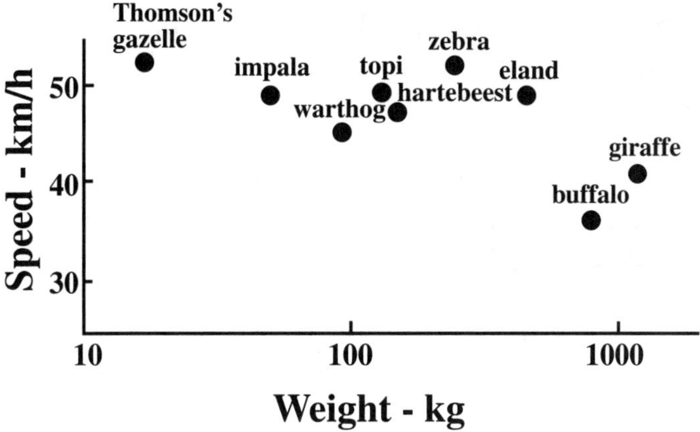

Figure 11.3. *Running speed vs. body size in African ungulates. Adapted from Ford (1984).*

It might be argued that it is inappropriate to compare animals of very different designs. The data for a 60-fold range of similar animals, African ungulates, is shown in Fig. 11.3. Again, there is a 2-fold range of speeds, with a slight tendency for larger animals to be slower.

It should also be emphasized that there is substantial variation in the data. A part of the variation arises because the observations have not been made in a standardized manner, and the animals may have had varying degrees of training and incentive. In addition, running speed and jumping ability are not the only qualities optimized by natural selection. The diversity of animal designs provides sufficient variation that many exceptions to these general rules apply. For example, Biewener et al. (1981) found that some large animals, such as kangaroos, store potential energy in their tendons when they land after a jump and use this energy to increase the size of the next jump. As with a person on a pogo-stick, they are able to jump substantially farther after they have made several preparatory bounces.

Muscle function

As described in Chapter 2, Hill (1938) showed that muscle force is inversely related to velocity by the rectangular hyperbola in Fig. 11.4 A. Muscle power, i.e. work rate, **PV**, is defined as the product of velocity times force. It is zero at each end of the curve, where either velocity or force is zero, and reaches a peak when velocity is about one-third of maximum. Muscle produces substantial heat when isometric, and the heat rate increases with shortening, so that total energy rate, heat rate plus work rate, also increases with velocity (Fig. 11.4 B). Chemomechanical efficiency, defined as work rate divided by total energy rate, reaches its maximum when velocity is about 20–25% of maximum (Fig. 11.4 C), the exact value being uncertain because of the uncertainties about the exact rates of heat production described in Chapter 2. The difference in the velocities as which peak power and efficiency are achieved may explain the observation described in the next chapter that power is optimized at higher velocities than efficiency.

In his lecture, Hill addressed the question of how force and velocity might vary among species. As mentioned, he reasoned that

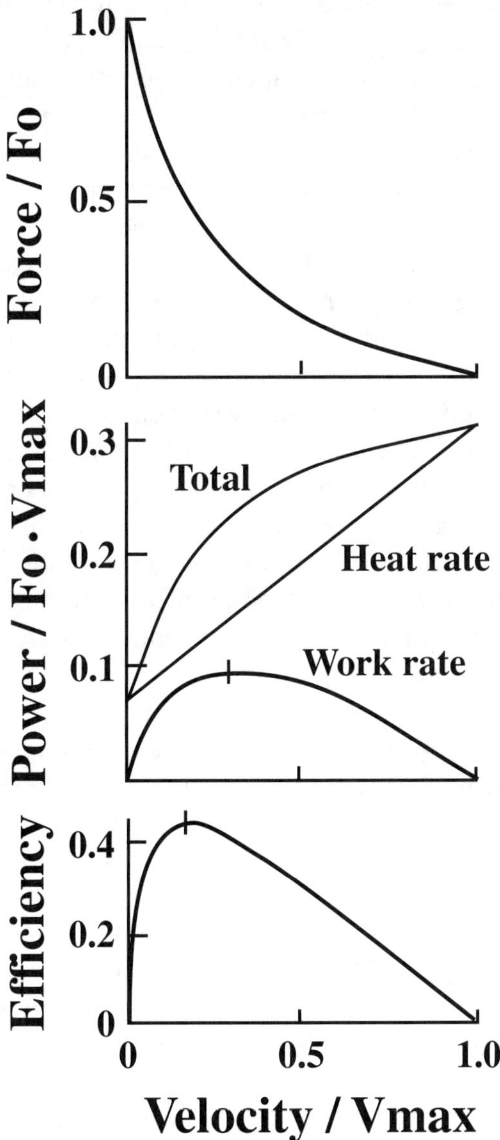

Figure 11.4. *Muscle function curves from the data of Hill (1938).*

force per cross-sectional area should be invariant among species, so that bone stresses would be the same. It is also likely that stresses on other structures, such as the myofilaments within the muscles, might be kept constant. Muscle velocity, however, should vary in a very specific way with size if animals are to operate near the peaks of their power and efficiency curves.

If the lever ratios of muscles and limbs do not vary with body size, then muscle speed should be proportional to animal speed. If animals of different size have the same velocity at the moment of toe-off during galloping or jumping, their muscles will then be shortening at the same overall velocity. By "overall velocity" is meant the speed at which the two ends of the muscle approach each other. When this overall velocity is the same in muscles of different length, the sarcomere velocity is faster in the shorter muscle, which has fewer sarcomeres in series. This shorter muscle has a greater **relative velocity**, i.e. the rate of fractional change is greater. Since power is the product of force times velocity, when overall velocity is the same, overall muscle power output is proportional to cross-sectional area (CSA) of the muscle (PV $\propto$ CSA). On the other hand, the power output per gram of muscle, called **specific power**, **SPV**, is inversely proportional to muscle length, because relative velocity is higher, i.e.

$$SPV \propto PV/mass \propto CSA/(CSAxL) = 1/L. \qquad (11.4)$$

It is not a large step from these considerations to conclude, as Hill did, that natural selection is likely to have provided that muscles operate near the peaks in their power and efficiency curves. This would require that the muscles from different animals have power-to-weight ratios that vary inversely with animal body length. He had the belief, then current, that animals scaled proportionately, so that both cross-sectional area and surface area were proportional to the 2/3 power of body mass. If this were true, both surface area and muscle power would be proportional to mass to the 2/3 power, and there would be a good match between maximum muscle power and the flow of heat and metabolites across body surfaces. More recent data suggest that this scaling does not occur and that there is a mismatch between surface area and muscle power. It is important when considering this newer finding to keep in mind that Hill's conclusion was that *animal muscle power varies inversely with body-length*. This conclusion is in agreement with the newer data and is generally regarded as the seminal work that helped to explain Kleiber's mouse-to-elephant series.

Hill also concluded that smaller animals' higher power-to-weight ratios make them faster in accelerating and hill climbing, even though they are only as fast as large animals running on level ground. This conclusion is used below to determine whether human power-to-weight ratios vary inversely with body size.

Cardiac output

Hill calculated the relationship between blood velocity and the pressure gradient along different sizes of aorta and concluded that the pressure required of large animal hearts would be impossibly high if blood flow per pound were the same as in small animals. On the other hand, if blood flow varied inversely with the length of the aorta, a constant pressure drop would result. Such a relationship between flow rate and aortic length would result if cardiac output is proportional to metabolic rate, metabolic rate is proportional to body cross-sectional area, and aortic length is proportional to body length. **Specific metabolic rate**, i.e. the metabolic rate per unit of body mass, would then be inversely proportional to body length.

In general, heart size is a constant fraction of body mass, and the fraction of blood ejected in each heart beat is size-independent. Thus, the volume of blood ejected in each beat, called the **stroke volume**, is proportional to body mass. If cardiac output is to be proportional to metabolic rate, heart rate must then vary in the same way as specific metabolic rate, i.e. in inverse proportion to body length. Two points can be made from these considerations.

First, differences in specific metabolic rates in different species can be estimated from a comparison of the resting heart rates in similarly conditioned animals. Second, Clark (1928) found resting heart rate in different species to vary with the −1/4 power of body mass. His observation would agree with Hill's calculations if body length varied with the 1/4 power of mass.

ELASTIC SCALING

It has been recognized for centuries that large animals are proportionately thicker than small animals. Galileo, for example, suggested that bones of larger animals had to become thicker to support the greater weight of the structures above them. A unifying theory for the exact relationship was developed by Thomas McMahon, a professor of Biomechanical Engineering at Harvard with many diverse talents and interests. In addition to his scientific work, he published several novels and designed the especially fast indoor running track at Harvard. His work on scaling includes the observation that tall species of tree are stouter in proportion to their height than shorter species (McMahon, 1973). In the 19*th* century, Greenhill derived equations to show that the diameters of self-supporting columns must vary with the 1.5 power of height if the columns are not to buckle, i.e. if they return to the upright position after being bent. McMahon found that dimensions of both mature trees and mature animals vary in the same way, their diameters vary with the 1.5 power of length. Double logarithmic plots of the heights against diameters of 576 species of tree were well fitted by a line with a slope of 1.5, as was a plot of chest circumference against height in the primates taken from the work of Stahl and Gummerson (1965) and length against diameter of long bones in ungulates (McMahon, 1975a). Thus, he appears to have found a general biological law of size, which he termed "elastic scaling" because it provided equal bending stresses per unit of cross-sectional area in structures.

The scaling of body dimensions with body mass are listed in Table 11.2. All follow from the Hill's calculation that power varies with cross-sectional area and the following proportionalities:

$$M \propto L{\cdot}CSA \propto L{\cdot}D^2 \propto L^4 \tag{11.4}$$

With geometric scaling, cross-sectional area and surface area vary in the same proportion and with the 2/3 power of body-mass. With elastic scaling, they do not vary in the same way. As indicated in Table 11.2, cross-sectional area varies with the 3/4 power of mass and surface area with the 5/8 power. The difference is sufficiently large to be distinguished in Kleiber's data in Fig. 11.1. Clearly, metabolic rate does not vary with the 5/8 power of mass. Thus, if animals scale "elastically," surface area is not a good indicator of metabolic rate.

The observation that both metabolic rate and Hill's predicted muscle power varied with cross-sectional area led McMahon to conclude that muscle power is a major determinant of metabolic rate. To some extent, this is expected because muscular activity is one of the determinants of basal metabolism. All physical activity in the body, including breathing, the heart beat, and gut peristalsis, as well as baseline motor movements, are the result of muscle con-

Table 11.2. *Elastic Scaling Relationships.*

Dimension	Symbol	Power	Derivation
Body mass	M	1	—
Length	L	1/4	McMahon
Cross-sectional area	CSA	3/4	M/L
Diameter	D	3/8	$\sqrt{CSA}$
Body surface area	BSA	5/8	LxD
Power	PV	3/4	M/L Hill
Specific power	SPV	−1/4	PV/M

traction. On the other hand, there is no *a priori* reason to expect that other functions, such as liver metabolism, should vary with cross-sectional area or muscle power. In addition, Hill's conclusions were for maximum power, not resting muscle activity. The concordance between McMahon's elastic scaling theory and Kleiber's metabolic data further suggests that basal muscle activity is a size-independent fraction of maximum muscle activity. This suggests that animals can increase their muscle activity by some constant, size-independent factor.

VARIATIONS ON THE THEORY OF ELASTIC SCALING

It should be stated that some data in more recent studies are described almost exactly by elastic scaling and other data are not. The closest agreement is obtained when animals within the same order are compared. For example, Alexander (1977) found that longitudinal dimensions of antelopes vary in almost exact proportion to 1/4 power of body weight. When looking across orders, it is sometimes found that longitudinal body dimensions vary with more than the 1/4 power of weight, but less than the 1/3 power predicted by geometric scaling. McMahon (1975b) was able to modify his theory slightly to accommodate the variations without substantially altering the predicted relationships between size and power. He did this by assuming that some larger animals compensate for added stresses not by increasing their diameters out of proportion to their lengths, but by an increased mechanical advantage. It was assumed in deriving the theories above that there is a constant lever ratio of the muscles acting on the bones, i.e. that the ratio of the length of the bones to the distance from the joint to the muscle insertions is size-independent. McMahon (1975b) also considered the case where this ratio might be reduced in larger animals. Less angular movement of the joints is then produced by the same fractional length change in the muscles of larger animals. Their muscles gain a force advantage, but in return must shorten more to produce the same movement. If the muscles grow so that they operate over the same range of sarcomere length, the range of joint motion would then be reduced in larger animals. He developed this variant of the theory to account for the observation that some dissimilar animals (mouse, rat, dog, horse) scaled nearly geometrically and had angular joint excursions that varied with body mass to the −0.1 power. As he emphasized, the reduced angular excursion reduces bone stresses in two ways, by a reduced lever ratio of the muscle acting on the bones of larger animals and by causing the larger animals to have a more upright posture, so that stresses are applied more parallel to the axis of their long bones.

Interestingly, McMahon calculated that the size-dependent variation of mechanical advantage in animals that scale geometrically keeps muscle power varying in approximately the same way that it does in animals that scale elastically. In this case, muscle velocity is determined as the product of mechanical advantage, that varies with the 0.1 power of body mass, multiplied by cross-sectional area, which varies with the 2/3 power of body mass ($V \propto M^{0.1} \times M^{0.67} = M^{0.77}$). The 0.77 power variation is very close to, and probably not distinguishable from, the 0.75 power predicted by elastic scaling.

Alexander et al. (1979) suggested that scaling in the animals selected by McMahon for this variant of his theory is due to differences in life-styles rather than to a general law of scaling. The larger animals, dog and horse, are **cursorial**, i.e. they roam widely in their daily lives, while the non-cursorial, smaller animals, mouse and rat, do not stray far from their homes. The more significant point, however, is that McMahon showed that the general law of muscle power scaling with the 3/4 power of body mass holds for different types of dimensional scaling. Furthermore, the relationship between power and body mass is linked closely to mechanisms that maintain size-independence of structural stresses.

TISSUE MEASUREMENTS

The forgoing discussion suggests that specific metabolic rates of tissues from animals of different species should vary with the −1/4 power average weight of the species. For unexplained reasons, such a finding has proven elusive, even in muscle. Close (1972) measured the shortening velocity in several animals having a 100-fold range of body size. He found that muscles from larger animals shortened at a slower relative rate than those from the smaller animals, but the range of velocity was about half that expected, i.e. velocity varied with body mass to the −1/8 power. Seow and Ford (1991) made similar measurements in skinned skeletal muscle fibers from animals having a 25,000-fold range of size (mouse,rat, rabbit, sheep, cow) and found the same relationship. A least-squares linear fit to the logarithms of their results (Fig. 11.5) had a slope of −1/8. Measurements were made in fibers from both fast and slow muscles, to be certain that there was not some size-dependent shift in the relative concentrations of fast and slow myosins. As indicated in Fig. 11.5, fast fibers shortened 1.7 times faster, and both had the same relationship to body size. The range of animal sizes in this more recent study was 25 times larger than the earlier study of Close, so that the body size dependence of velocity could be stated more confidently, but the conclusions in the two studies were nearly the same.

These results from skinned fibers are in keeping with the amended theory proposed by McMahon, and there is approximately the same mix of cursorial and non-cursorial animals as the group for which he developed the amended theory. If the measurements were made within the same order of animals, where elastic scaling appears to apply, perhaps a different result would be obtained. It should also be mentioned, however, that the experiments of Seow and Ford were done at a cold temperature, and it is possible that a body-size-dependent variation in the effects of temperature on velocity would produce a different relationship between body mass and velocity.

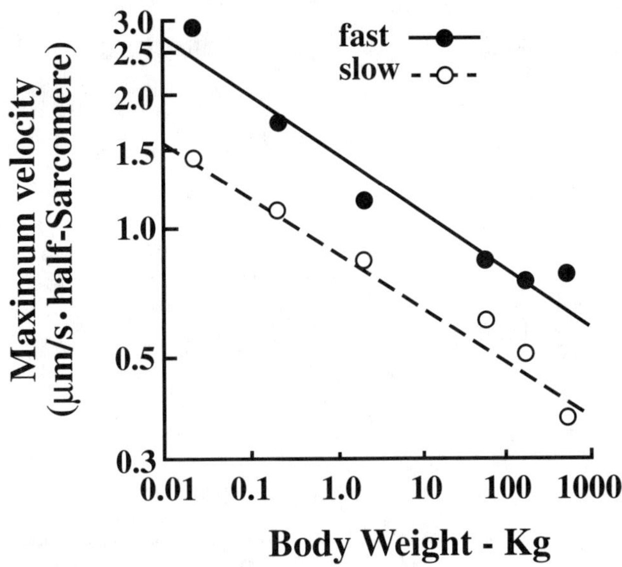

Figure 11.5. *Maximum shortening velocity in fast and slow skinned muscle fibers from animals having a 25,000-fold variation in body mass. From Seow and Ford (1991).*

OTHER RATES AND PERIODS

A number of bodily rates and periods have been found to vary with the 3/4 or −1/4 power of body mass. As mentioned, heart rate is one. Respiratory rate is another. Interestingly, life-span is yet another. Fig. 11.6 plots life-span for several animals against body size on logarithmic scales. The life span plotted is the mid-point between the longest recorded life and the mean life-span for the species. A line having a slope equivalent to the 3/4 power of body size is superimposed. As shown, there is a good fit. Kleiber, who first noted this relationship, suggested that, on average, animals live for approximately the same number of heart-beats or breaths. There are, of course, many exceptions to this generality. Yearly cycles, for example, may have caused smaller animals to evolve longer life-spans than predicted by their size, so that they can bear

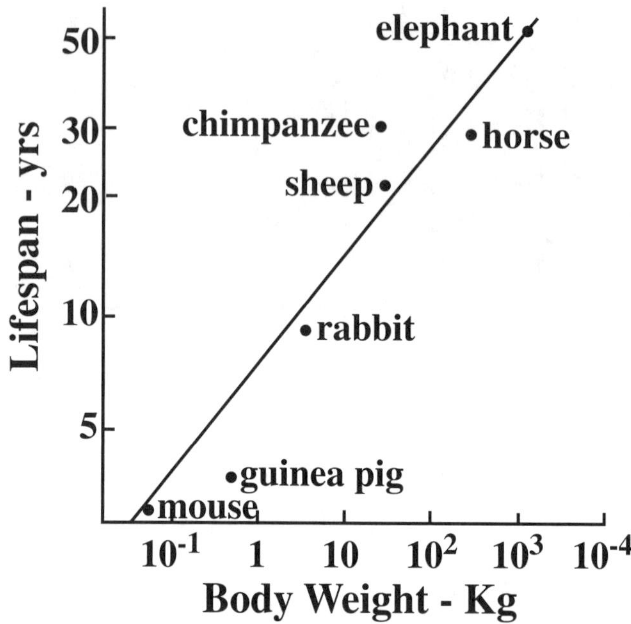

Figure 11.6. *Variation of life-span with body size. Value plotted is the average of mean and longest life-span for each species.*

and rear young in the most favorable seasons. Domestic breeding is another factor that is likely to have altered life-span as well as other periods, such as maturation. One of the most notable exceptions is the human life-span, which is off the scale of the plot in Fig. 11.6. The reasons are not known for our living very much longer than expected for our size and for having many more heartbeats or breaths. Perhaps it is because humans have adapted to use a much longer period of education and training in providing for survival of social groups.

METABOLIC DENOMINATORS

Before the genetic nature of proteins was known, it was not unreasonable to speculate that the body adapted its metabolism to match its final size. Thus, it was proposed that the same body-size dependence of metabolism in animals of different species should be applied when calculating the metabolic rates of humans. For example, when he found that metabolic rates in different species varied with the 3/4 power of body mass, Kleiber (1947) suggested that this same relationship should be used to estimate human metabolism. Now that the dependence of biochemical reaction rates on the inherited properties of enzymes is known, such a suggestion seems quaint, if not Lamarckian. But the refutation of the applicability of the laws for different species to humans begs the question of the proper metabolic denominator in a single species.

Metabolic denominators are needed for many physiological measurements, such as cardiac output and oxygen consumption, as well as for estimating the metabolism of food and drugs. The most frequently used denominator is body surface area, although most experts agree that this is both inappropriate and inaccurate.

If tissue metabolism in humans depends only on the mass of their organs, it might be expected that metabolic rate should vary with body weight. There are, however, several factors that confound such an expectation. To derive any hypothesis, it is necessary to measure metabolism in humans of very different sizes. The largest size range can be obtained by comparing infants and children with adults. It is also known, however, that children have higher metabolic rates than adults for reasons that are independent of their smaller size. Since children also have larger surface-to-volume ratios than adults, it has been easy to conclude that their metabolic rates are more nearly proportional to their surface areas than to their weights, even though the reasons for this may have little to do with body dimensions. In comparing adults, the size range can be extended substantially by including men and women, but such comparisons are likely to be confounded by gender differences. When examining adults of one gender, much of the range of body weights results from variations in obesity. Since fat has a substantially lower metabolic rate than other tissue, obese individuals are likely to have lower specific metabolic rates than non-obese individuals. Since they also have reduced surface to volume ratios, their lower rates create the false impression that surface area is a determinant of metabolic rate.

An extensive study of human metabolic rate by Harris and Benedict (1919) has served as the data base for several later analyses. The original authors concluded that metabolic rate correlated better with the 2/3 than the first power of body weight but that the range of body size was too small, and the scatter in the data too large to make a clear distinction between these two possible variables. Kleiber (1932) reanalyzed these data and concluded that metabolic rate correlated better with the 3/4 power than the 2/3 power of weight when allowances were made for differences in age, body build, and gender. However, Cunningham (1980) re-analyzed the data once more to show that metabolic rate is best correlated with lean body mass determined from a general formula applied to the published values of height and weight. He further concluded that correlations were not improved by corrections for age, gender, or body build.

In 1953, Miller and Blythe proposed lean body mass as a metabolic reference standard, and the superiority of this metabolic denominator has since been confirmed. In particular, the decline of specific metabolic rate with age has been shown to correlate with decreases in lean body mass and increases in adipose tissue (Keys et al. 1973; Tankoff and Norris, 1977, 1978). The best predictions of metabolic rate are obtained when lean body mass and fat mass were used as separate variables in a linear regression, and fat was assumed to have 0.1 to 0.3 times the metabolic rate of the rest of the body (Hoffman et al., 1979).

Muscle power

If allometric relationships between body mass and metabolic rate are due to differences in muscle power-to-weight ratios, as suggested by Hill and McMahon, and if the same laws apply to humans, then small humans should have higher power-to-weight ratios. These higher ratios would give shorter humans an advantage in some athletic events, such as sprinting and hill running. In his lecture, Hill argued that running speeds should be independent of body size, but only for the steady speed on level ground. He also marshalled evidence to show that the higher power-to-weight ratios in smaller animals give them advantages in hill-climbing and acceleration. Since marathon running requires hill-running, the heights of the winners of the Boston Marathon were examined to determine whether the event is dominated by small humans. The height of the winner in each year for the period from 1897 to 1977 is plotted in Fig. 11.7, which shows that elite marathon runners are

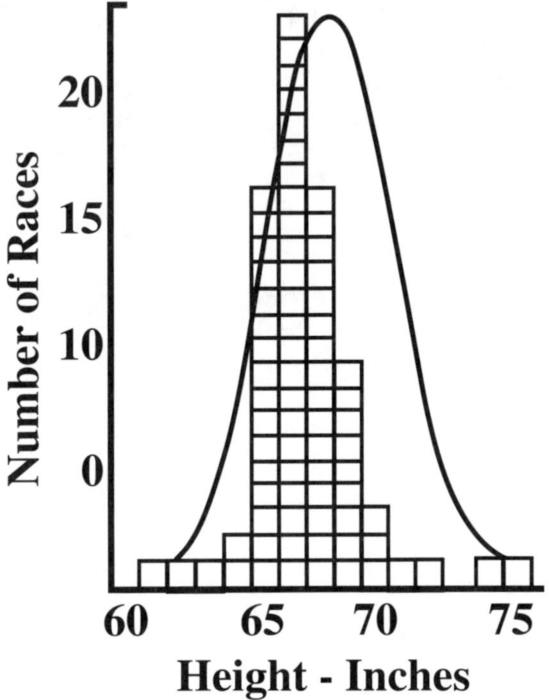

Figure 11.7. *Heights of the winner of the Boston Marathon, 1897 to 1976. Data from Falls (1977). Adapted from Ford, 1984.*

not inordinately short. The graph also shows that there is a narrower distribution than that for American men in tables compiled in the 1930s and plotted as the solid curve.

A similar distribution was found for sprinters (Ford, 1984), and these findings suggest that small humans do not have higher power-to-weight ratios. The relatively narrow distributions of heights in these events, however, led to the more extensive investigation of athlete sizes described in the next chapter.

SUGGESTED READING

HILL, A.V. (1949b) The dimensions of animals and their muscular dynamics. *Proceedings of the Royal Institution of Great Britain.* **186**: 450–471.
McMAHON, T.A. (1973) Size and shape in biology. *Science.* **179**: 1201–1204.
FORD, L.E. (1984) Some consequences of body size. *American Journal of Physiology.* **247**: H495-H507.

Chapter 12

OPTIMUM SIZES FOR SOME ATHLETICS

There is perhaps no better way to start an argument than to make a confident assertion about athletics. Everyone has an opinion. As mentioned in the last chapter, the data were collected initially to test a hypothesis that involved athletics only peripherally. As the data were examined, it became apparent that there are optimum body sizes for many different athletic events. There can be little argument with these observations; the heated discussion begins when explanations are advanced.

A cursory glance at body dimensions of athletes in different sports suggests that size optima for different events vary widely, but closer inspection shows that these optima are relatively near the mean values for humans. With the exception of basketball, no major sport is dominated by giants. Conversely, no sport selects for very small size, except for those athletes, such as jockeys and coxswains, who are transported by other participants. Instead, the different optimal sizes for each event are distributed fairly closely about the mean heights for humans, as might be expected if the human body evolved to optimize some forms of athletic prowess.

The data presented here suggest that the sizes of the athletes in a variety of events are determined by a variable balance among three factors: 1) an inertial advantage of short limbs; 2) limitations on the lateral growth of muscle; and 3) a frame size that brings muscle to its optimum operating point. Some of these principles are demonstrated by a comparison of men and women competing in the same event. Before discussing these three factors, some discussion of data selection is required.

DATA SELECTION

Only data for the most elite athletes are considered. This stringent selection is necessary because athletic ability is determined by many factors, and it is important that all factors be optimized to recognize the effects of body size. As will be shown, even with this extreme selection, there are still world class athletes whose dimensions are far from the mean, suggesting that training can overcome a size disadvantage. There are two reasons for making this point. The first is the scientific conclusion that body dimensions are relatively weak selection factors in determining outstanding athletes. The second, and perhaps more important reason is that individuals should not be discouraged from participating in a sport because they appear to be the wrong size.

BODY DIMENSIONS

The body dimensions considered here are height, weight, and weight divided by the third power of height, called the weight-height index. This index is a measure of proportional thickness, i.e. it will be the same in similarly proportioned individuals of different heights. It is used to assess muscularity in minimally obese athletes of different heights.

These dimensions for the selection of athletes discussed here are plotted in Fig. 12.1. Fig. 12.2 is an artists' rendition of a male athlete scaled to match the dimensions plotted in Fig. 12.1.

Heights of modern Americans

Heights of athletes can be compared with the mean heights for present-day 18-24 year old Americans. These are 176.9 cm ± 6.9 S.D. (69.7" ± 2.8") for men and 7.7% less, 163.3 ± 6.5 cm (64.3"

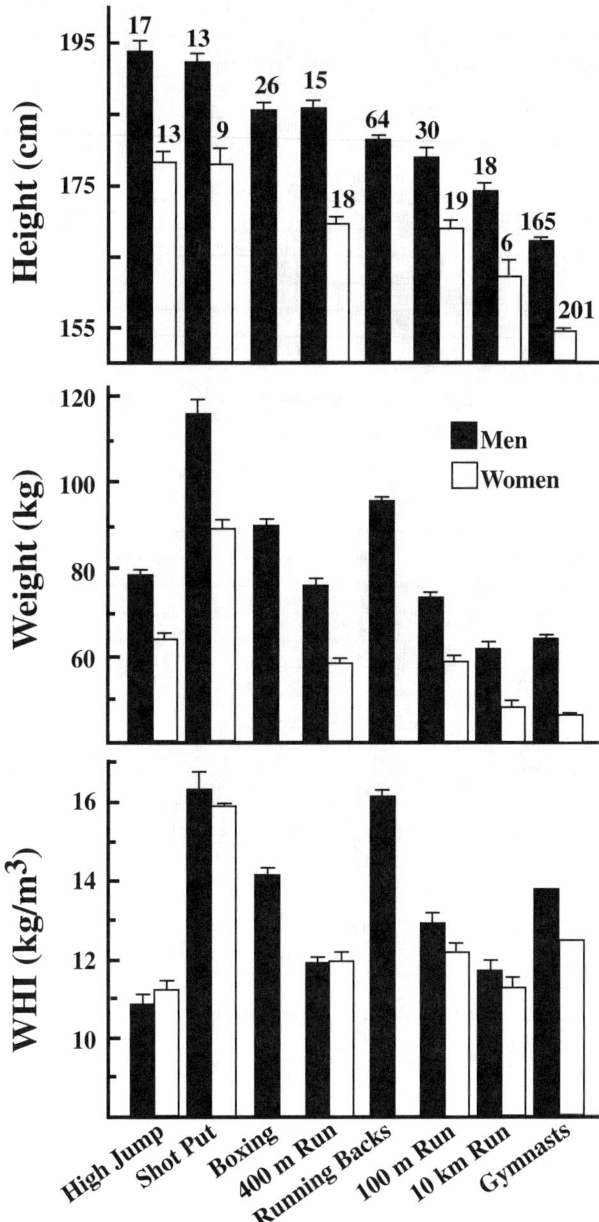

Figure 12.1. *Average body dimensions of elite athletes.*

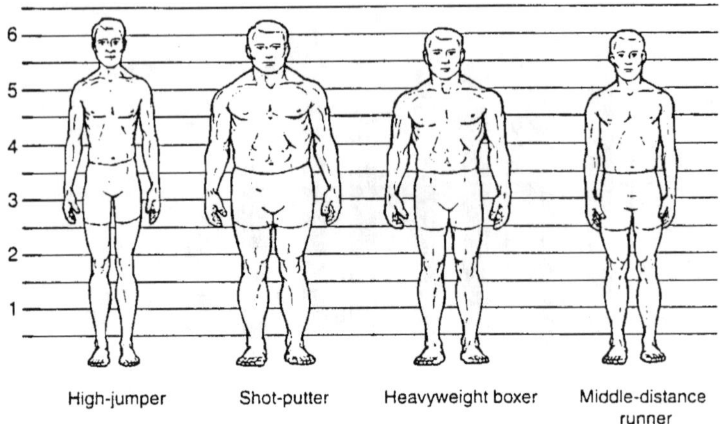

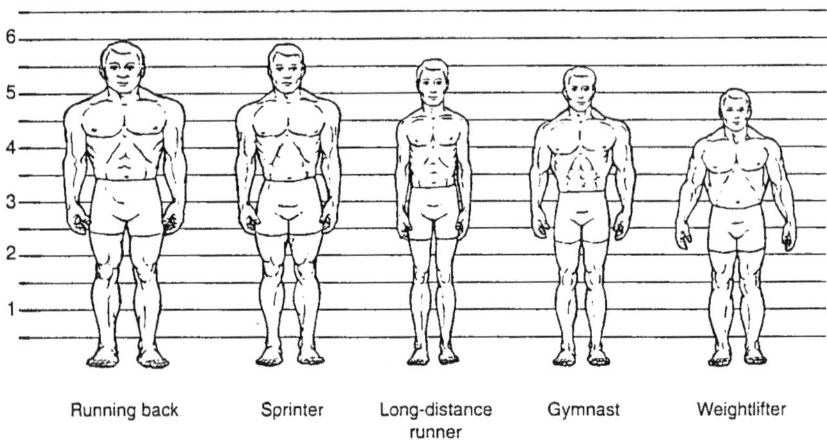

Figure 12.2. *Athlete scaled to match the body dimensions Fig. 12.1. Drawn by Craig Gosling and Tim Yates.*

± 2.6") for women (Naijer and Rowland, 1998). The mean heights of the athletes plotted in Fig. 12.1 range from −1.4 to +2.4 standard deviations from the means for young American men and women. It should also be noted that the average heights of present day American men used here is about 2" taller than the distribution for American men compiled for pediatric tables in the 1930s and plotted in Fig. 11.7 of the preceding chapter. The standard deviations for the two distributions are, however, nearly identical.

INERTIAL ADVANTAGE OF SHORT LIMBS

Hill's (1949b) calculation, derived in the last chapter (eq. 11.3), that acceleration is inversely proportional to limb length, i.e

$$\text{acceleration} \propto L^{-1}, \tag{12.1}$$

is essential for understanding the relationship between size and athletic ability. As described in the previous chapter, acceleration is independent of muscle cross-sectional area because limb mass and accelerating force are both proportional to muscle cross-sectional area. As emphasized in the last chapter, this proportionality only holds if muscle mass comprises a constant fraction of limb or body mass. One of the objects of athletic training is to cause selective hypertrophy of the relevant muscles, so that proportionality 12.1 is valid only for athletes of the same body build. The data discussed below, indicates that taller athletes carry less of their body mass as muscle, so that their accelerations are reduced out of proportion to their increased limb length.

The validity of Hill's conclusion can be seen from the absence of very large athletes in contact sports requiring both strength and speed. The "unlimited" weight category in boxing, for example, has not led to very tall world heavyweight champions. Fig. 12.3 plots the sizes of the 26 boxers who held the undisputed championship since its inception in 1899 to 1978, when more than one champion became recognized. The average height and weight

                                             Part II: Whole Body Function

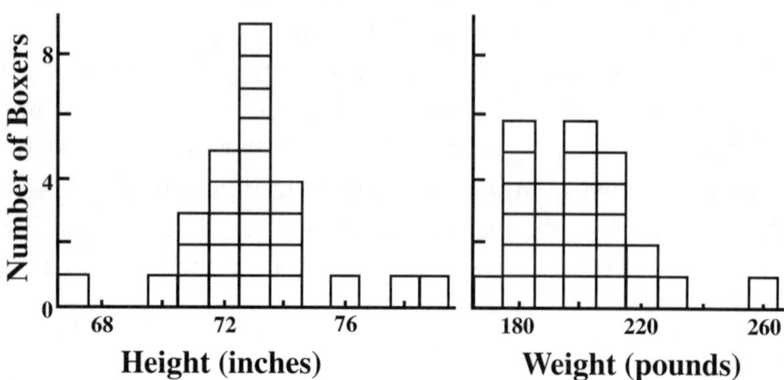

Figure 12.3. *Distributions of heights and weights of world heavyweight boxing champions 1899 to 1978. Weight is the value when the boxer first won the championship.*

were 6'1" (185 cm) and 198 lbs (90 kg). Only three of the 26 were taller than 6'2.5", and there was only one successful title defense among the three. These data suggest that the longer reach and greater punching power of larger athletes are insufficient to overcome their reduced acceleration.

A similar absence of large athletes is seen among elite running backs of professional American football. The backs who led the National and American football conferences in rushing for the 32 years from 1966 to 1997 averaged 5'11" (181 cm) and 210 lbs (95 kg). The size distributions of these yearly conference leaders are plotted in Fig. 12.4. The data also show that while these athletes were not exceptionally tall, they are very muscular, having weight-height indices comparable to shot-putters and body-weight restricted weightlifters.

The inertial advantage of short stature is also recognized in gymnasts, who are among the shortest athletes considered here (Fig. 12.1). They must perform demanding rotational movements while their bodies are falling through limited distances, and their short limbs and bodies enable more rapid rotational accelerations and therefore more movement while they are in the air.

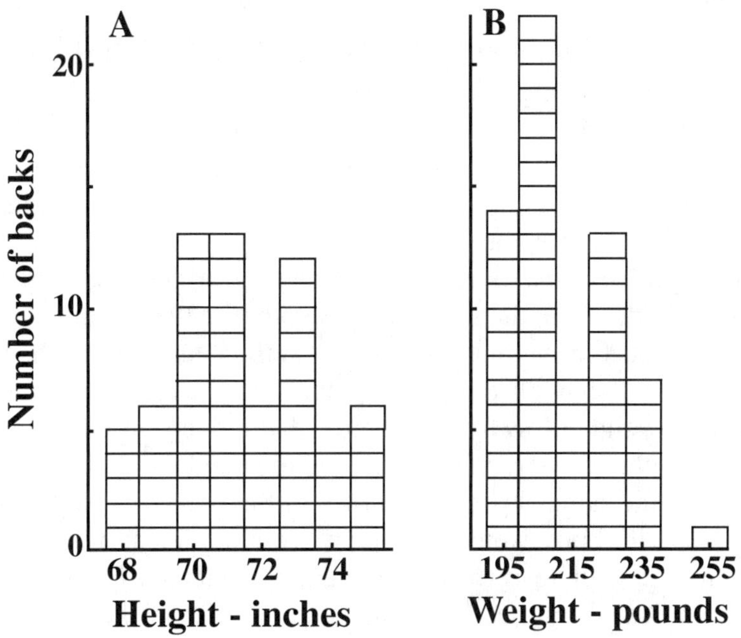

Figure 12.4. *Heights and weights of professional American football backs who led their conference in rushing each year from 1966 to 1997.*

LIMITS ON THE GROWTH OF MUSCLE

Absolute upper limits of strength and muscle growth are indicated by the absence of very tall athletes in sports where great height would be an advantage in the absence of such limits. We have not studied basketball, but the frequent reports of players taller than 7' (213 cm) suggest that very tall athletes are available to compete in other sports. Their absence from the ranks of elite competitors in events where height might be expected to provide an advantage, such as high jumping, shot-putting, and weightlifting suggests a height-related limit to muscle growth.

Weightlifting

Weightlifting provides an opportunity to assess how large and how strong muscles can grow. For the body-weight-limited classes, short stature is an advantage because a shorter athlete's muscles can have a larger cross-sectional area and therefore generate more force. The unlimited class provides insight into the absolute upper limits of muscle growth.

The goal in this sport is to achieve the highest sum of two weights lifted, one in the snatch and the other in the clean-and-jerk. Fig. 12.5 plots the average sums that won the international championships, together with the body dimensions of the champions in each body-weight class from 1993 to 1997, a five year period when the weight classes were unchanged and men and women competed in some of the same classes. As shown in Fig. 12.5 C, weight-lifted increased with body weight but approached a plateau for the men and achieved a plateau for the women. The plateau for women is further indicated by the observation that in four of the five years, the limited-body-weight class women's champion lifted more than the unlimited-class champion. The approach to the men's plateau is suggested by the observation that the men's unlimited-class champions lifted only 6% more weight than the heaviest limited-class lifters, in spite of weighing 42% more.

As shown in Fig 12.5 A, the heights of the lifters also reached a plateau. Only one of the male champions was taller than 183 cm (6'), and none of the female champions was taller than 175 cm (5'9"). As expected, the unlimited class champions were stouter and therefore had larger weight-height indices than the body-weight restricted lifters (Fig. 12.5 B). This expectation arises because these lifters have no need to minimize their weight, and if they find that more body weight provides greater strength, they will work to achieve an excess. The unexpected finding, however, was that the weight-height index began to increase at body weights less than the unlimited class, >83 kg for men and > 64 kg for women.

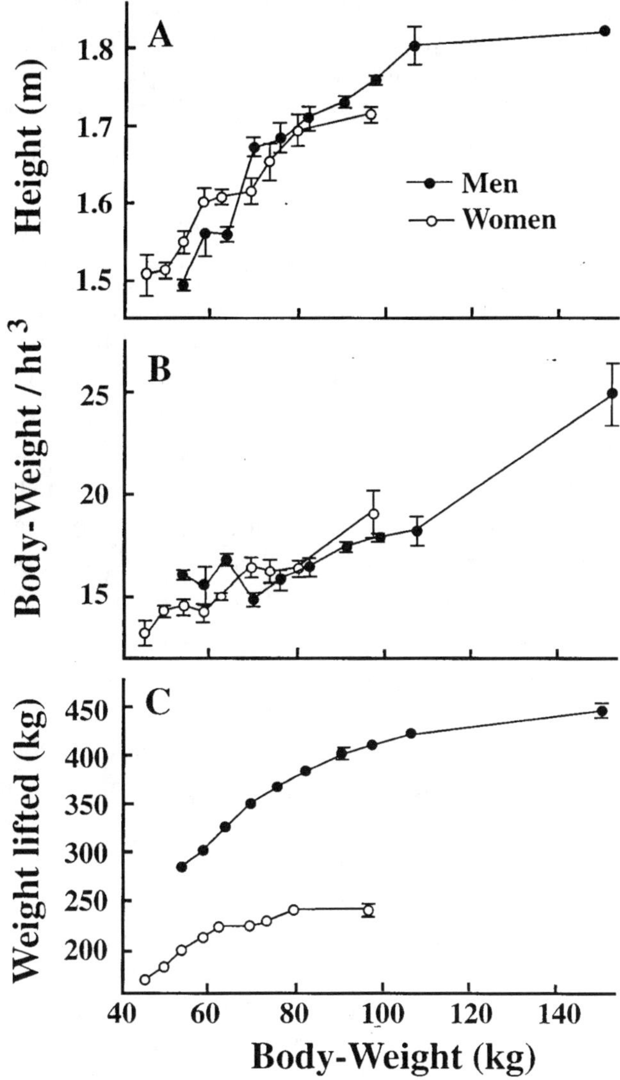

Figure 12.5. *Body dimensions and weight lifted by world weightlifting champions 1993 to 1997, adapted from Ford et al. (2000) with permission.*

Since muscle force is proportional to muscle cross-sectional area, it might also be expected that the weight lifted by body weight restricted athletes would vary in proportion to their mean cross-sectional areas, calculated as body weight divided by height. Fig. 12.6 A shows that this expectation is met, but only up to body weights of 64 kg in women and 83 kg. in men. At higher body weights, the ratio of weight lifted to mean cross-sectional area declines in a manner that is nearly continuous with the decrease in the ulimited class lifters.

The observation that the ratio of weight lifted to body cross-sectional area declined over the same range that the weight-height index increased raised the question of whether the two relationships were related. If a decline in one were proportional to an increase in the other, the product of the two would be constant. To test for such a proportionality, the product of the two was therefore calculated for each weight class. This calculation is computationally the same as dividing weight lifted by height squared. This quotient was was found to be very nearly constant over the entire range of body weights, including the unlimited weight class. The qualifying phrase "very nearly" is used because a least squares method showed the exponent for the relationship to be 2.16, slightly greater than the squared relationship. A plot of (weight lifted)$^{2.16}$ against height in Fig. 12.6 B shows very little variation over the entire range of body sizes.

The data shown in Figs. 12.5 and 12.6 suggest several major conclusions. First, there are well defined limits to muscle strength and growth. Second, these upper limits are related in some way to height. Third, an upper limit of muscle strength and growth is achieved at a body height of about 183 cm in men and 175 cm in women. Finally, if the weight lifted by these champions is determined by the cross-sectional areas of their muscles, their muscle mass scales in slightly greater proportion than the third power of their height, i.e to height$^{3.16}$.

These findings raise the question of how muscle strength and growth are related to height. A possible explanation is that both

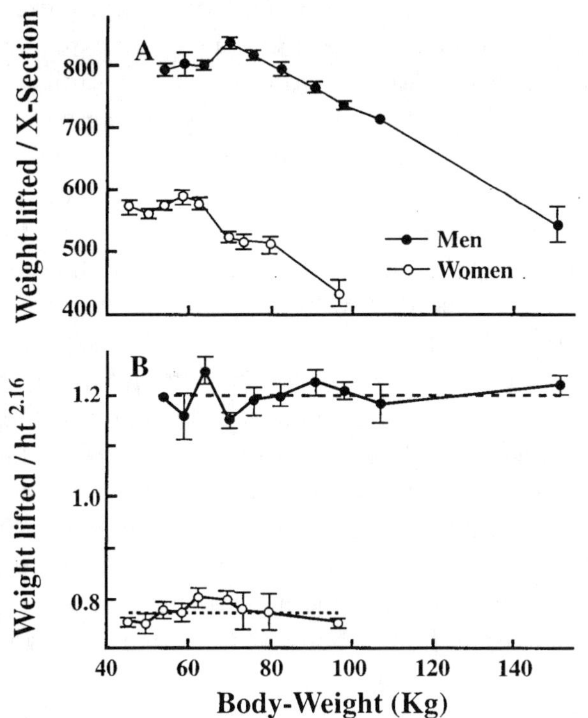

Figure 12.6. *Relationships between between body weight and weight lifted by world champions lifters. Adapted from Ford et al. (2000) with permission.*

the number of muscle cells in cross-section and height are deter-mined by a common factor during maturation. It is known that in-dividual muscle cells achieve an upper limit of cross-sectional area (Alway et al., 1992), possibly because of diffusion limitations in larger cells. There is also the controversial evidence that there are limits to the number of muscle cells in cross-section (Taylor and Wilkinson, 1986), probably determined by maturational fac-tors. A common factor determining both height and the number of muscle cells in cross-section would explain all of the data here.

The absolute height limits also raise interesting questions. An absolute limit to muscle growth would not, by itself, explain the height limits, since taller athletes with the same muscle cross-sectional area should develop as much force, and therefore lift as much weight as shorter athletes. One possible explanation is that the best weight lifters are individuals who were initially destined to grow much taller, but whose bone lengthening was arrested, so that their muscles are larger in proportion to their height. Another possible explanation is that there is some physical disadvantage of height, such as having to lift weights through a greater distance. Such a disadvantage would provide that shorter athletes would lift the most weight, once they were of sufficient size to achieve the upper limit of strength.

A final point to be made about the data in Figs. 12.5 and 12.6 is a comparison of the performances of men and women. Women do not lift as much weight as men of the same size, and when other factors are the same, their cross-sectional areas are not as great. Their more slender forms are seen in a comparison of the lighter lifters in Fig. 12.5 A, where women are taller. The comparison is obscured, however, by the tendency for women's weight-height indices to increase at lighter body weights than men, so that in the heavier body weight classes, women are shorter than men. A similar effect is seen in female/male ratio of quotient of weight-lifted to body cross-sectional area in Fig. 12.7. The ratio declines by about 14% at body weights above 64 kg (Fig. 12.7 A). A much more constant ratio of about 0.7 is found when the ratio is calculated after the weight classes have been shifted to account for the increase in weight-height index beginning at lighter body weights in women (Fig. 12.7 B). A shift by three weight classes brings into coincidence both the heaviest limited body weight class and the weight where the weight-height index increases. It also results in a nearly constant ratio. This finding suggests that the body dimensions of men and women weightlifters are governed by the same factors, and not surprisingly, that some of the size dependent shifts in these factors occur at lighter weights and shorter stature in women than men. Finally, the constant ratio in Fig. 12.7 B indicates that women lift about 30% less than men of the same proportions.

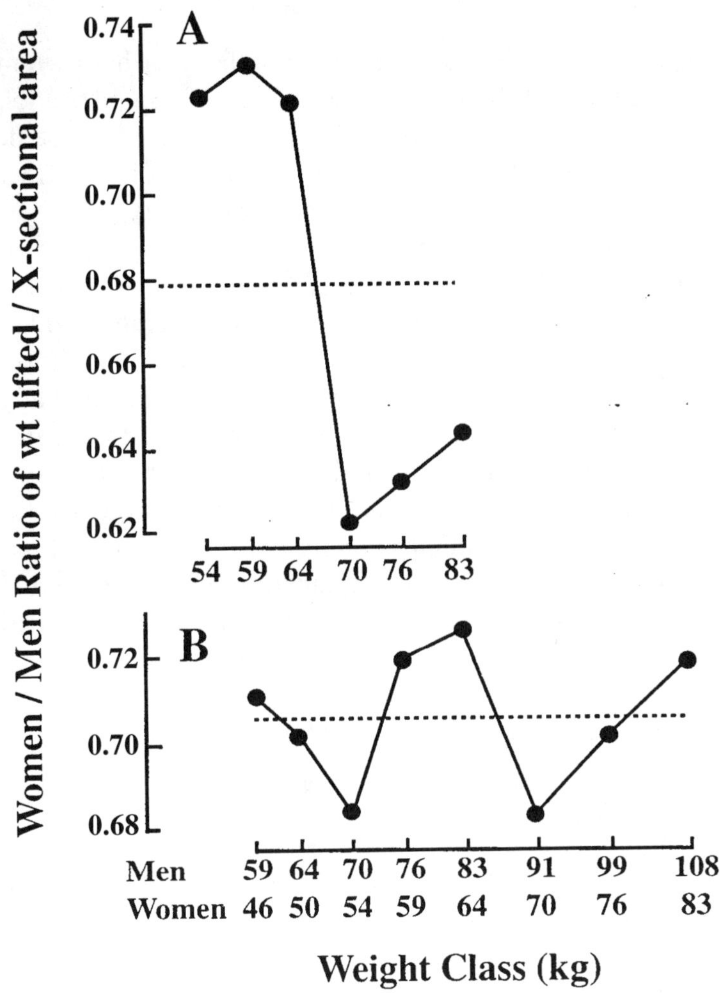

Figure 12.7. *Female/male ratios of weight lifted per cross-sectional area. A) Ratios calculated for same weight class. B) Ratios calculated after the women were shifted up by 3 weight classes. Redrawn from Ford et al. (2000) with permission.*

Shot-putting

In this event, the athlete must accelerate both his or her own mass plus the fixed mass of the shot. In the absence of any restriction on muscle growth, it might be expected that the event would be dominated by very large athletes in whom the shot is a lesser fraction of the total mass to be accelerated. The data in Fig. 12.1, for the world record holders since World War II, show that while these athletes had some of the largest weight-height indices considered here, they were not exceptionally tall. The men averaged 192.2 cm (6' 3.7") and the women 8% less, 178.1 cm (5'10"). In addition, as explained on p. 250 for runners, taller athletes have a small advantage in accelerating their limbs and bodies because their sarcomere velocities are slightly slower than those of shorter athletes when they are moving at the same limb and body velocity. Thus, the absence of very tall athletes among the ranks of world record holding shot-putters requires an explanation. One likely explanation, supported by the data for weightlifters, is that there is a height-related upper limit to muscle growth, and that when some critical height is exceeded, increasing muscle strength cannot overcome the inertial disadvantage of longer limbs.

High-jumping

A tall high-jumper has the advantage of beginning his jump with a high center of gravity. This advantage is shown by a comparison of the holder of the world record for the traditional competition with the holder of the *Guinness Book of World Records* title for the greatest height jumped above his own head (Young, 1998). The former is 193 cm (6'4") tall and jumped 7'11.5" (242.6 cm), 49.6 cm above his own height. The latter is 172.7 cm (5'8") tall and jumped 7'7.25" (231.8 cm), 59.1 cm above his height. Each jumped nearly the same height above his own half-height, 146.1 and 145.4, respectively, for the tall and short jumper. If the physiologically relevant achievement is the height they raised their centers of mass, it is likely that the shorter athlete achieved a little more; the center of mass of the body is probably substantially above the half-height.

For the medical purpose of establishing the normal lengths of limbs, an extensive study was made of the limb-length to body-length ratio (McKusick, 1972, pp. 70–74). This study showed body half-height to be 3–4% of body height below the pubic symphysis. Three percent of the 20 cm height difference in the two jumpers is 0.6 cm, the exact difference in the heights they jumped above their center of height. Thus, each raised his pubic symphysis to the same extent. For each 5% of body height the center of mass is above the pubic symphysis, the shorter athlete raised his center of mass by 1 cm more than the taller jumper. It seems quite likely that the center of mass is substantially above the pubic symphysis. Thus, the shorter jumper is likely to have raised his center of mass by an extra 1 to 2 cm. In addition, the more precisely relevant accomplishment is the extent to which the center of mass was raised after toe-off. The taller athlete undoubtedly had a longer foot that raised his center of mass another few mm before toe-off. A final reason for believing that the shorter jumper had a slightly better performance is the much greater competition for the world record in the traditional high jump. If the Guinness record were contested as vigorously, it seems likely that shorter jumpers would set higher records.

Although the taller jumper raised his center of gravity to a slightly lesser extent than the shorter jumper, he jumped 10 cm higher. His greater performance in the traditional contest occurs because he begins with a higher center of gravity. If there were no limit to ability in taller jumpers, this advantage would increase indefinitely, so that the records would be held by very tall jumpers. The observation that the world record holders were substantially shorter than basketball players suggests that very tall athletes have a disadvantage. The likely explanation is that the fraction of body cross-sectional area devoted to muscle is reduced in very tall athletes.

A final point about the high jumpers is that their heights are almost the same as the shot-putters, but they weigh substantially less. The heights of the men's world record holding high jumpers since World War II (Fig. 12.1) averaged about 1.5 cm taller than men's record holding shot-putters (193.7 vs. 192.2 cm) and the women's heights were almost the same height (178.3 vs. 178.1 cm). By contrast, the high jumpers had the smallest weight-height indices among the athletes considered while shot-putters had nearly the highest. A facile explanation might be that the lighter body weights of the high jumpers imposed lesser loads on their legs, but this explanation ignores the consideration that the muscles of a lighter jumper are also smaller. A more complete explanation is that less non-muscle tissue is required in athletes who are both tall and slender, such that their strength-to-weight ratios are greater than in more muscular athletes of the same height.

MATCHING FRAME SIZE TO MUSCLE CHARACTERISTICS

Bicycle racing

Bicycle racing is an excellent example of how mechanical linkages can be altered to match muscles to their task. The length of the bicycle crank, for example, can be adjusted to obtain the optimum output from the cyclist. Not surprisingly, taller individuals require longer cranks to compete successfully. A constant ratio of crank length to muscle length keeps the muscle working over the same, optimal sarcomere length in athletes with different leg lengths.

Cyclists have also found empirically that they must maintain a high, constant pedalling frequency, called **cadence**. When they encounter resistance from a head-wind or a hill, they gear down to keep the legs moving at the same rate, irrespective of the bicycle speed. The plots of muscle power and efficiency in the preceding chapter (Fig. 11.4), as well as the discussion Chapter 2, indicate a reason for this empirical finding. Muscle power and efficiency reach distinct peaks at specific velocities. Constant cadence keeps the muscles operating on these peaks when the resistance to bicycle movement changes. The observation that power peaks at higher velocity than efficiency can also explain why a short race is won with a higher cadence than a long race.

Running

Unlike cycling, where mechanical linkages adjust the muscles of the athlete to his task, the optimum match of muscle to frame for other sports must be obtained by selection of the correct body height. This selection is made by competition, and the optimum sizes are determined from the sizes of the winners. The body dimensions of runners who set or tied world-records for the outdoor distances and the 50 & 60 meter indoor sprints are plotted against distance in Figs. 12.8. The data for outdoor events are for holders of world records set after World War II and before 1997. The data for indoor events are from 1971, when electronic timing began, to 1997. The starting date for the indoor events is more recent because body dimensions were not available for many of the earlier record holders.

There are 468 records set by 280 runners. Of the 68 runners who held multiple records, only seven men and five women held records for distances that differed more than 2-fold, indicating that elite running is highly specialized by distance. Two related reasons are suggested by Figs. 12.8 and 12.9. First, sprinting and long distances favor shorter stature while middle distances favor taller stature. Thus, runners who attempt a much different distance may find themselves at a size disadvantage.

Specialization by height is even more extreme than indicated by the data in Figure 12.8. There is a very sharp division at a height of 183 cm (6'). Of the 44 men who held records in the 200, 400 or 800 m distances, 32 were at least 183 cm. By contrast, only 6 of the 50 record holders for the 50, 60 or 100 m sprints and none of the 26 record holders in the 10 and 20 km distances were taller than 183 cm.

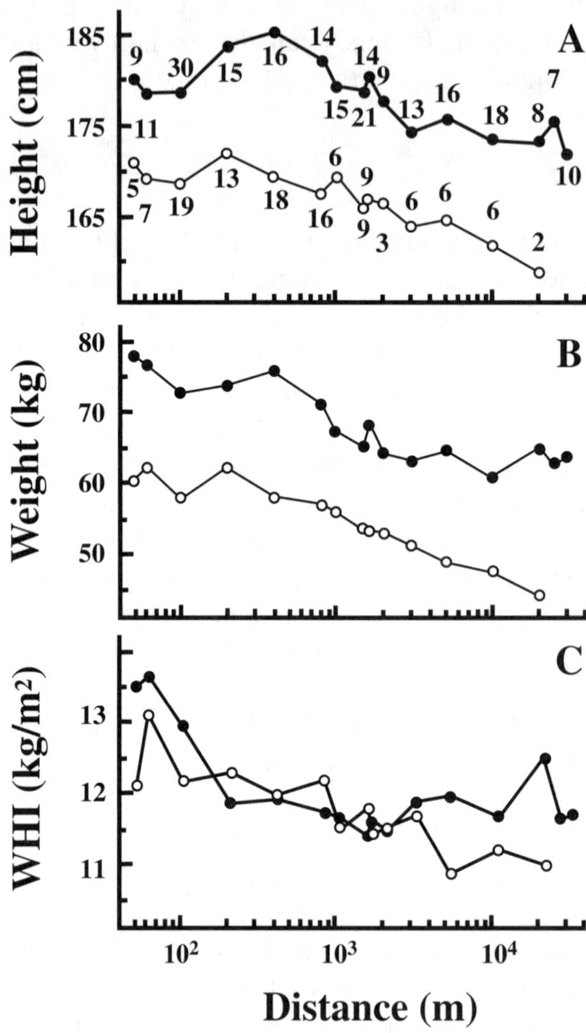

Figure 12.8. *Average body dimensions for holders of world-records in running. Numbers above symbols in A indicate the number of runners averaged, both here and in Fig. 12.9.*

The second reason for the specialization is shown by the plot of mean velocity in the current world records (Fig. 12.9 A). The highest mean velocities are achieved in the 100–200 m sprints, indicating that rapid acceleration is a major part of successful sprinting, and probably a very important component of middle distance running as well. In addition, velocities fall off sharply in longer distances, suggesting different requirements for middle and long distances. These data are consistent with the hypothesis that shorter stature optimizes power for acceleration in sprints and efficiency in long distances, while taller stature optimizes top speed in middle distances. Thus, runners who attempt multiple distances

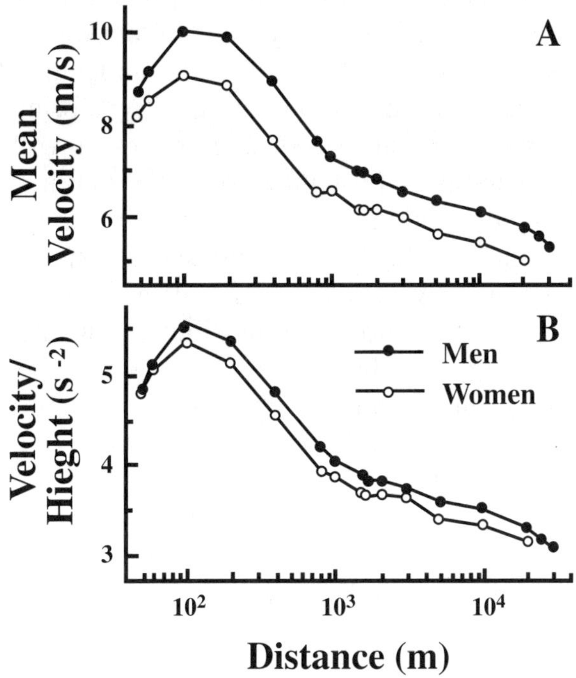

Figure 12.9. *Variation with running distance of A) average speeds of world records and B) average of speed divided by height, indicating relative muscle velocity.*

may have to adapt to very different running styles which are determined partly by height and partly by other factors.

The likely reason for different heights in different distances is again provided by the relationships between the several muscle characteristics, power, efficiency, force, and velocity. As with bicycle racing, the optimum skeletal frame size is one that brings the muscle to its optimum for each task. If lever ratios for muscle and bone are the same in tall and short runners, overall muscle velocity will be the same when the runners are moving at the same speed. Sarcomere shortening velocity will be higher in shorter runners, however, because their muscles have fewer sarcomeres in series. Fig. 11.4 of the last chapter indicates that there are optimum sarcomere velocities for maximum efficiency and power. For the same running speed, velocity is at its optimum only when frame size is correctly matched to the muscle length. Taller runners will not accelerate their limbs sufficiently to bring their muscles to the optimal velocity while shorter runners will accelerate so rapid that velocity is brought beyond the optimum. Interestingly, the average speed in current record for the longest distance in Fig. 12.8 is 60% of the speed in the shortest sprint. This ratio might be explained by maximum efficiency occurring at about 60% of the sarcomere velocity as maximum power, as described with Fig. 11.4 in the last chapter.

As also described in the last chapter, Hill (1949b) calculated that while acceleration is diminished in proportion to limb length, the reduced acceleration is applied over a longer stride in a taller animal, so that at the end of the stride, tall and short animals move at the same speed. This relationship is only approximately true, however, when the muscle force-velocity characteristics are also considered. Taller runners' reduced sarcomere velocities provide slightly greater force to accelerate the limbs. Thus, a taller runner is likely to do better in races where top speed is most important. This may explain why middle distance runners are taller than the others; they drive their limbs at higher velocities, but at the expense of reduced efficiency and power. They are not very tall, however, perhaps because they must develop sufficient power to accelerate rapidly and operate with enough efficiency to maintain speed for the duration of the race. If this explanation is correct, there may be size related differences in runners of longer sprint and shorter middle distances, with taller individuals having higher maximum speeds and shorter athletes accelerating more rapidly.

An analogous relationship may be seen in boxing. Heavier fighters move more ponderously and throw fewer punches than lighter boxers. One reason is that they accelerate less rapidly, and therefore appear slower. If their muscles move less efficiently in their taller bodies, they may also require more energy to accelerate their limbs. Fewer punches would then result from their greater tendency to exhaustion with less efficient activity.

COMPARISON OF MEN AND WOMEN

In all sports except weightlifting where men and women compete in the same event, women are 7–8% shorter than men (Fig. 12.1). For sports where calculations could be made, women's perfomances were reduced by an amount suggesting that they develop 70–81% less muscle force per unit of body cross-sectional area than men. The 70% factor for weight lifting was calculated above. An 81% factor for running and high jumping is calculated here.

As shown in Fig. 12.1, the weight-height indices of men and women competing in the same running event are approximately the same, especially over the middle distances. Calculations for both runners and high jumpers suggest that women in these sports develop about 81% of the force per unit of mean body cross-sectional area as men.

For all running distances plotted above, women's heights averaged 93% of men's and their speeds averaged 90%. If muscle lengths vary in the same proportion as their heights, the numbers of sarcomeres in series in women's muscles should have been 93% of the men's. When running at 90% of the men's speed, the sarcomere velocity would be 97% (=90/93), close to men's sarcomere velocity for each of the distances. This calculation suggests that muscles of men and women operate at very nearly the same point when running the same distance.

Calculation of the relationship between muscle force and speed depend on the estimations of the force required to accelerate the limbs. The calculations made here are for accelerations of the mass of the limb, or body on the end of the limb, accelerating from zero. For running, the assumption of beginning from zero velocity is certainly valid for muscles moving the legs, where the foot is stationary during part of the stride. It is probably also valid for high jumping, where the upward acceleration occurs in a nearly continuous movement.

Calculations of the relative strengths of men and women require the familiar physical equations relating acceleration, a, velocity, V, distance, D, and time, t, over which the acceleration is applied:

$$V = a{\cdot}t \tag{12.2}$$

$$D = {}^1\!/_2 \cdot a{\cdot}t^2 \tag{12.3}$$

which can be rearranged to

$$V^2 = 2{\cdot}a{\cdot}D \tag{12.4}$$

For running, the value D is stride length, which is proportional limb length, L

$$D \propto L. \tag{12.5}$$

Equation 11.3 in the previous chapter indicates that limb acceleration is directly proportional to muscle force, F, and inversely proportional to limb length, i.e.

$$a \propto F/L \tag{12.6}$$

Substituting proportionalities 12.5 and 12.6 in eq. 12.4 yields

$$F \propto V^2 \tag{12.7}$$

Thus, if the women's velocities are 90% of the men's, their accelerative forces were 0.9^2, equal to 81%.

Similar calculations of the relative strengths of men and women high-jumpers yield an identical result. For these calculations, the height of their jumps above their half-height is used as a surrogate for the amount by which they raised their centers of gravity. The current holder of the men's record is 193 cm (6'4") tall and jumped 243 cm, 146.5 cm above his half-height. The current holder of the women's record is 180 cm (5'11") tall and jumped 209 cm, 119 cm above her half-height, equivalent to 81% of the extent to which the man raised his half-height. At the moment of toe-off, the jumpers began to decelerate under the same force of gravity, and reached zero as they crossed the bar. Substituting 81% for the D in equation 12.4 suggests that the woman's velocity was 90% of the man's at toe-off. Proportionality 12.7 then indicates that her accelerative forces were 81% of the man's, identical for that of the runners.

EVOLUTION OF THE HUMAN BODY

It seems very likely that the qualities selected by many athletic events are the same that led to the physical characteristics of the human body, i.e humans have evolved as fighters, runners, and jumpers, but not basketball players. The evolutionary selection of heights that optimizes various athletic capabilities can thus explain the observation that the mean sizes of very elite athletes are close to the mean for Americans.

EVOLUTION OF ATHLETICS

While it might be imagined that the ideal athlete for an event would set a record which could not be broken, such a chance event has not yet occurred, at least in a major sport. All of the world records in these sports have been broken multiple times in the past half-century. The continued improvement in performance suggests an improvement in training techniques, possibly combined with more competitive selection of athletes.

The possibility of more competitive selection raises the question of whether the variation about the mean heights of world record holders in the various events would become smaller in more recent years. We therefore examined the deviations from the mean in the height data described above, to determine whether the limits converged over a period of decades. No such narrowing was found. We did, however, find a few examples where the heights of the athletes increased significantly over the period of examined.

Tennis

The average height of the winners of the four "Grand Slam" tournaments of the World Professional Tennis Tour (Wimbledon, plus the French, Australian, and U.S. Open Tournaments) are plotted for the years 1960 to 1998 in Fig. 12.10. As shown, there was a 7 cm (3") increase in the average heights of the athletes between 1983 and 1985, without a significant increase in height during the periods before 1983 and after 1985. The cause of the large increase in height has not been identified. The tennis racquet has evolved progressively in the last three decades, and it is possible that the increase in height resulted from an advantage provided by one of the racquet changes. It is also said that the serve has achieved a greater importance recently, and perhaps an edge has been given to taller servers, possibly as the result of racquet changes.

Quarterbacks

The heights of quarterbacks of professional American football who led their conference in passing increased by about 2 inches over the period from 1966 to 1997, while the heights of running

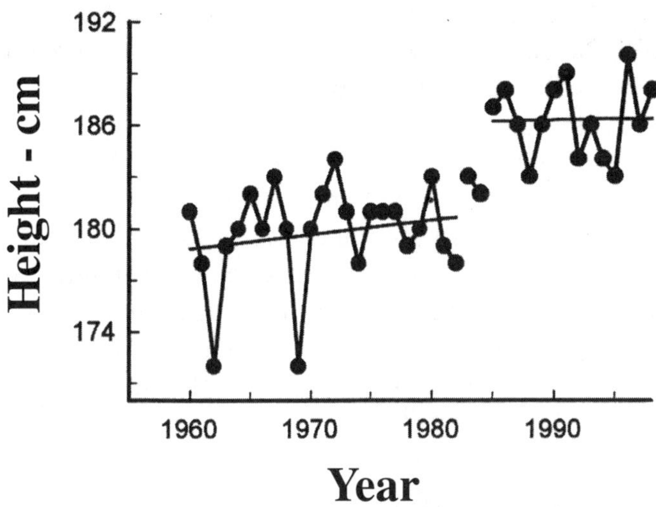

Figure 12.10. *Mean heights of the winners of the four Grand Slam tournaments for each year.*

backs and wide-receivers did not change significantly. A possible explanation is that the heights of the linemen increased as well, and their greater heights required that quarterbacks be taller, to see the pass receivers. The heights of the linemen have not yet been examined, at least by us, but it seems possible that more competitive selection and improved weight-training in tall athletes has led to taller linemen.

High-jumping

Three techniques have been used to set the men's world high jump record since World War II. The first record holder and the only one to use the western roll was, by far, the tallest athlete considered here. He was 204 cm (6'8") tall, 3 cm taller than the next tallest high jumper and 4 cm taller than the tallest shot-putter. The change to the straddle was associated with a substantial reduction in the heights of the record holders and a commensurately large increase in the extent to which they elevated their half-height. The first 4 of the 6 record holders to use this technique were either 185 or 186 cm (6'1") tall. The "Fosbury flop," introduced at the 1968 Olympic games and made possible by softer landing cushions, resulted in a substantial increase in the height of the jumpers. The immediate effect of the change in the new technique was to enable taller jumpers to be more competitive. The last jumper to set a record using the straddle elevated his center of height by 5.3 cm more than the first jumper to use the flop. It should also be emphasized, however, that the greatest increases in records have occurred by progression using the same technique and not as the result of new techniques.

EXCEPTIONS TO SIZE RULES

Although the distribution of heights about the mean value for each sport is fairly small, there are also outliers in many sports, and some lie far out. One of the most notable examples is a successful National Basketball Association player who is only 5'3" (160 cm) tall. Undoubtedly his rapid acceleration gives him an advantage in some aspects of the game. It might also be mentioned that one of the greatest basketball players of all time, who was especially known for his jumping ability, was also at the short end of the spectrum of these tall players, and well within the distribution of heights for high-jumpers.

Although most heavyweight boxing champions are clustered in a 4 inch band from 5'11" to 6'2", the entire spectrum spans 13 inches, from 5'7" (170 cm) to 6'7" (201 cm) (Fig. 12.3).

As mentioned at the outset, one reason for describing these outliers is to avoid discouraging aspirants to a sport from participating because they might have a size disadvantage. Another reason is to avoid discouraging coaches from considering promising athletes whose size is not at the optimum. There is one story of a basketball player, now in the National Basketball Association Hall of Fame, who was not allowed to try out for his local professional team because the coach believed he was too short, even though he had been a star player in college. He was only given an opportunity when his name was drawn randomly, as part of a promotional event, to play in an exhibition scrimmage with the professional team.

A final reason for making these points is that the data presented here are for the most elite athletes in the world. When less stringent selection is made, wider size ranges are found. In addition, the balance among various factors may change when the training level is different. A cautionary example is boxing. The advantage of greater acceleration in a smaller frame accrues mainly to well trained fighters. Anyone who enters an impromptu match with courage derived from the belief that a larger opponent cannot move quickly is liable to have a surprise.

AN IDEAL SPORT

An ideal sport might be one without body-size advantages, so as to include all athletes who might like to participate. A sport that comes close to meeting this criteria is, arguably, the world's most popular, namely soccer. The mean heights of the non-goaltenders on each of the top eight teams in the 1998 World Cup ranged from 176 cm (Argentina) to 183 cm (Croatia), with an overall mean of 180.5 ± 6.2 cm S.D. and a range of 170 to 192 cm. There was a distinct tendency for goaltenders to be taller, averaging 187.5 ± 3.5 cm S.D., and having a range of 180 to 197 cm.

From the perspective of encouraging wider participation, soccer has the added advantage of requiring very little equipment, other than a ball and a reasonably level field. From the health perspective, it provides a great deal of outdoor exercise.

PART III

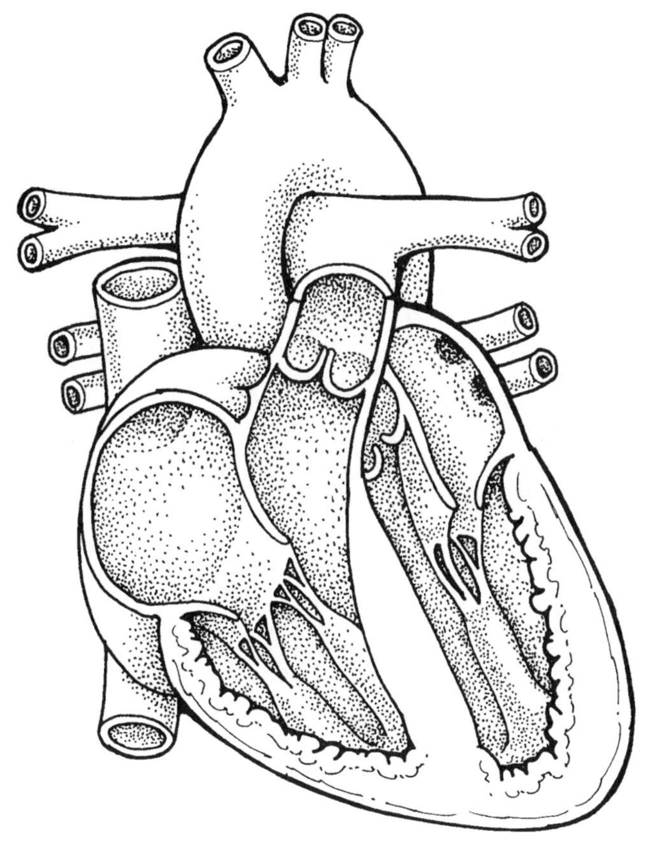

CARDIAC FUNCTION

Chapter 13

HEMODYNAMICS

Understanding cardiac function requires a knowledge of the task the heart must accomplish: pumping blood through the circulation. This task is known as **hemodynamics**. The word evolved from **hemastatics**, a term coined by the Rev. Stephen Hale (1733), who made the first reliable measurements of blood pressure. He inserted tubes into the arteries of horses to see how high their blood would rise. These seminal measurements of blood pressure were an extension of Hale's earlier studies of the fluid pressures in plants, which he called **vegetable statics**. The advance from hemastatics to hemodynamics is more than semantic; seemingly static parameters, such as blood pressure, result from a dynamic balance among multiple variables. A description of this balance provides a good summary of some of the subjects presented in this chapter.

The arterial pressure, as measured by Hale, is an average determined by the distensibility of the aorta and the volume of blood in it. This volume is a dynamic quantity that varies with the inflow rate from the heart and the outflow rate to the capillaries. Superimposed on the mean pressure are two sets of fluctuations, variations in time due to the pulsatile heart beat and local fluctuations due to the interaction of the inertia of the blood with the compliance of the arteries. Thus, the instantaneous pressure loading the heart is determined by: 1) the compliance of the arteries, 2) the resistance to flow in the vessels, 3) the inertia of the blood, and 4) the recent history of the heart's own activity. Finally, the volume ejected from the heart is determined by how much blood is returned to it from the circulation. This **cardiac return** is in turn dependent on venous tone, which determines the pressure that dilates the heart. This chapter describes how the properties of the vasculature affect the pumping ability of the heart. The remainder of the book describes how linear muscle is wrapped into a hollow viscus that pumps blood through the vessels.

OVERVIEW

The circulation supplies blood to separate organs that require different flow rates at different times. Adequate flow is ensured by maintaining a high perfusion pressure in **arteries**, thick-walled, high-pressure vessels that carry blood from the heart to individual organs. Flow to separate organs is controlled by muscle tone in **arterioles**, small branches of the arteries supplying the organs, and **pre-capillary sphincters**, bands of smooth muscle encircling the ends of the arterioles. Arteriolar tone is regulated both by local feedback from tissues and by the executive functions of the central nervous system. Blood pressure is maintained within a narrow range both by an intrinsic cardiac mechanism and by central nervous system control. The individual organs possess mechanisms that allow them to vary the amount of blood they receive, and in general, they operate with an enviable efficiency, taking only as much flow as needed for their immediate function.

The exchange of material between blood and tissue occurs in **capillaries**, very small vessels with diameters only slightly larger than a red blood cell. Blood from capillaries perfusing tissues is carried to the heart in thin-walled, low-pressure **veins**.

For simplicity, the circulation is described as if it were a single circuit, i.e. as if the heart and lungs were a single, central unit, but in fact, it is two distinct circuits in series. The **pulmonary circuit** perfuses the lungs and the **systemic circuit** perfuses the rest of the body. Ease of description often justifies the simplification, but the potential for error should be remembered.

The circulation can be regarded as a complex servo-system with multiple controls for ensuring adequate blood flow to all organs. The heart supplies the energy, and control mechanisms distribute blood in accord with various demands. This chapter describes the functions of blood vessels in meeting these demands, leaving the heart for later chapters. To fully comprehend the circulation, however, it is essential to be aware of the intrinsic, Frank-Starling mechanism that enables the heart to maintain an adequate pumping strength under very different conditions. This mechanism is described briefly here and at greater length in Chapter 16.

FRANK-STARLING MECHANISM

A brief description of this mechanism is given here because it is so essential in compensating for variations in other parts of the circulation. Simply stated, it provides that the strength of the heart beat, and therefore the volume of blood ejected, increases when the heart is distended. The importance of this mechanism can be seen in three simple examples. The first is the case in which the heart skips a beat. The continuing flow of blood returning to the non-beating heart causes it to distend. This distention strengthens the heart beat over several contractions until the extra blood is expelled and the heart returns to its normal volume. The second example is the case of increased vascular resistance increasing arterial pressure. These pressures initially prevent the heart from expelling as much blood as previously, so that the heart dilates. This dilation increases the strength of contraction to match the resistance, and outflow from the dilated heart returns to its original value. The final example is the maintenance of the equality of flow in the systemic and pulmonic circuits. Since the two circuits are in series, blood flow must be the same in both. If the flow in either circuit transiently exceeds that in the other, more blood is returned to the circuit with the lesser flow, and the ventricle providing power to that circuit is distended. This distention increases the pumping capacity of the ventricle so that its output is increased to match that of the other ventricle.

THE CARDIAC CYCLE

The pulse felt in peripheral arteries is due to rhythmic pumping of the heart produced by contraction-relaxation cycles of heart muscle. A single cycle is illustrated by the plots of pressures in the left heart chambers and aorta in Fig. 13.1. These pressure plots are accompanied by an **electrocardiogram (ECG)** tracing showing the electrical activity that initiates the processes.

The cardiac cycle is divided into two unequal parts, called systole and diastole. **Systole** is the period when the muscle is contracting. **Diastole** is all the rest, when the muscle is relaxed. The cycle is initiated by an electrical **pacemaker** impulse arising in the **sinoatrial (SA) node** of the right atrium. This impulse spreads throughout both right and left atria, creating the **P-wave** in the electrocardiogram and causing the atria to contract. The impulse is then transmitted to the ventricles through the **atrioventricular (AV) node**, a band of cardiac muscle tissue with very slow electrical conduction. The slow conduction delays the arrival of the impulse at the ventricles, allowing time for atrial contraction to finish filling the ventricles. Once the electrical impulse reaches the ventricles, it spreads rapidly to depolarize all of the cells of both ventricles, causing the **QRS** complex in the ECG. Because the action potential of cardiac muscle has a long plateau, when the cells are depolarized, there is a delay before repolarization, which causes the **T-wave** in the electrocardiogram.

As illustrated, the venous side of the circulation, including the atrium, operates at low pressures, while the arterial side operates at high pressures. The ventricle cycles between the two pressures, falling during diastole so that blood can enter from the atrium and rising to eject blood into the aorta during systole.

Atrial contraction causes a transient rise in atrial pressure which forces blood into the ventricles. When the ventricles contract, the one-way valves between atria and ventricles (**AV valves**) close as pressures rise in the ventricles. There is a period of **isovolumic contraction** while ventricular pressures rise until they

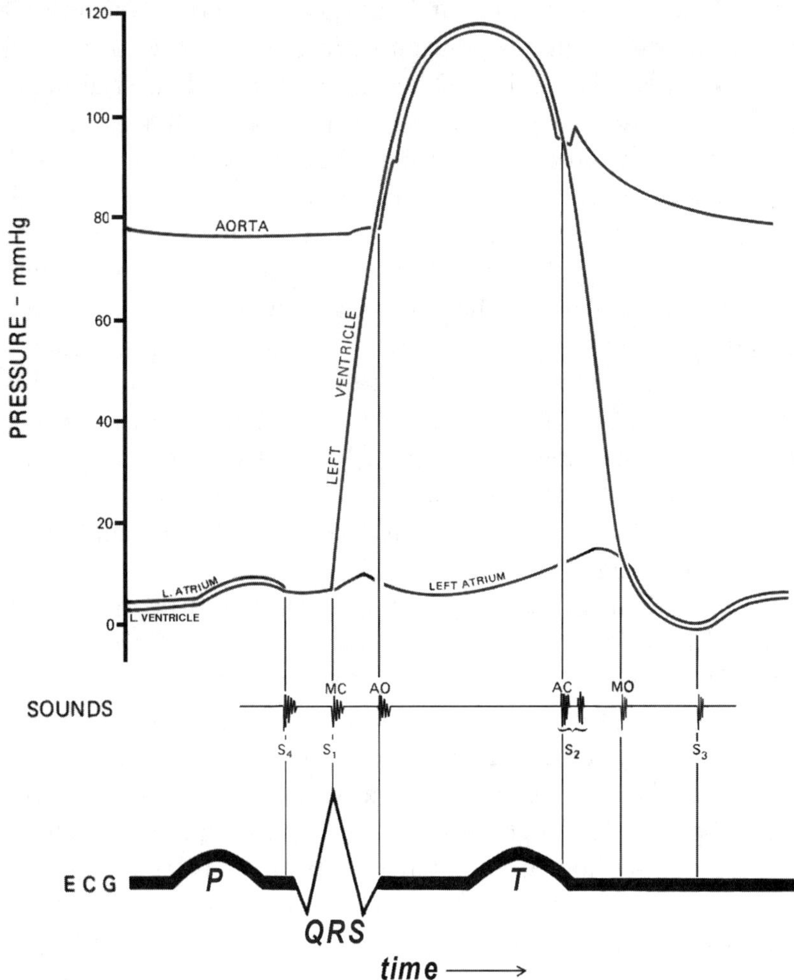

Figure 13.1. *The cardiac cycle. Upper traces: pressure in aorta, left ventricle, and left atrium. Lower trace: ECG. Middle trace: phonocardiogram of heart sounds associated with valve motions.*

reach the levels in the aorta and pulmonary arteries. The aortic and pulmonic valves then open, and pressures in the arteries rise. When the heart is nearly finished ejecting blood, arterial pressure declines due to the run-off of blood into the capillaries, and when it falls below ventricular pressure, the aortic and pulmonic valves close, ending systole. There is then a period of **isovolumic relaxation** while the muscle relaxes and ventricular pressures fall to the level in the atria. The AV valves open and blood flows from the atria to the ventricles. During systole, when the AV valves are closed, atrial pressures rise because of the unremitting return of blood from the circulation, so that there is a large inrush of blood early in diastole. Following the early inrush, there is a period of **diastasis**, when flow across the AV valves is very nearly zero. Flow to the ventricles resumes when the atria contract in response to the electrical activity associated with the P wave, and the next cycle is initiated.

Also shown in Fig. 13.1 are the **heart sounds**, heard with a stethoscope. The most prominent of these result from valve closure. The **first heart sound, S1**, occurs with closure of the AV valves. The **second heart sound, S2**, is produced by closure of the aortic and pulmonic valves. Frequently, a slight delay between closure of the two valves splits the second heart sound. In young and old people, it is normal to hear a **fourth heart sound, S4**, due to atrial contraction stretching the ventricle. A **third heart sound, S3**, caused by rapid stretch of the ventricle when blood from the atrium rushes in, is distinctly abnormal, and heard only in severe heart failure, when diastolic pressures in the ventricle are high. Also shown are **opening clicks**, MO and AO, for mitral and aortic opening. These sounds are audible only when valves are diseased and stiff, but are included here to indicate valve timing.

ARTERIAL PRESSURE

Rhythmic contraction of the heart produces pulsatile blood flow. Pulsations are reduced and flow is made more constant by arterial compliance. These vessels distend during systole and re-coil during diastole to maintain pressure and flow to organs. Arterial compliance also reduces the work of the heart by lowering pressure fluctuations and by spreading the work over the entire cardiac cycle, as discussed below. The recoil of the arteries also ensures that blood flow is maintained if the heart skips a beat.

Arterial compliance

It might be expected that arterial pressure would follow the time course shown in Fig. 13.2, declining quasi-exponentially during diastole, and rising from its minimum, diastolic value at the onset of systole to reach its peak, systolic value shortly before the end of systole. This expectation arises because blood flows continuously from arteries into capillary beds. When no blood is being pumped

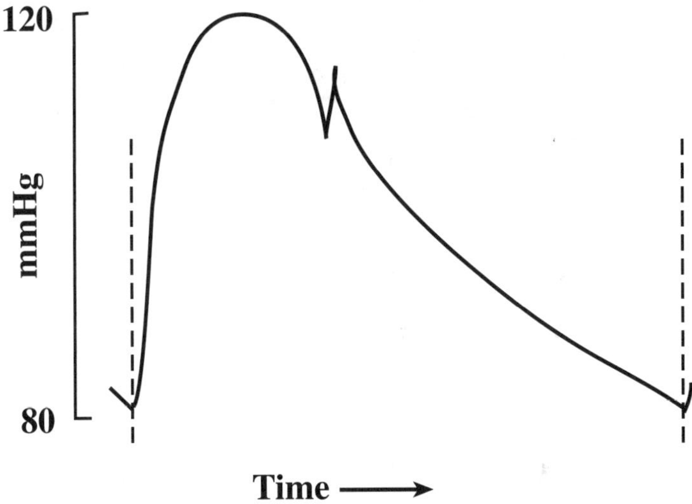

Figure. 13.2. *Length averaged aortic pressure.*

from the heart, the pressure in the aorta should therefore decline with a rate proportional to the pressure. Similarly, pressure is expected to rise monotonically during systole, when the volume of blood in the aorta is increasing. In fact, the arterial pressure traces rarely resemble that in Fig. 13.2 because of pulse waves travelling up and down the arterial tree.

Reflected arterial waves

The aortic pressure profile illustrated in Fig. 13.2 is probably valid for pressure averaged along the length of the aorta, but it cannot be recorded at any point in the arterial tree because inertial pulses distort the local waveforms. Typically distorted waveforms at evenly spaced intervals along the aorta are plotted in Fig. 13.3. As shown, the wavefront steepens and becomes taller further along

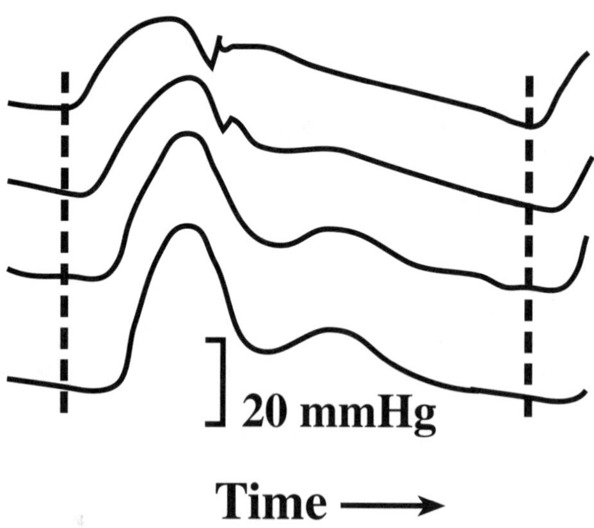

Figure 13.3. *Pressure profiles at evenly spaced intervals along the aorta. Adapted from O'Rourke (1967).*

the aorta. This phenomenon is due to reflections of the wavefronts from both ends of the vessel. If the wave were reflected perfectly from the distal end, its height would be twice that of the original wave, the factor of two arising from summation of the energies needed to decelerate the forward wave and to re-accelerate it in the opposite direction. The later bumps in the record usually occur after aortic valve closure, as indicated by the **incisura**, a notch in the record at the end of systole. These are due to reflected waves traveling in both directions.

The cause of the pressure waves can be summarized as follows. At the onset of systole, when flow through the aortic valve begins, there is a tendency for the entire column of blood in the aorta to be accelerated at once. This tendency is resisted by the inertia of the blood. The incompressible blood cannot absorb the shock wave of this acceleration, but the compliant aorta can, and a small quantity of blood at the root of the aorta moves sideways, distending the root of the aorta slightly, causing a locally higher pressure there. This locally higher pressure accelerates the local blood, and the wave of distention passes down the aorta. The onset of the distal waveform is therefore delayed and made steeper, until it meets the end of the vessel, where it is reflected backward. When the front of the reflected wave meets the later parts of the forward wave, they sum to produce a local pressure that is substantially higher than the original driving pressure in the heart.

Although the waveforms seem very complex and require sophisticated instruments to record accurately, the basic principles, including the relevant mathematical derivations, were first worked out by Thomas Young two centuries ago and were reviewed in his Croonian Lecture of 1809. In addition, the pulse characteristics were known to cardiovascular scientists, such as Frank, more than a century ago.

As is obvious from Fig. 13.3, reflected waves greatly complicate the interpretation of arterial pressures. On the other hand, they contain information about the state of the arteries. With the advent of better methods of recording pulses non-invasively and computers for analyzing the waveforms, it will be possible to learn increasingly more about the arteries without actually looking inside them. Interestingly, clinical assessment of the pulse was a skill learned by most clinicians in previous centuries, and is now largely forgotten. As we learn how to assess the pulse more quantitatively, interest in this area may be revived.

It is well known that arteries stiffen as they age. This stiffening increases systolic pressure and steepens the wave fronts, as shown in Fig. 13.4. Thus, a recording of the pulse profile gives some indication of the physiological age of the arteries. Such profiles can be used to measure the premature aging of the vessels caused by such diseases as diabetes mellitus and hypertension.

The stiffening of the arteries that occurs with age and causes the changes shown in Fig. 13.4 increases cardiac work in two ways. First, since blood is incompressible, a larger column of blood must be accelerated at the onset of the heart beat when the arteries are stiffer. In addition, the reflected waves increase the ventricular pressure when they arrive back at the heart before aortic valve closure.

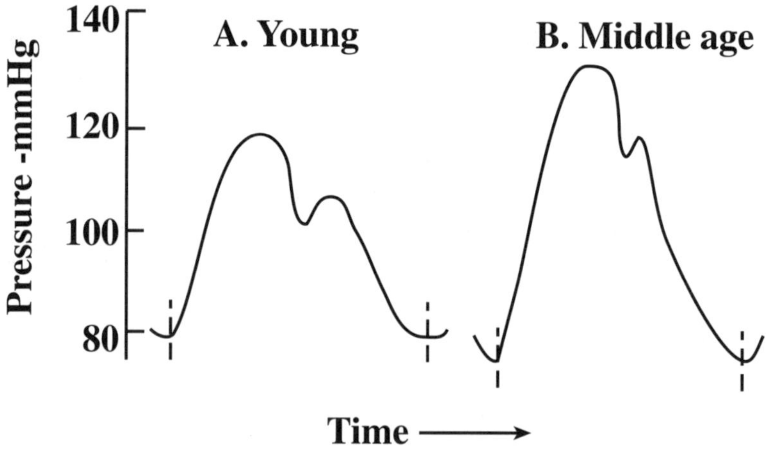

Figure 13.4. *Changes in arterial pulse wave with age.*

VASCULAR ANATOMY

As blood passes through tissue, it flows first in the arteries, then in small branches of arteries, called **arterioles**, before entering the thin-walled **capillaries**, where the exchange of chemical substances between blood and tissue occurs. Upon leaving the capillaries, it passes first through small tributaries of the veins, called **venules**, before reaching the larger veins that carry blood from tissues back to the heart. The ends of the arterioles are encircled by bands of smooth muscle, called **pre-capillary sphincters**, that regulate blood flow to individual capillaries.

As explained below, most of the pressure drop between arteries and veins occurs in the arterioles, which are sometimes called **resistance vessels**, because they are used to regulate the resistance to flow and therefore the amount of blood a tissue bed receives. This term distinguishes them from the larger arteries, called **conductance vessels** that carry blood between heart and arterioles, and the veins, called **capacitance vessels**, which can accommodate changes in blood volume without large changes in pressure.

The arteries and arterioles have thick muscular walls that enable them to withstand the high arterial pressures without dilating. Because the pressure drop occurs upstream, the capillaries and veins are not subject to high pressure and therefore do not require thick muscular walls. The thinner muscular walls of the veins are, however, used to great effect in regulating the venous pressure, which is a major determinant of cardiac output.

Two physical laws govern the architecture of the circulatory system. The first is the Law of Laplace, derived at the beginning of Chapter 15, which dictates that the tension in the wall of a vessel be proportional both to the pressure within the vessel and to its radius. This law provides that the smaller arterioles can withstand the high arterial pressures with thinner walls simply because they have smaller radii.

Poiseuille's law dictates that laminar flow along a vessel be proportional to the pressure drop along the vessel and to the fourth power of the vessel radius. This fourth power relationship has several important consequences. For example, it dictates that there be a much higher pressure drop along the length of an arteriole than along the length of a larger artery. This large pressure drop in an individual arteriole further requires that there be a large number of arterioles to carry blood to the tissues. If as the blood vessels became smaller, their number increased in inverse proportion to their cross-sectional areas, there would be too much resistance to flow in the small vessels. To allow adequate flow, the number of vessels increases out of proportion to the reduction in their area. This increased number means that the total cross-sectional area of the arterioles and venules is substantially larger than the total cross-sectional area of the arteries and veins.

HEMODYNAMIC RESISTANCE

When fluid is moving in a tube, flow rate (F) and pressure drop (P) across the tube vary directly with each other. It is often assumed that these two dynamic quantities vary in direct proportion to each other, so that their relationship is defined by a proportionality constant (R), i.e.

$$R = P/F. \tag{13.1}$$

In this case, the proportionality constant is called resistance, and it is analogous to electrical resistance defined by Ohm's Law as R=E/I. Although there is often no experimental validation for the assumption of a linear proportionality, it is justified by the simplification that it affords. It should be remembered, however, that it is an often unstated assumption which may be incorrect. To understand the importance of the concept of resistance, consider that the flow of blood out of the heart, called **cardiac output**, increases several-fold when an animal goes from rest to strenuous activity. This increase occurs because the vascular beds supplying the muscles dilate, allowing more blood to flow through them.

The effect of this dilation is quantified in terms of the ratio of pressure drop between arteries and veins divided by the flow rate, i.e. resistance. As an example of the simplicity of assuming that resistance does not vary with flow rate, consider that when this is true, and when the pressure drop between arteries and veins is constant, cardiac output and resistance are inversely proportional to each other.

In the discussions that follow, resistances across several different structures are recognized. One is resistance across the peripheral circulation, called the **systemic vascular resistance**, and abbreviated **SVR**. Another is resistance across the pulmonary circulation, called **pulmonary vascular resistance**, and abbreviated **PVR**. The third is resistance across a cardiac valve, called **valvular resistance, VR**.

REGULATION OF CARDIAC OUTPUT

Adequate blood flow to the body is maintained by keeping arterial pressure within a narrow, normal range. This constancy of arterial pressure results from the Frank-Starling mechanism acting in concert with variations in venous filling pressure and variations in inotropic state. If, for example, resistance across large tissue beds were to diminish without a change in cardiac output, arterial pressure would fall, and blood flow to the entire body would be compromised. This does not happen because the more rapid flow of blood through the tissue beds increases venous return to the heart and increases cardiac filling pressures, dilating the heart. The more dilated heart is then able to eject more blood in each beat and expel the increased amount of blood returned to it.

In addition to the Frank-Starling mechanism, any fall in blood pressure resulting from a decline in peripheral vascular resistance immediately stimulates sympathetic nervous reflexes. These reflexes increases both heart rate and the strength of the heart beat, so that the volume of blood expelled from the heart, called the **cardiac output**, is increased and arterial pressure is restored.

Maintaining venous pressure

An extremely important, but often overlooked aspect of these regulatory mechanisms is the central role of venous tone in maintaining hemodynamic homeostasis. To understand this, consider the case of a fall in systemic vascular resistance, as might occur during exercise. If arterial pressure is to be maintained constant, cardiac output must increase to match exactly the increase in blood flow through the systemic circuit. This will happen only if filling pressure in the heart is exactly correct. If, for example, the more rapid return of blood causes the veins to distend, the return of blood to the heart will be compromised and cardiac output will fall. In this case, more blood would be returned to the venous side of the circulation than is pumped by the heart, the volume of blood in the arteries would decrease, and arterial pressure would fall.

There are at least three reasons that the importance of venous tone goes unrecognized. The first is that it is very difficult to measure under conditions where cardiac output is made to vary in a physiological way. The second is that under most physiological conditions, venous pressure does vary much, leading to the incorrect impression that it is not important in maintaining hemodynamic stability. The third is that more attention has been directed to the heart because its active role in maintaining blood flow is more obvious. One major effort to explain the integration of the various components of the circulation, including venous tone, is explained at the end of this chapter. Chapters 18 and 19 describe a medical condition called **acute pulmonary edema** that results from a sudden and severe increase in venous tone.

REGULATION OF BLOOD FLOW TO ORGANS

Adequate blood flow to each tissue bed is ensured by the maintenance of arterial pressure, as explained above. In the presence of adequate pressure, flow to an organ can be regulated by adjusting the hemodynamic resistance of its blood vessels. In general, the different tissue beds receive only as much blood flow as they need. This economy is achieved and demonstrated by a phenomenon called autoregulation.

Autoregulation

Blood flow to many organs is independent of pressure over the physiological range. This independence is illustrated in Fig. 13.5, plotting flow against pressure. At very low pressures there is a proportionality between flow and pressure. The slope of the line, i.e. the ratio of flow to pressure, is the inverse of vascular resistance. The ratio is constant because the vascular bed is maximally dilated, and there is no change in resistance with changes in pressure. A similar proportionality is observed at very high pressures. Again, the ratio of pressure to flow is constant, now because the arterioles are maximally constricted, and there is no variation in

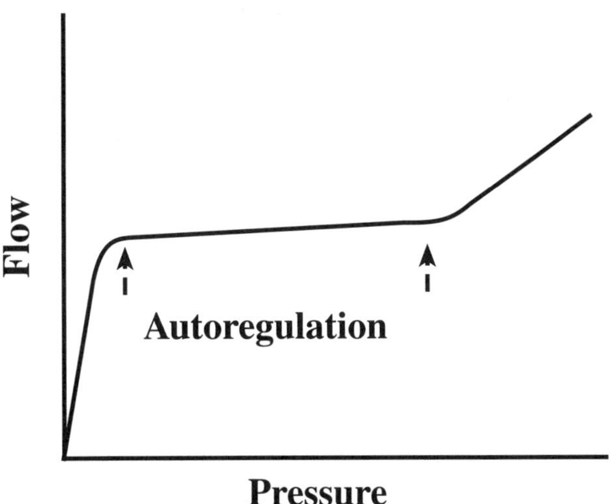

Figure 13.5. *Flow-pressure relations to an organ.*

resistance with pressure. The slope of the relationship is much less steep because the resistance is substantially higher. In the middle of the range between these two extremes, resistance increases almost in proportion to pressure, so that flow remains nearly constant. The change in local resistance to maintain a constant flow is called **autoregulation**. It appears to be regulated locally because it occurs even when the organ is denervated.

Local control of blood flow

The finding that autoregulation is controlled locally leads immediately to the suspicion that metabolites affect the opening and closing of the pre-capillary sphincters. This suspicion has been confirmed. Blood flow to skeletal muscle, for example, increases enormously when the muscle does work, and similar increases can be produced by raising the CO_2 in the arterial blood or making it more acidic. These findings suggest that the local mechanisms increase blood flow to remove metabolic wastes and to deliver more oxygen, which minimizes lactic acid production. Similar increases in blood flow to the brain have been described in subjects asked to carry out mental arithmetic. In addition to their dilatory effects, some of the same metabolites may constrict other arterioles, thereby shunting blood to regions where it is most effective. This is not the complete explanation, however, because blood flow to different organs must be coordinated by executive functions of the central nervous system, as described below.

Regulation of resistance

The pressure gradient across an organ or tissue bed is not uniform along the path of the blood, but shows a sharp drop in the arterioles. Thus, most of the resistance to flow occurs in these resistance vessels. It is important to recognize, however, that there must be some flow in the arterioles if they are to create a pressure drop. If flow is stopped elsewhere, there will be no pressure drop across the open arterioles and their regulatory power will end. This issue is important because flow in individual capillaries often ceases completely, but the pressure drop across the entire tissue bed still occurs upstream, in the arterioles, because flow continues in other capillaries and because the arterioles constrict sufficiently to maintain the upstream pressure drop.

Critical closing

The consideration that flow in a blood vessel is proportional to the fourth power of vessel radius suggests that the small, pre-capillary sphincters could cause very fine regulation of flow rates in capillary beds. If one observes blood flow in capillaries through a microscope, a very different picture is presented. Blood flow in individual capillaries is largely an "all-or-none" phenomenon, either stopped or proceeding at its maximum rate. This is caused by a phenomenon called **critical closing**.

As mentioned above and derived in Chapter 15, the Law of Laplace dictates that pressure in a vessel be proportional to both wall tension and vessel radius. When the radius of a pre-capillary sphincter diminishes sufficiently, wall tension overcomes pressure, and radius is reduced to its minimum. This minimum size is less than a single red cell, which plugs the vessel and stops flow entirely. Opening the vessel requires that the muscle relax substantially below its previous level. Thus, the capillaries resemble computer bits, having only two states, open and closed.

As long as flow continues in other capillaries supplied by the same resistance arteriole, the pressure drop in the arteriole will maintain a low pressure in the pre-capillary sphincter. If flow should stop completely, however, the pressure in these sphincters will rise substantially, and some of them will open. Thus, the flow to a tissue bed is maintained by a cooperation between the tone in the arterioles and the pre-capillary sphincters.

Central control of blood pressure

If venous pressure is constant, blood flow to a capillary bed depends only on arterial pressure and the resistance in the bed. When local beds require more blood, as do skeletal muscles during exercise and gastrointestinal capillaries after a meal, the increased demand is met by a combination of increased cardiac output and decreased flow to other organs. The competition for blood flow among organs must be adjudicated by the executive functions of the central nervous system. The regulatory mechanisms include both direct alterations in nervous tone and **neurohumoral** effects mediated by hormones, such as adrenaline, in response to central nervous stimulation and peptides secreted directly from the atria. A full exposition of these mechanisms is not needed for an understanding of cardiac loading. Only the simplest descriptions will be given here, to show how the load on the heart is regulated to maintain the necessary blood flow to all organs.

Arterial pressure is sensed by **baroreceptors** located in the aorta and the **carotid sinus**, a slight bulge in the carotid artery that carries blood up the neck. When arterial pressure increases, the baroreceptor nerves increase their firing rate. These nervous impulses are transmitted to the brain to initiate reflexes that lower pressure. The opposite happens when blood pressure falls; the baroreceptor nerve firing rate diminishes, initiating reflexes that support blood pressure. One effect of increased baroreceptor tone is a reflex increase in the firing rate of the cardiac branch of the vagus nerve, which slows the heart rate and thereby diminishes cardiac output and arterial pressure. Since there is always some vagal tone slowing the heart at rest, a fall in blood pressure decreases baroreceptor tone and vagal suppression of heart rate, so that heart rate increases. The changes in vagal tone affecting heart rate are a well understood part of the overall reflex that maintains a nearly constant blood pressure. Another major effect of a fall in blood pressure is to restrict of blood flow to non-vital organs, such as the skin. Thus, another effect of low arterial pressure is cool skin. The coolness is produced both by direct nervous stimulation of pre-capillary sphincters in the skin and by reflex stimulation of the adrenal gland which increases circulating adrenaline. The adrenaline causes both a decrease in blood flow and sweating, which makes the skin feel damp as well as cold.

Additional mechanisms for supporting blood pressure are initiated by atrial stretch receptors, which detect venous pressure and therefore circulating blood volume. The reflexes initiated by these receptors include fluid retention by the kidneys and thirst, which stimulates volume replacement.

To appreciate the importance of these reflexes in maintaining blood pressure it is only necessary to lie recumbent in a hot bath for many minutes and then stand up quickly. The dizziness caused by this maneuver results from pooling of blood in the legs, causing a fall in cardiac output and less perfusion of the brain. The tendency for blood to pool in the legs on standing is normally counteracted by fast reflex constriction of veins and shunting of blood away from non-vital centers. These reflexes are defeated by the heat, which relaxes the veins of the legs and the arterioles of the skin. The dizziness is usually only transient because the reflexes are only blunted, not abolished. It should also be emphasized that the response can be variable and unpredictable, so that anyone attempting such an experiment should always secure a firm hand-hold and be prepared to sit down quickly.

HEMODYNAMIC WORK

The power required to force blood through the circulation is the pressure drop between arteries and veins multiplied by the flow rate (power=P·F). Power is an instantaneous quantity, and the work done is the integral of the power over time. The work done in perfusing the body is often calculated as the mean pressure drop between the arteries and veins multiplied by the cardiac output. This simplified calculation contains an inherent error because the pressure and flow rise together during the cardiac cycle. Since the power is the product of the two, the variations in power are proportionately greater than variations in either flow or pressure alone. The difference between the work calculated from mean values and that calculated by integration of the instantaneous pressure-flow values is sometimes called the "pulse work," but this concept is flawed because the difference arises from using mean values, not from extra work. It does, however, make the point that the greater the pulsations, the greater the work.

A major advantage of compliant arteries is the smoothing of pulsations arising from the cyclical heart beat. The elastic aorta absorbs energy when blood flows out of the heart, and thereby reduces pressure, so that the power needed to eject blood is reduced. Thus, a compliant aorta can reduce cardiac work, even when cardiac output and mean arterial pressures are the same.

RESISTANCE

An important determinant of cardiac work is the resistance to blood flow, defined by equation 13.1, above. As mentioned, this formulation assumes that pressure and flow vary linearly with each other, a condition that may not always obtain. In spite of this potential inaccuracy, the concept of resistance can be very useful, both in assessing clinical states and in making hemodynamic calculations. Some examples of the clinical use of resistance follow.

Pulmonary vascular resistance

In general, systemic and pulmonic pressures, both venous and arterial, remain relatively constant during exercise, in spite of several-fold increases in blood flow through both circuits, at least until the limits of exertional tolerance are reached. This constancy of pressures must result from a decrease in resistance that is inversely proportional to cardiac output. The mechanisms responsible for the decreased peripheral resistance are relatively well understood and have been outlined above. The mechanisms which maintain a constant pulmonary resistance are less well understood. They do not appear to be due to the local or global metabolic changes that cause peripheral dilation; pulmonary resistance decreases long before global metabolic changes occur, the local effects do not reach the pulmonary circuit, and the metabolic effects that dilate peripheral arterioles usually increase pulmonary vascular resistance. It is certain, however, that the pulmonary vascular resistance can decline several-fold during exercise because pulmonary artery pressure rises very little as blood flow through the lungs increases several-fold. The best current explanation of this dilation is that the small pulmonary arterioles are extremely compliant, so that very small increases in pressure cause them to open.

Because all the cardiac output must pass through the lungs, the heart's ability to increase output is strongly dependent on the ability of the pulmonary vasculature to dilate. A clinical state called **pulmonary hypertension,** defined by an elevated pulmonary artery pressure at rest, is associated with marked shortness of breath during exercise and a poor prognosis. The probable reason for the poor prognosis is that most of the capacity of the pulmonary vasculature to dilate is lost before the resting pressures rise, so that the disease is in an advanced state when it is diagnosed. In the early stages of the disease, when the arterioles have a diminished but not absent ability to dilate, arterial pressure measurements made at rest are likely to be normal, and the criteria for diagnosis will not be met. The pathology can then be diagnosed only by showing that resistance does not decrease as much as expected when cardiac output increases. The shortness of breath during exercise is due to an inadequate cardiac output caused by the inability of the right ventricle to increase blood flow through the inelastic pulmonary circuit.

Valvular resistance

The placement of a catheter in the heart was first done by Forssman, on himself! This dramatic demonstration that the heart could be catheterized soon led to the widespread clinical use of the technique. Among the first useful catheter measurements were pressure differences across **stenotic**, i.e. narrowed, valves, that often result from rheumatic fever. There is almost no pressure difference when blood flows forward across a normal valve, but as the disease progresses, the obstruction to flow increases, and the pressure difference rises. Thus, the pressure drop across a diseased valve can be used to indicate the severity of stenosis. However, a difficulty occurs because the pressure difference depends both on the degree of stenosis and the blood flow. Some method must be employed to compensate for differences in flow when assessing the severity of stenosis.

The earliest proposal for assessing stenosis was to calculate resistance (VR) as the ratio of pressure drop (P) to flow (F)

$$VR = P/F \tag{13.2}$$

(Dow et al., 1950; Silber et al., 1951), analogous to electrical resistance calculated as voltage drop divided by current flow. In a seminal effort to understand more about stenosis, Richard Gorlin, a cardiologist with his father S.G. Gorlin, an engineer, in 1951 developed a formula for estimating valve area based on the Torricelli principle of laminar flow through a flat orifice, shown in Fig. 13.5 A. Their formula indicated that area (A) is proportional to the flow divided by the square root of pressure

$$A = k_A \cdot F/\sqrt{P} \tag{13.3}$$

There is a major difference between these two quantities, resistance and calculated area, used to assess the severity of valvular stenosis. The difference arises because one assumes that pressure drop is proportional to the first power of flow, and the other assumes that it is proportional to flow squared.

In proposing their formula for area, the Gorlins referred to the resistance formulation as the Poiseuille relationship because this relationship for laminar flow through a uniform tube (Fig. 13.5 B) predicts that flow and pressure are linearly proportional. Neither the Torricelli principle, on which the area calculation is based, nor the Poiseuille relationship is appropriate to the stenotic valve because the valves often taper (Fig. 13.5 C) and the murmurs heard across stenotic valves suggest that flow is turbulent. Furthermore, a general relationship cannot be developed from physical principles because most of the relevant parameters are likely to vary among patients. These parameters include the shape and distensibility of the valve and the viscosity of blood. Thus, the utility of either index depends, in part, on the extent to which it is found to be constant at different flow rates.

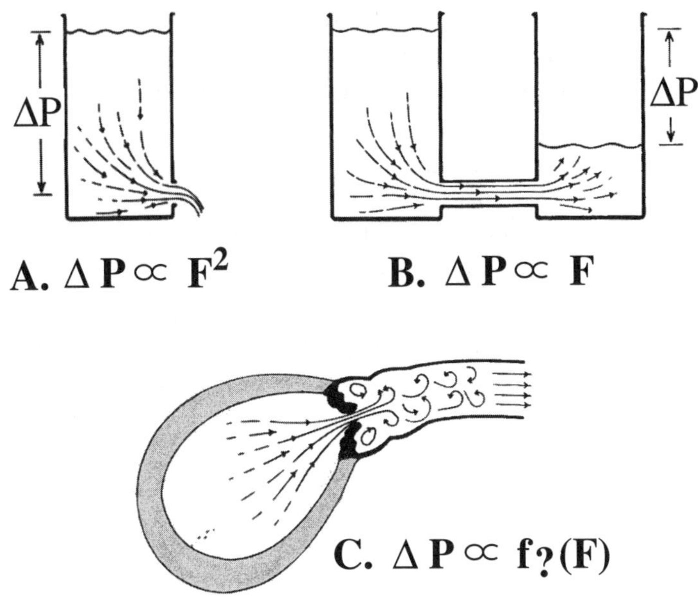

Figure 13.6. *A) Laminar flow through a flat orifice, the basis for the Torricelli principle. B) Laminar flow through a uniform tube. C) Turbulent flow through the tapered orifice of a stenotic valve. Adapted from Feldman et al. (1992), with permission.*

Perhaps because of the simplicity of its units, cm^2, and its evocation of an anatomic image, valve area came into common clinical usage, and resistance, which is often reported in the arcane units of dyne·s·cm^{-5}, fell into disuse. The question of whether to use area or resistance was revisited recently because it was found that area calculated from the Gorlin formula appeared to vary directly with flow. Sometimes, area appeared to vary with the square root of the pressure drop (Cannon et al., 1985), as would occur if resistance remained constant at different flow rates. On re-examining this data, it was found that resistance remained much more constant than area (Ford et al., 1990). This finding is surprising because the sounds caused by a stenotic valve, **murmurs**, imply turbulence, and turbulence usually increases pressure out of proportion to flow. Possible explanations for pressure being proportional to the first power of flow are 1) that the valve distends slightly at the higher pressures and flows and 2) that the viscosity of blood decreases with increasing turbulence. Blood viscosity is a dynamic quantity that depends on the aggregation of red blood cells. When sheer forces are low, as when flow is slow, the viscosity is markedly increased. When sheer forces are high, as in rapid and turbulent flow, the relatively weak bonds between the red cells are broken, and viscosity decreases.

It should be emphasized that there is no reason to expect flow and pressure to be related by a specific function or the relationship to be the same in different valves. It is convenient that they appear empirically to be linearly proportional to each other, but it should also be kept in mind that the use of this simple relationship may obscure variations in clinical states that cause the relationship to deviate from linearity. More studies are needed to test whether the same relationship obtains in all clinical settings.

While the constancy of resistance might be sufficient reason for preferring it to area as an index of stenosis, a more compelling reason is its utility in calculations. Resistance is a statement of the relationship between flow and pressure. When one is known, the other can be estimated. In addition, resistance facilitates calculations of other hemodynamic parameters, such as work. By contrast, calculated area gives only an estimate of anatomic deformity and does not allow such calculations.

INTEGRATION OF THE CIRCULATION

Arthur Guyton, known to generations of students for his concise textbooks of Physiology, is perhaps best remembered scientifically for his effort to develop a comprehensive scheme for explaining the regulation of the circulation. His explanation attempts to include the properties of the individual components of the circuit in an integrated whole. In particular, he called attention to the importance of venous tone by developing the concept of mean vascular pressure. This construct regards the circulation as akin to a refrigeration system which has a mean pressure that can be measured when the compressor is at rest. Similarly, if the heart could be stopped with no other change in the circulation, arterial and venous pressures would be the same, and equal to mean vascular pressure. This concept is illustrated in Fig. 13.7, which shows that when the pump becomes active, and cardiac output increases from zero, arterial and venous pressures diverge from the mean value. As indicated, the decline in venous pressure is less than the rise in arterial pressure because of the greater venous compliance. It should also be emphasized that, as drawn here, the figure represents a major simplification described at the beginning of this chapter, namely that it ignores the important complication of having two sides to the circulation.

There are three major determinants of the cardiac output: the venous filling pressure, the arterial pressure against which the heart must work, and the inherent strength of the heart, determined by its inotropic state. For a given inotropic state, the cardiac output therefore settles at a specific operating point, determined by the balance among these three factors and indicated by the vertical arrows in Fig. 13.7.

The importance of venous tone is illustrated by the dashed

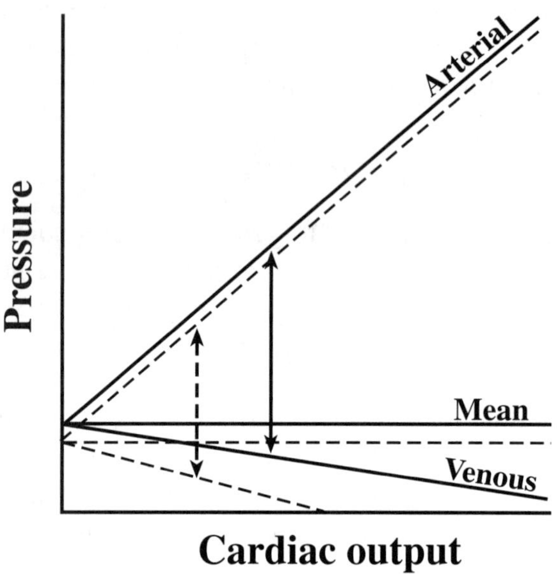

Figure 13.7. *Relationship between vascular pressures and cardiac output. Dashed lines indicate changes caused by increased venous compliance. Adapted from Levy (1979).*

lines representing the changes that occur when venuos tone is diminished, as might result from lying recumbent in a hot bath and then standing quickly. The result shown is venous dilation which decreases cardiac filling pressure and leads to diminished cardiac output. The consequent fall in arterial pressure diminishes blood flow to the brain and dizziness. The dizziness is only transient, however, because reflexes from the brain stimulate both arterial and venous constriction to restore arterial perfusion pressures.

Similar types of dizziness and low blood pressure on standing, respectively called **orthostatic dizziness** and **orthostatic hypotension**, are frequently seen in patients taking medicines to lower blood pressure. These medicines work by partially inhibiting the mechanisms that support blood pressure when gravity diminishes cardiac return, and are a necessary consequence of adequate treatment.

The major contribution of this construct, and the reason for presenting it here, is that it calls attention to the importance of venous tone in maintaining hemodynamic homeostasis

SUGGESTED READING

LEVY, M. N. (1979) The cardiac and vascular factors that determine systemic blood flow. *Circulation Res.* **44**: 739–747.

FORD, L. E., FELDMAN, T., CARROLL, J., and CHIU, Y. C. (1990) Hemodynamic resistance as a measure of functional impairment in aortic valvular stenosis. *Circulation Res.* **66**: 1–7.

Chapter 14

CARDIAC ANATOMY

The irregular shape of the heart is so similar in such a wide variety of animals as to suggest that natural selection has proven the design optimal. The task in this chapter is to describe the little that is known about the advantages of this unique shape. More attention in subsequent chapters is devoted to the left ventricle because it provides most of the power to the circulation and has thus been studied more extensively. The atria usually receive less attention, mainly because they have been studied less, and less is known about their structure-function relationships.

A major point to be made in later chapters is that the heart rarely uses its full capacity and has substantial reserves for increased body demands. This point is first mentioned here, however, because some cardiac structures, such as the atria and the right ventricle, can be removed or inactivated without causing much apparent disability. While these observations are sometimes taken to indicate that the inoperative structure was relatively unimportant, they are probably more valid evidence that the luxury of the modern life does not require the animal to use the full reserve provided by evolution.

GENERAL ARCHITECTURE

In a remarkable scientific accomplishment, William Harvey (1628) showed that the function of the heart is to pump blood in a circuit through the body. The heart and circulation are divided into two parts. The **left side** of the heart receives blood from the lungs and pumps it out to the rest of the body, called the **systemic cir-**

cuit. The **right side** receives blood from the systemic circuit and pumps it to the lungs. Each side of the heart has two chambers, an **atrium**, sometimes called an **upper chamber** because it is upstream in both the blood flow and the electrical circuit, and a ventricle, sometimes called the **lower chamber**. The electrical impulse that initiates the heart beat originates spontaneously in the **sinoatrial (SA) node**, an area of specialized tissue high in the right atrium, and travels down the atrium to the **atrioventricular (AV) node**, another band of specialized tissue that conducts the impulse to the ventricles with a delay that allows the contracting atria to fill the relaxed ventricles before they contract.

Fig 14.1 shows the heart cut away to reveal the four chambers and four valves. There are two valves that separate the atria from

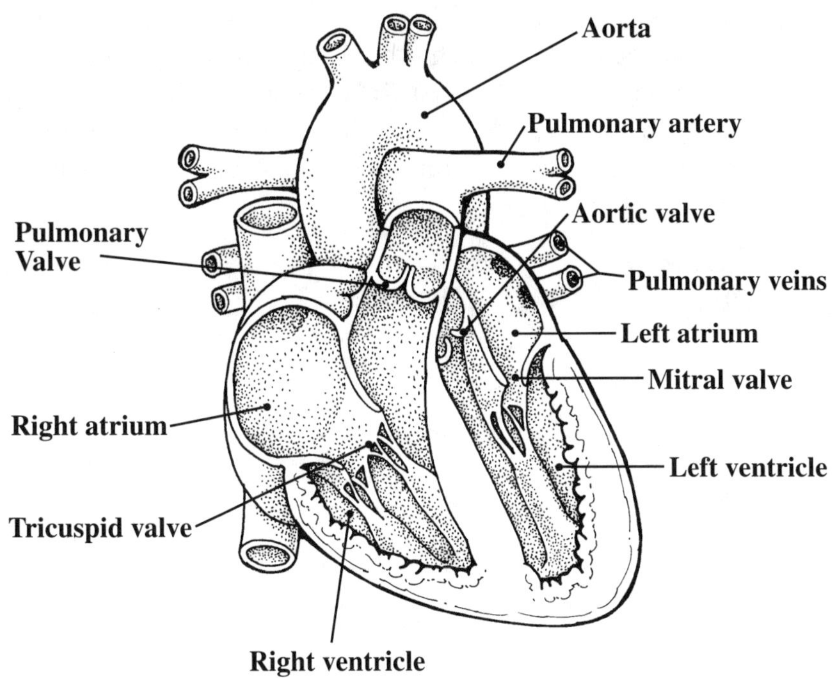

Figure 14.1. *The heart cut away to show the four chambers and the valves.*

the ventricles and another two that separates the ventricles from the aorta and pulmonary artery. The four valves are located in an irregular plane of fibrous tissue, called the **cardiac skeleton** (Fig. 14.2), that makes up the circumferential valve rings and separates the atria from the ventricles. This plane is normally traversed by only one bundle of muscle, the atrioventricular node. The fibrous tissue provides an electrical "insulator" which ensures that the atrial electrical activity is not transmitted to the ventricles prematurely. The benefits of this insulation are well recognized because in the **Wolf-Parkinson-White syndrome** a band of normally conducting tissue, the **Kent bundle**, also traverses the cardiac skeleton and frequently leads to abnormalities of the cardiac rhythm.

The relevance of the term "cardiac skeleton" is that, unlike skeletal muscles that originate and insert in the bony skeleton, ventricular heart muscle is anchored only by the fibrous tissue surrounding the valves. At all other points, the individual muscle cells are connected only to each other longitudinally, so that force is transmitted axially from the myofilaments of one cell to the myofilaments of the next. Long bundles of muscle cells originate and insert in the cardiac skeleton and spiral around each other to

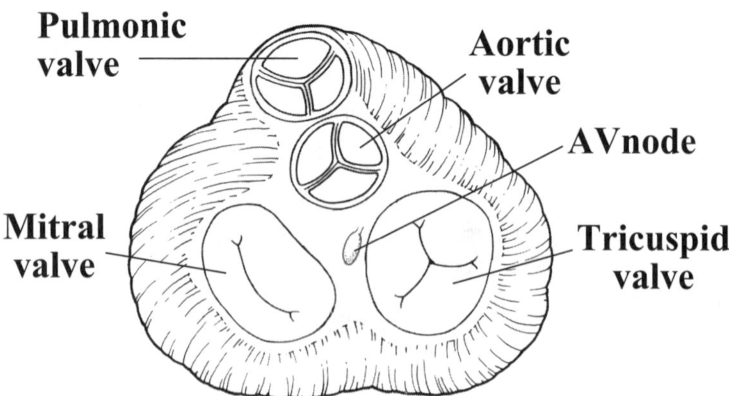

Figure 14.2. *Arrangement of valves in the cardiac skeleton.*

form the ventricular walls. The crossings of these spirals provide lateral support so that contraction does not separate adjacent cells and cause the ventricle to leak. Similar bundles comprise the atria, and most of these originate in the fibrous tissue near the place where the main vein from the upper systemic circulation, the **superior vena cava**, meets the right atrium.

Whether contraction in a spiral direction also provides a functional benefit other than offering lateral support remains to be discovered. As explained in Chapter 15, a recent theory suggests that the left ventricle twists as it contracts. If it occurs, this twisting might be facilitated by the spiral arrangement. It is also true, however, that interruption of the continuity of the spirals does not appear, by itself, to interfere with cardiac function to a greater extent than expected from the loss of muscle that produced the interruption.

Valves

The heart has four one-way, "flap" valves that open freely when blood flows across them in the normal direction, but close immediately when the flow attempts to reverse. The two **atrioventricular (AV)** valves prevent blood from returning to the atria during **systole**, the period when the ventricles contract. The valve between the left atrium and ventricle has two cusps and is called the **mitral valve,** because of a loose, and not easily seen, resemblance to a bishop's hat, called a mitre. The right sided AV valve is called the **tricuspid** because it has three cusps. The valves separating the ventricles from the arteries are named for the arteries, i.e. the **aortic valve** sits at the root of the **aorta**, the main arterial trunk, and the **pulmonic valve** is at the root of the **main pulmonary artery** that carries blood to the lungs. These valves prevent blood returning from the arteries during **diastole**, when the ventricles are relaxed. Each valve normally has three cusps. Occasionally the aortic valve has only two valves, and these **bicuspid** valves often deteriorate later in life, perhaps demonstrating an advantage of a three cusp valve.

Fig. 14.2 shows the cardiac skeleton and the relative positions of the four valves. The irregularity of the plane of the skeleton is indicated by the cutaway view of the four chambers in Fig 14.1, showing that the long outflow tracts of the ventricles causes the plane of the aortic and pulmonic valve to be displaced from the planes of the AV valves.

The free margins of the cusps of the AV valves are supported in part by tendons, **chordae tendinae**, attached to **papillary muscles** arising in the ventricular wall, as illustrated Fig. 12.1. This muscular support helps to prevent edges of the valves from prolapsing into the atria during systole.

Heart shape

Some lower forms of life, such as the sea cucumber (tunicate), have a rudimentary circulation powered by a heart consisting of a contractile segment in a straight length of blood vessel. A wave of constriction propagated along the tube propels blood. In an example of ontogeny recapitulating phylogeny, the embryonic heart begins as a nearly straight tube in the circulation and forms the four chambers by developing the bulges that form the atria and a prominent bend that folds the ventricular structure back on itself so that the valves are nearly in the same plane. In the fully developed heart the flow of blood makes U-turns in the ventricles. This shape causes the heart to eject blood in the same general direction whence it came and provides a more compact structure that fits conveniently within the chest cavity, the **thorax**, where it is protected by the bony skeleton.

Blood supply

Blood is delivered to the heart by the **coronary arteries**. There are two main coronary arteries that arise from the root of the aorta behind two of the three aortic valve cusps. The larger of the two, the **left main coronary artery**, divides almost immediately to form the **left anterior descending artery** and the **circumflex artery**. The left anterior descending artery passes over the front of the heart and supplies the anterior portion of the left ventricle including the **septum** between the right and left ventricles. The circumflex artery passes over the back of the heart and supplies the posterior part of the heart. The **right coronary artery** travels over the right side of the heart and supplies the atria, the inferior portion of the heart, and much of the right ventricular free wall. In general, the main arteries travel over the outer, **epicardial** surface where they are not constricted by contracting muscle. There are anatomical variations where a band of muscle "bridges" over a coronary artery, and sometimes this **bridging** is associated with the clinical picture of insufficient coronary blood flow.

A "heart attack," called a **myocardial infarction**, occurs when a coronary artery becomes occluded, usually as the result of atherosclerosis. The muscle supplied by the occluded artery dies and becomes a fibrous scar. This disease process is described here because the resulting damage may alter the architecture of the heart. These alterations interfere with normal cardiac function in several ways, described briefly in Chapters 15 and 19.

The arrangement of arteries, and their anatomical variations, were mainly of academic interest until the health consequences of myocardial infarctions became recognized. The exact anatomy has taken on a new importance with the development of techniques for visualizing the arteries using X-ray cinematography and additional techniques for bypassing or opening stenotic vessels.

Venous blood flows from the muscle into the right atrium via the **coronary sinus**, which is sometimes catheterized to measure metabolites. In general, it is not possible to isolate the venous return from a specific chamber, such as the left ventricle. In addition, the individual chambers of the heart share their blood supply with other chambers, so that it is difficult to isolate the blood flowing to a single chamber. This sharing of blood supply complicates estimates of the ratios of work done by a single chamber, such as the left ventricle, to the amount of blood flow it receives or the amount of oxygen it consumes.

CARDIAC CHAMBERS

Left ventricle

The normal left ventricle is shaped like a tapered tube. Its short axis normally changes length to a relatively greater extent than its long axis, so that it behaves more like a tube than a sphere. Failure of the heart muscle causes the ventricle to dilate, and the increased expansion of the short axis causes the ventricle to become more spherical. The dilated, spherical ventricle contracts more uniformly in all dimensions.

The left ventricle has a thick wall, which enables it to generate greater pressures for a given amount of force per unit of muscle cross-sectional area than the thinner walled right ventricle. The thick wall also enables it to achieve more complete emptying, as described in the next chapter. The nearly complete emptying avoids stagnation of blood that might lead to clotting and systemic emboli. The inner, **endocardial**, surface of the ventricle is traversed by bands of muscle that form an irregular network of ridges and bridges. The functions of these **trabeculae** are unknown, but at least three possible benefits of an irregular endocardial surface have been suggested. One benefit, suggested by comparative physiology, is that the irregular shape provides a larger surface area for the diffusion of oxygen and other metabolites between muscle and blood in the chambers. This benefit is exaggerated in amphibian hearts, which have only a very rudiment coronary system but very deep clefts in their endocardial surfaces. The muscles of these "reptilian" hearts receive most of their nutrient supply directly from the blood within their chambers. A second benefit, described in the next chapter, is that the irregular shape facilitates more complete ventricular emptying. As ventricular volume diminishes, the endocardial layers heap up, and the irregular clefts provide a place to accommodate this tissue accumulation. Finally, the contractile movement of the individual trabeculae might also avoid the adherence of clots. **Mural thrombi**, clots on the walls of the ventricle, which sometimes break off to form systemic **emboli**, are vanishingly rare in normal hearts but occur with distressing frequency when cardiac emptying is impaired. Thus, it might be expected that natural selection has provided the irregular surface to accommodate more movement of the inner layers of muscle, and thereby avoid the adherence of endocardial clots.

Right ventricle

The right ventricle has a thin wall, as expected from the lower pressures that it generates, and it is wrapped around the left ventricle. Under normal circumstances, the septum between the ventricles has a thickness and curvature similar to the remainder of the left ventricular wall. Thus, the septum and left ventricle bulge into the normal right ventricular cavity, as illustrated in the cross-sectional view of ventricles in Fig. 14.3.

The point of these two observations is that the right ventricle achieves its large fractional volume changes not from the thick-

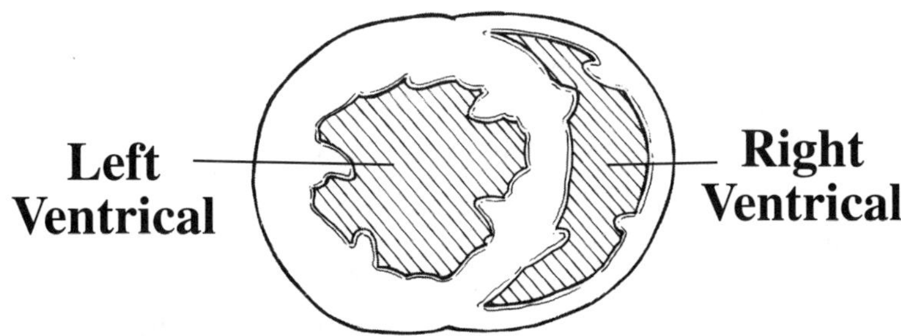

Figure 14.3. *Cross-section through the heart at the level of the two ventricles.*

ness of its wall, as does the left ventricle, but from the septum and left ventricle partially filling its cavity. As with the left ventricle, the walls of the right ventricle are made irregular by a network of trabeculae.

The right ventricles of experimental animals have been completely inactivated with little obvious clinical consequences, at least when the animals enjoy a sedentary life-style. While this has sometimes been martialled as evidence that the right ventricle is not needed to power the circulation, it is probably better evidence that the domesticated animals are not required to use the full capability of their hearts.

Atria

The atria are wide places in the rivers of blood returning to the heart. One of their functions is to serve as reservoirs that absorb the nearly continuous flow of blood during systole, when the AV valves are closed, and thereby minimize interruptions of venous flow. To augment this capacitative function, they have *cul-de-sac*-like **appendages** that increase their volume. Both atria, including their appendages, contract immediately before the ventricles. This contraction greatly increases the flow of blood across the AV valves, and thereby augments the ejection of blood from the ventricles during systole. The importance of the atrial contraction is readily illustrated in patients with **atrial fibrillation**, a condition in which separate portions of the atria contract rapidly and without coordination. The capacitative function of the atria is retained, but the coordinated contraction is lost. The cardiac output falls immediately when the atria begin to fibrillate, but ventricular filling and output are returned to near normal by an increase in venous pressure. The effects are variable, with some patients developing symptoms of breathlessness, particularly with exertion, and others being unaware of the change in rhythm. As with inactivation of the right ventricle, the ability of some persons to tolerate the absence of co-ordinated atrial contraction is sometimes taken to indicate that the atria are relatively unimportant, but a more likely interpretation is that such persons are not required to perform at maximum levels.

Atrial fibrillation usually occurs because of increased atrial pressures and is frequently associated with conditions that cause atrial dilation. Occasionally it occurs for no demonstrable reason. Atrial fibrillation from any cause is deleterious for two reasons. As mentioned, it is associated with a reduced cardiac output and higher venous pressures. In addition, it is associated with the formation of clots, **thrombi**, that break loose and float downstream as **emboli** that occlude arteries. When a systemic artery becomes occluded, the tissue supplied by that artery frequently dies. A frequently observed consequence of such tissue death is a **stroke** due to death of brain tissue. It is probably observed more frequently than other embolic phenomena both because the brain receives such a large fraction of the cardiac output, about 25%, and because the clinical consequences are so immediately obvious.

Clotting of blood in the normal atrium is avoided in part by vigorous contraction of the muscle that prevents the adhesion of clots to the wall and in part by emptying of the chambers that prevents the stagnation of blood. Individuals with atrial fibrillation and without an anatomical demonstrable cause have a nearly tenfold increase in risk of strokes resulting from emboli to the brain. Patients with atrial fibrillation due to a condition that increases dilation may have a 20–30 times greater stroke risk.

CONCLUSION

As with other biological structures, the heart has a very irregular shape that has been preserved over a large number of species, and only some of the advantages of this shape have been discerned so far. It seems likely that more advantages will be recognized as knowledge of cardiac function accumulates. As with the several possible advantages of an irregular endocardial surface, described above, it will probably be found that the benefits of some aspects of this highly conserved shape will be found.

Chapter 15

GEOMETRIC PRINCIPLES APPLIED TO THE HEART

Understanding the application of muscle physiology to cardiac function requires a firm grasp of the geometric principles that relate linear muscle contraction to 3-dimensional, volume work done by the heart. The necessary geometric relationships are derived here from first principles to provide a precise explanation of the mechanisms. The importance of such an understanding will be obvious when the calculations of muscle tension in the heart derived below are compared with those derived by Woods over a century ago, before the physiological basis of muscle contraction was understood. The main reason for these derivations, however, is to demonstrate the utility of asking physiological questions in terms of parameters measured in the ventricle, work and power, rather than in calculating linear muscle performance, force and shortening velocity, from 3-dimensional ventricular parameters.

THE LAW OF LAPLACE

The issue here is to define the relationship between pressure in a cardiac chamber and muscle tension in its wall. This relationship has been shown empirically to be governed by the same physical principle that explains such diverse phenomena as surface tension in soap bubbles and the greater ease in expanding a balloon as its size increases. This principle was first elucidated by the French mathematician Laplace (1749–1827) and bears his name. It relates the wall tension in a hollow cavity to the cavity radius and its transmural pressure. It is derived here for a thin-walled sphere because an exact understanding of the principles is required to distinguish between theoretical possibilities.

Fig. 15.1 shows a hollow sphere transected by a plane through its center. The force tending to separate the two halves of the sphere is equal to the pressure in the sphere multiplied by the area of the plane, which is in turn equal to $\pi \cdot r^2$, i.e.

$$F = P \cdot \pi \cdot r^2. \qquad (15.1)$$

A complication arises in normalizing tension, T, from this force because there are several denominators that might be used to define tension in terms of a unit quantity of muscle. The denominator used by Laplace was the force per unit length of circumference. When wall tension, T_W is defined in this way, both sides of eq. 15.1 are divided by circumference, $C = 2 \cdot \pi \cdot r$, to give

$$T_w = F/C = P \cdot r/2 \qquad (15.2)$$

which is the usual statement of the Law of Laplace.

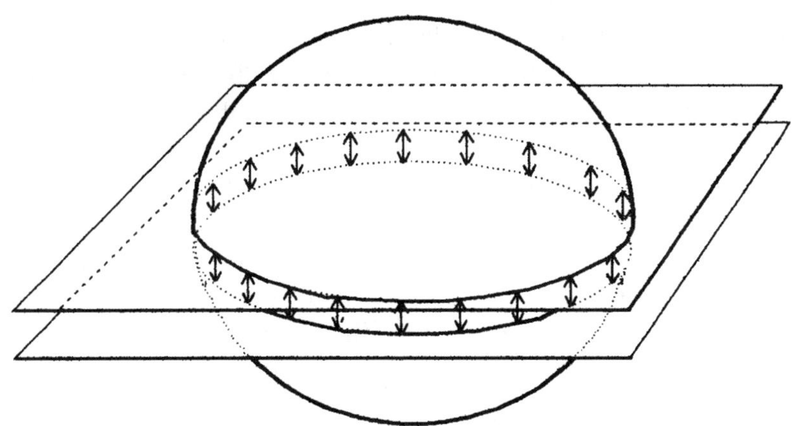

Figure 15.1. *Hollow sphere bisected by a plane through its center.*

To express chamber size in terms of muscle length, circumference, $C = 2 \cdot \pi \cdot r$, is substituted for radius,

$$T_w = C \cdot P / 4 \cdot \pi \tag{15.3}$$

Relationship between muscle and ventricle

To be applicable physiologically, wall tension, T_w, must be replaced by the muscle cell tension, T_m. Since a constant number of cells, N, crosses the circumference, the number of cells per unit length of circumference varies inversely with the chamber dilation, and the tension per muscle cell varies with the ratio N/C, i.e. $T_m = T_w \cdot N/C$, and equation 15.3 becomes

$$T_m = C^2 \cdot P / 4 \cdot \pi \cdot N \tag{15.4}$$

This relationship shows that tension per cell increases parabolically as the heart dilates, so that a much higher cell tension is required to develop ventricular pressure in a dilated heart.

It should be re-iterated that this derivation is for the special case of a spherical, thin-walled sphere. In this case, the Law of Laplace dictates that tension on a muscle cell perpendicular to the plane in Fig. 14.1 be directly proportional both to the intra-cardiac pressure and to the square of the circumference of the sphere. One final correction is needed before moving on to the next derivation, and that is for the random orientation of cells. For the case considered here, the thin-walled sphere is assumed to have an equal distribution of cells oriented equally in all directions from perpendicular to parallel to the plane. The mean force of these cells is equivalent to the force exerted by the perpendicular cell divided by $\pi/2$, so that mean force per muscle cell becomes

$$T_c = C^2 \cdot P / 2 \cdot \pi^2 \cdot N \tag{15.5}$$

The reason for this correction is to calculate the work per cell, which is derived by multiplying the force per cell by the fractional shortening velocity of the cells. To do this, it is convenient to re-arrange eq. 15.5 to obtain an expression for pressure

$$P = 2 \cdot N \cdot \pi^2 \cdot T_c / C^2 \qquad (15.6)$$

Ejection rate

As with the law of Laplace, the relationship between the rate of ejection of blood from the heart is determined both by muscle characteristics and chamber size. This relationship is derived here, again using the simple case of thin-walled sphere.

The volume of a sphere, Vol, is defined by

$$\mathrm{Vol} = 4 \cdot \pi \cdot r^3 / 3 = C^3 / 6 \cdot \pi^2 \qquad (15.7)$$

Ejection rate, defined as the rate of change of volume with time, dVol/dt, is determined by differentiating equation 15.7, to obtain

$$d\mathrm{Vol}/dt = C^2 \cdot (dC/dt) / 2 \cdot \pi^2. \qquad (15.8)$$

which indicates that ejection rate is directly proportional both to the shortening velocity of the circumferential muscle (dC/dt) and to the square of the circumference. This equation shows that a much lower muscle velocity is required to produce a given ejection rate from a dilated chamber, exactly balancing the increased force required to produce a given pressure in a dilated heart.

Pressure-volume work from force-length work

The rate of work done by the ventricle, power, PV, is calculated as the product of pressure times ejection rate, i.e., multiplying eq. 15.6 by eq. 15.8,

$$\begin{aligned} PV = P \cdot d\mathrm{Vol}/dt &= (2 \cdot N \cdot \pi^2 \cdot T_c / C^2) \cdot (C^2 \cdot (dC/dt) / 2 \cdot \pi^2) \\ &= N \cdot T_c \cdot (dC/dt) \end{aligned} \qquad (15.9)$$

Eq. 15.9 simply states the irrefutable fact that ventricular work rate must equal the muscle work rate defined as the work rate per cell, $T_c \cdot (dC/dt)$ multiplied by the number of cells, N. This relationship is independent of the size of the ventricular chamber or any other geometric relations. The observation that the preceding equations lead to this expression is not proof that the eq. 15.9 is correct, because it must be correct, but it can be taken as a suggestion that the preceding expressions are formulated correctly.

Woods' observation

In 1892 Woods, a physician in Dublin, published a paper entitled "A few applications of physical theorem to membranes in the human body in a state of tension," in which he defined the relationship between wall tension and cavity radius in three organs, an adult heart, an infant's heart, and a urinary bladder. This seminal paper firmly established the importance of the Laplace phenomenon in organ physiology. His main observation was that wall thickness at different locations in a single heart was proportional to the average radius of curvature at that location. His work is mentioned here because it established the importance of the regulatory mechanisms that increase cardiac muscle strength at long lengths. It is also described here because one of his calculations yielded a different result from the derivation above, and the difference between them shows the importance of understanding how the calculations are applied.

The equations derived above indicate that wall tension per unit length of circumference increases with the first power of circumference while the tension on the individual cells increases with the square of circumference. Woods calculated that the wall tension per unit of cross-sectional area of the muscle increased inversely with the third power of circumference. This greater inverse dependence arose because he also accounted for the thinning of the wall that must occur if the muscle maintains a constant volume as it spreads over a larger ventricular volume.

All of these calculations are correct, and the issue is to decide which of them is relevant to the problem at hand. The tension per muscle cell is used here because estimates of muscle performance are made in the same way. Cross-sectional area is measured at a standard length, and force at all lengths is normalized to this standard cross-sectional area rather than to the cross-sectional area specific to each length. As explained in Chapter 2, this use of a standard cross-sectional area provides an estimate of the force per filament, because the number of filaments in the cross-section does not vary with length. Cross-sectional area varies inversely with length but the smaller areas at longer lengths are associated with decreased spacing between filaments and not changes in their number. Neither the existence of filaments nor their importance was known to Woods. His conclusion that force per unit of muscle cross-sectional area increased enormously with cardiac dilation was correct, but he might have used a different denominator if he had worked at a later time.

The difference between the two expressions for tension is explained by one expressing tension per unit cross-sectional area of muscle and the other expressing it in units that are proportional to force per cell or per myofilament. The higher force which Woods estimated is borne by an increased density of filaments in the cross-section of a stretched cell.

The primary importance of Woods' observations was that they demonstrated very convincingly that the muscle in the wall of the heart must operate at a severe mechanical disadvantage when the heart expands. From this consideration he concluded, before Frank or Starling, that the muscle must possess a mechanism for increasing its strength at longer lengths.

Geometric complexities

The overly simplified case of a thin-walled, spherical ventricle with cells oriented in every direction normal to the radius can be used to build a more complicated, thick-walled, non-spherical ventricle with cells in different layers having different orientations. This would be done by calculating the force and velocity of individual cells in fractional, thin-walled spheres and summing the lot using a modern computer. The reader will be spared a description of such a complicated process, however, because a much simpler conclusion can be made from the calculations above. A major point of these derivations is that they show how calculations relating ventricular performance to muscle physiology are very much simpler when the relevant measurements are work and power rather than force and velocity.

Calculations of pressure from muscle force and ejection rate from muscle shortening require a detailed knowledge of ventricular size and shape. Furthermore, since the left ventricle has a thick wall, both the force and the rate of circumferential change is different in the different layers of the wall, further complicating exact estimations. By contrast, muscle work and power are the same as ventricular work and power, irrespective of any geometrical considerations. Thus, it is extremely advantageous to ask questions that can be answered in terms of these parameters.

HEART SIZE

The title of this section, heart size, may be a misnomer because the data are for the left ventricle. On the other hand, it seems likely that the biological principles derived for the left ventricle apply equally well to the other chambers, so that the overall heart size is determined by the same factors.

The main point of this discussion is that heart muscle grows to match the load imposed by the circulation. There are two distinct types of load, volume overload and pressure overload, and the growth response to the two types are different.

Wall thickness

As described above, the Law of Laplace dictates that tension per unit length of circumference in a hollow vessel varies in proportion to both the pressure across the wall and to the radius of curvature of the wall, as first recognized by Woods (1892). He related wall thickness at multiple locations in a single ventricle to the average radius of curvature at the same location. Both thickness and radius varied by a factor of 5 but their ratio was constant. Since the pressure experienced by each segment of the ventricle was the same, Woods' findings suggest that the muscle thickness varies to maintain a constant tension per unit of muscle cross-sectional area. Thus, the muscle appears to grow to match the force it is required to generate. Subsequent observations by others confirmed and extended Woods' measurements to show that the wall thickness increases to maintain a constant force per unit of cross-sectional area when tension is increased, either because of increased pressure or because of an increased radius of curvature.

Pressure overload

An intraventricular pressure greater than a normal human's is pathological except in the giraffe, whose heart must generate an higher blood pressure to perfuse its elevated brain. The conditions that chronically increase the ventricular pressure include high blood pressure, a narrowing of the aortic valve orifice, called **aortic stenosis**, and obstruction due to the growth of muscle into the outflow tract of the ventricle. All of these conditions, including the elevated blood pressure in the giraffe, lead to an increase in wall thickness without an increase in cavity size unless there is an additional pathological condition. In fact, the increased wall thickness is achieved by the growth of the wall both inward as well as outward, so that the diameter of the compensated and hypertrophied ventricular cavity is reduced slightly.

Volume overload

Conditions requiring a chronically increased cardiac output will cause heart growth and increased heart size, called volume overload hypertrophy. Normal body growth, for example, demands an increased cardiac output and is therefore a form of volume over-load hypertrophy. Other forms of volume overload include: 1) **arteriovenous fistulae** that shunt blood from the arteries directly to the veins; 2) incompetent valves that allow blood to flow backward so that stroke volume must be increased to accomplish the necessary forward flow; 3) **septal defects** in the heart that allow blood to flow directly from left to right chambers without passing through the systemic circulation, 4) chronic increases in muscular activity, either from hard labor or strenuous athletics; 5) **anemia**, a decrease in the number of hemoglobin-containing red blood cells that requires an increased cardiac output to achieve the needed oxygen delivery; 6) pregnancy; and 7) increased metabolic activity, usually due to over-activity of the thyroid gland. When these conditions occur in a setting where cardiac function is not impaired, the heart grows in a way that chamber volume and stroke volume increase while ejection fraction remains the same. Thus, the circumference of the wall increases while the percentage change of circumference in each beat is unaltered. As might be expected, these changes are accomplished by the growth of new sarcomeres in series around the circumference, so that the muscle operates over the same range of sarcomere length.

The increased cavity radius in volume-overload hypertrophy increases wall tension. To compensate, the wall grows thicker, in proportion to the increased radius, with the ratios of all dimensions being unchanged. This is called "magnified hypertrophy" because all dimensions increase in the same proportion.

Thickness/radius ratio

The foregoing discussion suggests that, at a constant pressure, ventricular wall thickness and radius of curvature should increase in the same proportion as the ventricle grows, maintaining a constant thickness/radius (t/r) ratio, and that the ratio should vary in proportion to systolic blood pressure. Ford (1976) tested this hypothesis by analyzing published data from hearts having a 2,000-

fold variation in size. The linear dimensions of thickness and radius in human hearts were obtained from a study by Grossman et al. (1975) using **echocardiography**, a technique that uses high frequency sound waves to distinguish tissue density borders in internal body organs. End-diastolic thickness and radius were determined from these borders and the ratios are plotted against systolic blood pressure in Fig. 15.2 A. The average values are shown for four conditions in which the hearts remained compensated, including normal controls (circles), volume overload due to mitral regurgitation (squares), volume overload due to aortic regurgitation (inverted triangles), and pressure overload due to aortic stenosis (upright triangles). The average intraventricular volume of patients with aortic insufficiency was three times that of the normal controls, and as indicted, the wall thickness in the hearts with aortic stenosis was nearly twice that of the controls.

The straight line passing near the points in the graph shows a nearly exact proportionality between the thickness/radius ratio and systolic blood pressure. The qualification "nearly" is used because it is also seen that the data points for hearts with valve lesions all lie below the line while the normal control value lies above the line. There are several possible explanations, most of which suggest that the patients with valvular heart disease were not fully compensated. The value for another group of patients with heart failure (X) lie substantially below the line. If the heart fails, it maintains cardiac output by dilating, using the Frank-Starling mechanism. The ventricular function curves, explained in Chapter 16, indicate that heart failure causes the heart to move to a lower function curve and that dilation returns the stroke volume towards normal. The thickness/radius ratio is a doubly sensitive index of dilation because the wall thins as the radius increases. Thus, a possible explanation of the data points for patients with diseased valves lying below the line is that they had a mild degree of heart failure, either because of underlying disease or because the valvular lesions altered coronary blood flow. Another, similar explanation is that their valvular lesions were not static, but pro-

gressive so that their hearts were not fully adapted to their overloads.

A much wider range of animal heart sizes are shown in Fig
15.2 B. Thickness and radius measurements were not available,
but estimates of ventricular mass and heart volumes were. Different symbols represent different groups of animals. Open symbols

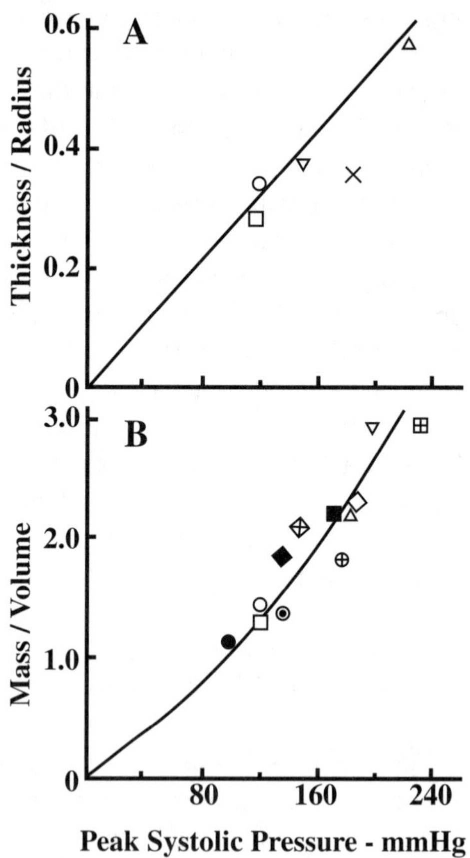

Figure 15.2. *Thickness/radius ratio (A) and mass/volume ratio (B) plotted
against pressure for hearts having a 2,000-fold variation in size. From Ford
(1976) with permission.*

show data for spontaneously hypertensive rats at different ages as triangles and diamonds with circle and square representing normotensive controls at the same age. Open symbols with cross are for dogs, with circle as control, diamond as hypertension due to coarctation of aorta, and square as arterio-venous fistula. Closed symbols are for human children, with circle as control, diamond as coarctation of aorta, and square as aortic stenosis. The circle with central dot indicates value for human adults.

The t/r ratios for these ventricles were calculated from the M/V ratios.

$$M/V = [(t+r)^3 - r^3] \cdot [1.05 \text{ g/cm}^3]. \qquad (15.10)$$

All of the data were pooled and fitted with the straight line through the origin. This is the line drawn in Fig. 15.6 A. The line was then converted to the curve in Fig. 15.6 B, using eq. 15.10.

There was some scatter in the data about the line and curve in Fig 15.6 B, as might be expected from studies obtained for different purposes using different methods, but the data is sufficiently close to the line and curve to suggest that mammalian hearts grow to maintain a constant relationship between t/r and systolic pressure.

EXTENT OF VENTRICULAR EMPTYING

A puzzle presents itself when the functioning length range of isolated cardiac muscle is compared with the volume changes the ventricle can undergo. Typical isometric length-force curves for isolated cardiac muscle, Fig. 15.3, show that the total length range is 25% when the muscle is maximally stimulated in the presence of unphysiologically high levels of calcium. When the inotropic state is closer to its baseline, the slope of the relationship is steeper over much of the range where it normally operates. For a given change of circumference, the greatest volume change in a cavity will occur when all dimensions change equally, so that changes in all three dimensions contribute to volume changes. In this case, when muscle length shortens from Lmax to 75% Lmax, volume will be reduced from its optimum to 0.75^3 of its optimum, i.e. to 42% of its optimum volume or a 58% volume excursion.

The puzzle is shown schematically in Fig. 15.4 where the 25% functioning range of isolated muscle is shown in panel A and the pressure volume curve with a 58% volume excursion is shown in panel B.

Not only is the volume change in this spherical ventricle much less than the capability of the normal ventricle, but the ventricles do not diminish equally in all dimensions. Under normal circumstances, the ventricular short axis changes more than the long axis. These considerations suggest that the maximum volume reduction expected of the ventricle would be 58% if the cavity were to shrink uniformly in all directions and less than 50% if most of the reduction occurred in the short axis.

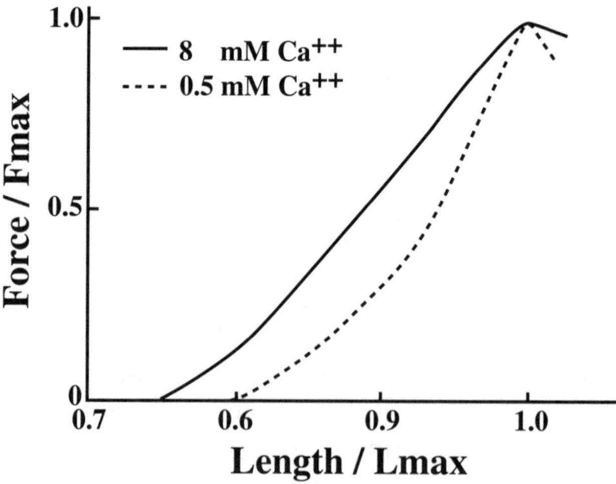

Figure 15.3. *Length-force relations in cardiac muscle at two calcium concentrations. Adapted from Allen et al. (1974).*

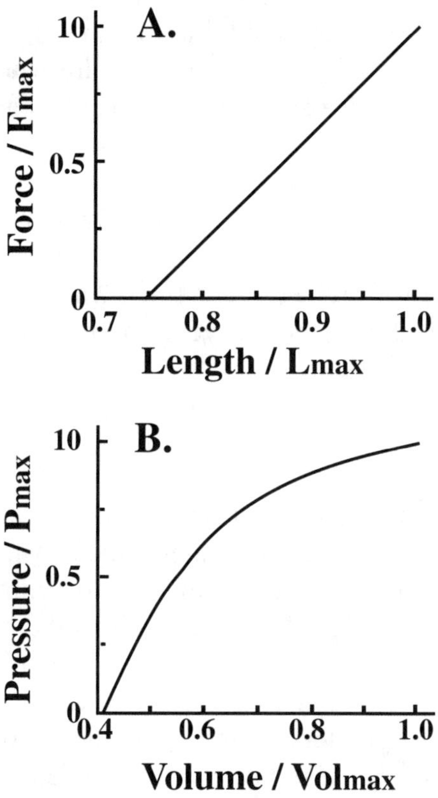

Figure 15.4. *A) Length-force relation. B) Volume-pressure relationship for a thin walled spherical ventricle enclosed by muscle with the length-force properties in A.*

In contrast to this expectation, nearly complete emptying of the ventricle is often seen and yet the ventricle maintains normal systolic pressures. During exercise in the upright position the left ventricle ejects more than 90% of its end-diastolic volume and with rest and recumbency the end-diastolic volume dilates up to 40% in the same subjects (Poliner et al., 1980). Thus, in going from rest and recumbency, where the ventricle is 140% of its up-

right value, to upright exercise, the volume of the ventricle can decline by more than 93% and continue to maintain perfusion pressures. Right ventricular ejection fractions are usually somewhat
less than those of the left ventricle, but typically can be more than
the 58% maximum calculated.

The answer to this puzzle appears to be that each ventricle has
a filler that displaces volume and provides more complete emptying, and that the filler is different for the two ventricles. In the left
ventricle, the filler is provided by its thick wall. In the thin-walled
right ventricle the filler is the left ventricular septum bulging into
the cavity.

Left ventricle

Muscle volume remains constant during contraction, as expected
from tissue comprised largely of water. This constant muscle volume causes the endocardial layers of the left ventricle to shorten
more that the outer, **pericardial** layers so that the wall thickens as
the ventricle contracts. As explained in Chapter 6, the steep
length-force curves for cardiac muscle result from inactivation at
shorter lengths. The greater shortening of the endocardial muscle
therefore suggests that these layers are completely inactivated at
small volumes. At these small volumes, the inactivated muscle
serves to fill the volume enclosed by the epicardial layers and
allow nearly complete emptying of the ventricular cavity. Since
these layers are also stretched more when the ventricle dilates,
they become activated to a greater extent than the epicardial layers. This increased activation of the inner layers with stretch is a
form of recruitment that augments the Frank-Starling relationship.

The wall thickness and diameters of different muscle layers of
a thick-walled spherical ventricle are shown schematically for
three different volumes in Fig. 15.5. For ease of representation,
muscle volume is chosen to be equal to the intracavity volume at
the largest volume. The data in Fig. 15.2 B indicates that the ratio
of muscle volume to cavity volume in normal, non-dilated ventri-

cles at rest is about 1.4. It seems likely that these normal ventricles can dilate by at least 40% before reaching their maximum volumes, so that the ratio of 1.0 chosen for simplicity in Fig. 15.4 is approximately correct, or perhaps a little high.

Each of the six rings in Fig. 15.5 represents a thin-walled shell of muscle, with an equal volume of tissue enclosed between consecutive pairs of rings. As shown, the radial distances between inner rings varies more with volume changes than those between outer rings, so that the inner layers shorten more. The reduction in cavity volume to 42% of its optimum brings the muscle of the inner layer to the length where it would just reach a state of total inactivation if it had the length-force characteristics shown in Fig. 15.4. When the volume falls below this length, further shortening cannot occur, and the wall heaps up. This observation suggests that the irregular inner surfaces of the ventricles evolved to facilitate this inward movement of inactive muscle that increases ventricular emptying.

The isovolumic volume-pressure relationships for the theoretical ventricle in Fig. 15.5 is shown in Fig. 15.6. Panel A shows

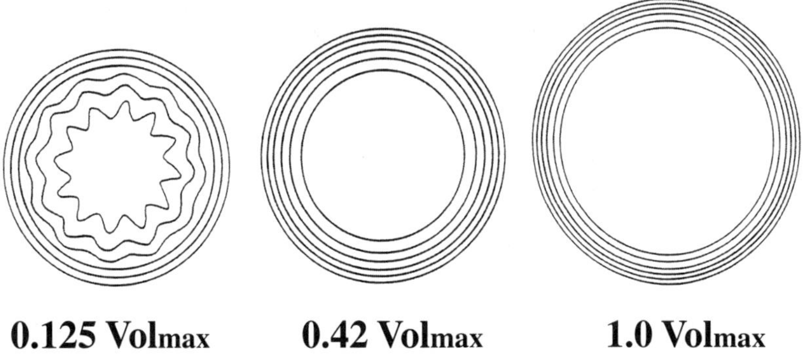

Figure 15.5. *Circumferential differences in muscle layers of a spherical ventricle at the 3 volumes indicated.*

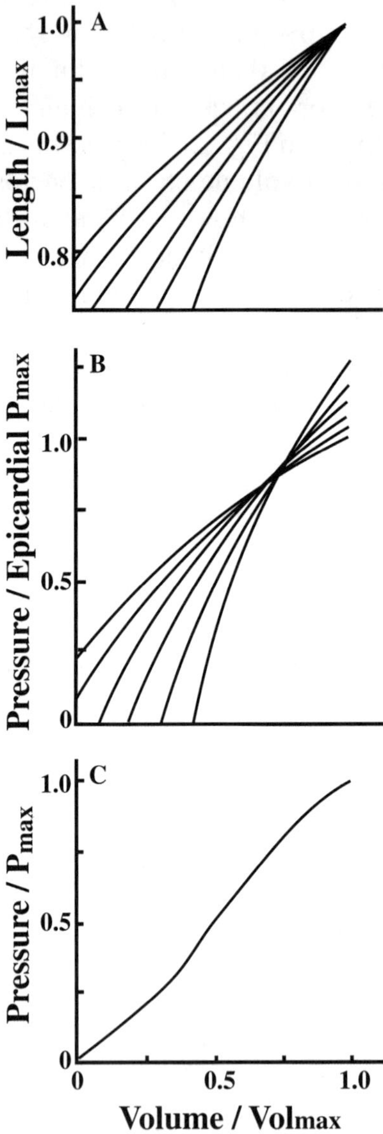

Figure 15.6. *A) Sarcomere length in each muscle layer, B) pressure produced by each muscle rings, and C) sum of pressure generated by the ventricle depicted in Fig. 15.5.*

how sarcomere length in each ring of muscle varies with ventricular volume while panel B shows the pressure developed by each ring. Fig. 15.6 B shows that the inner layers of muscle develop higher pressures at the largest volumes, but their pressure declines to zero with smaller volume excursions of the ventricle. The sum of the pressures in all layers (Fig. 15.6 C) shows that finite pressures are still developed when the volume of the thick walled ventricle is reduced to zero.

Comparison of theory and experiment

Fig 15.6 A indicates a steeper relationship between sarcomere length and ventricular volume in the endocardial as compared with the epicardial layers of muscle, so that there is a substantial difference in sarcomere length across the wall at small volumes. The expected sarcomere length in different layers at each volume is plotted against volume in Fig. 15.7 A, with lengths less than some minimum relaxed value plotted as dotted lines. Sarcomere lengths measured in relaxed ventricles fixed at different filling pressures (Fig. 15.7 B) have been found to be nearly uniform across the wall (Grimm et al., 1988). The important difference between the typical experimental finding and the theory is that the sarcomere lengths have been measured in relaxed muscle.

Muscle has a mechanism for restoring sarcomere length to some minimum relaxed value as it relaxes. The solid curves in Fig. 15.7 A show the muscle coming to rest on this minimum value at the smaller volumes, in a manner very similar to the experimental findings in Fig. 15.7 B. Thus, the difference between the experimentally measured sarcomere lengths and those expected of the simple, thick-walled ventricle may be due entirely to the measurements being made in relaxed muscle that has achieved its minimum resting length.

Another prediction of the model is that at smaller volumes, most of the ventricular pressure will be developed by the outer layers of muscle. Pressure within the muscular wall will rise steeply from atmospheric at the epicardial surface and achieve the full intraventricular value at a fraction of the distance through the wall, as shown in Fig. 15.7 C. This is found experimentally (Fig. 15.7 D).

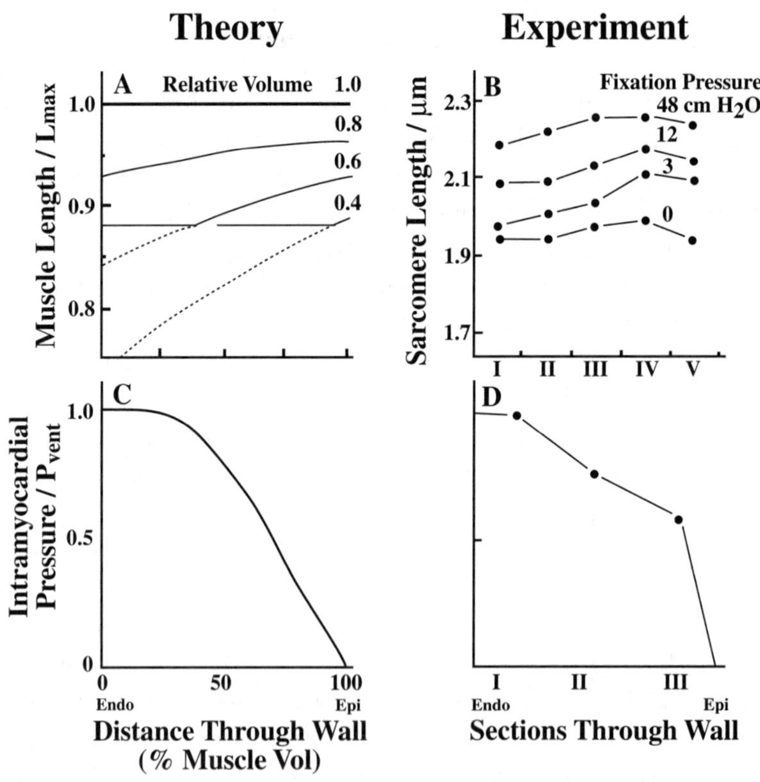

Figure 15.7. *Comparison of theory developed in Fig. 15.6 (A & C) with experiment (B&D). A) Muscle length as fraction of L_{max}. B) Sarcomere lengths measured by Grimm et al. (1988) in five different layers. C) Intramyocarial pressure as a fraction of ventricular pressure in the model and D) in three different layers measured by Baird et al. (1970).*

Twisting theory

It should be mentioned that more complex explanations of the completeness of ventricular emptying have been advanced. One such suggestion is that the left ventricle twists as it contracts (Arts, 1979). This hypothesis was proposed to explain the constancy of sarcomere length across the wall and the reduced sarcomere length excursions shown in Fig. 15.7 B. The essence of the theory lies in the consideration that when a thick-walled tube is twisted, its volume is reduced and the inner layers of material are stretched more than the outer. Thus, if a tubular left ventricle is twisted as its sarcomeres shorten, a greater volume change and more uniform sarcomere change will result than if the ventricle does not twist.

It isn't clear that such a complex geometric relationship is needed, however, because the measurements of sarcomere length in a contracting ventricle have not been made. Such measurements pose a major technical challenge, but until they are available, there is no reason to invoke twisting to explain the observations.

Right ventricle

Since the right ventricle does not have a thick wall to act as a filler at small volumes, the questions arise as to whether its volume excursions are as large as those of the left ventricle and whether other geometric factors facilitate more complete emptying. The answer to the first question appears to be that the right ventricle generally empties less completely. The stroke volume of the right and left ventricles must be identical to keep blood from accumulating on one side of the circulation, but the right ventricle usually has a larger end-diastolic volume, so that its ejection fraction is less (Arcilla et al., 1971). The ejection fraction is sufficiently large, however, as to require some geometric factor to explain its large volume excursion. This factor appears to be the bulging of the left ventricle and septum into its cavity in a way that facilitates emptying. This conclusion leads to the interesting speculation that derangements of the left ventricle can cause a mechanical impairment of the right ventricle. It is well known clinically that right heart failure most frequently follows from left heart failure. This finding is usually attributed to the increased left ventricular filling pressure increasing the load on the right ventricle. Geometric considerations suggest an additional explanation; that left ventricular dilation or reduced contraction reduces the bulging of the septum into the right ventricle, and thereby interferes with its emptying.

GEOMETRIC REARRANGEMENTS

A **myocardial infarction** occurs when a **coronary artery** supplying blood to a segment of heart muscle becomes occluded. The muscle dies and is replaced by scar. Two possible results of such an injury are shown in Figs. 15.6 & 15.7. In the first, the scar simply forms in place of the original muscle. In the second, systolic pressures stretch the dying tissue, producing further dilation, called **remodeling**, as the scar is forming. The scar may be inextensible, or it may form an **aneurysm** that bulges every time the heart contracts. The effects of an aneurysm are not discussed here, except to say that it is the worst possible outcome because it absorbs some of the contractile energy of the remaining muscle during systole, without doing any useful work.

The immediate effect of myocardial infarction is decreased cardiac output secondary to reduction of muscle mass. The fall in output is compensated, at least partially, by two mechanisms. The body may compensate by increasing adrenaline and similar hormones that increase the strength of contraction. The muscle may also compensate by stretching and "moving up the Frank-Starling curve." What happens next may depend upon the relative contributions of these two compensatory mechanism. As discussed in Chapters 19, the best outcomes appear to occur when the adrenaline effects are minimized.

The desired outcome is that the remaining muscle grow, and in so doing, replace some of the contractile elements lost to the infarction. As with any muscle, the increased work is a stimulus to growth. Cell growth in the lateral direction is restricted to a 1.5-fold increase in diameter (Linsbach, 1960), possibly due to diffusion limitations within the cells. There have, as yet, been no limits described to cell growth in the lengthening direction. Tissue growth by an increase in cell number through cell division is very slow, if it occurs at all.

The type of cell growth is likely to depend upon whether remodelling has occurred. If none has occurred, as in Fig. 15.8, the remaining normal muscle must shorten more to achieve the same stroke volume. The effect is similar to volume overload, and if the cells respond normally, they will grow new sarcomeres in series. This lengthening increases cavity diameter to a small extent, and the resulting increase in tension will stimulate a small increase in muscle thickness.

When the scar stretches (Fig. 15.9), the resulting dilation increases force on the remaining wall, and it decreases the shortening required to produce a given stroke volume. Thus, the work produced as shortening is traded for work done as force generation when the pressure and volume work of the ventricle are unchanged. The immediate effect of these changes is to move the remaining muscle to a higher-force, lower-shortening state where it is less efficient. But if the muscle grows thicker, it will generate sufficient force to overcome the increased load imposed by the stretch. There is also the possibility that the muscle will decrease its length by reducing the number of sarcomeres in series, to keep sarcomere shortening near normal in the recovered heart. Thus, thickening of normal ventricular wall can be a normal and desirable response to injury of another part of the heart.

Between the infarction and the compensatory hypertrophy, when the same amount of work is done by less muscle, the heart has less reserve because some normal reserve is used to maintain cardiac output in the resting state. If hypertrophy causes sufficient growth that the muscle mass is the same as in the original ventricle, the work capability of the recovered heart will be restored and

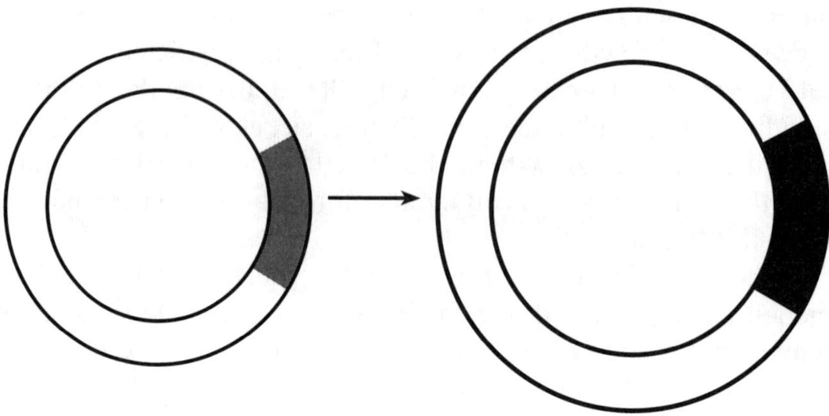

Figure 15.8. *Myocardial infarction without remodelling.*

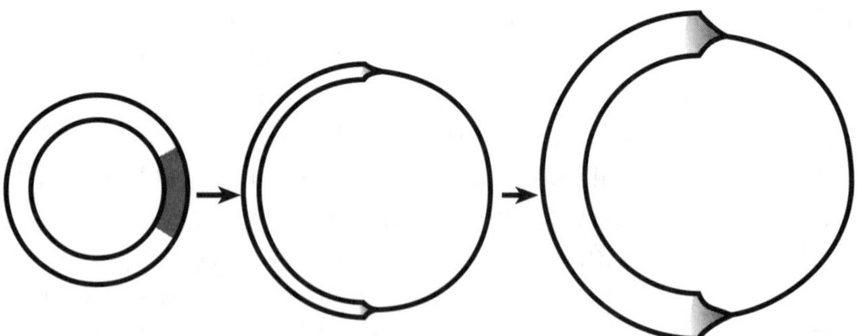

Figure 15.9. *Myocardial infarction followed by stretch of the scar producing ventricular dilation and remodelling.*

cardiac reserve may be very near normal. These hearts would be expected to function in a manner similar to that of the original, uninjured heart. Indeed, some hearts recover sufficiently following myocardial infarction that their function appears normal by the usual measures. Other hearts never recover fully. It is now coming to be recognized that the extent of recovery may depend on the extent of remodelling and the prevalence of conditions that promote cell growth.

When the recovered heart is dilated, as in Fig. 15.9, tension is increased to a greater extent at the end of systole because the cavity radius is not diminished to the same extent by muscle shortening. In addition, the inner layers of muscle are not as deactivated at end-systole as in the normal ventricle because the gradient in shortening across the wall is not as great. Finally, the dilated ventricle does not empty as completely, and is more prone to form thrombi which can embolize. As the effects of these geometric differences are coming to be recognized, greater clinical efforts are being made to avoid remodelling, as discussed in Chapter 19.

SUGGESTED READING

FORD, L.E. (1976) Heart Size. *Circulation Res.* **39**: 297–303.

Chapter 16

THE FRANK-STARLING LAW OF THE HEART

Ernst Starling (1918) summarized his experiments done with colleagues at University College London in the 1915 Linacre Lecture entitled "The Law of the Heart." They showed that the strength of the heartbeat was augmented when the heart was larger. The experiments were done on heart-lung preparations isolated from dogs, the lungs being necessary to oxygenate the blood used to perfuse the heart. Although technically complex, these experiments made the simple but important point that the heart possessed an autoregulatory mechanism for adapting to different loads. Because of Starling's prominence, the principle elucidated soon became known as "Starling's Law of the Heart." But as Starling and his colleagues had pointed out (Patterson et al., 1914), the general principle had been described in 1895 by Frank. Frank's contribution was later emphasized by others, and Frank's original paper, which is very difficult to read, was translated from German to English by Chapman and Wasserman (1959). With this "re-discovery" of Frank's work, it has become customary to call the cardiac autoregulatory phenomenon the "Frank-Starling mechanism" or "principle." The form of words is important because there is another "Starling hypothesis" or "phenomenon," relating transcapillary fluid flow to a balance between hydrostatic and oncotic pressures, and it is easy to confuse these terms.

The cellular mechanisms responsible for the Frank-Starling Law were described in Chapter 6. The present chapter explains the experimental observations and the implications of the work, together with the later experiments of Sarnoff and Berglund (1954) that put this work into a more general context. A brief summary of the mechanisms described here were also outlined in Chapter 13, at the outset of this section of the book on cardiac function, because it is impossible to discuss the circulation or cardiovascular physiology without referring to the mechanism discovered by Frank and Starling.

STARLING'S EXPERIMENTS

Starling made several major improvements over Frank that led to his receiving the larger share of the credit for the principle first described by Frank. He and his colleagues measured volume changes, whereas Frank had inferred volume changes from pressure measurements in the relaxed heart. He also studied mammalian heart rather than the frog heart which Frank had studied because it did not require oxygenated blood. Thus, Starling's experiments seemed more applicable to human physiology. Most importantly, his papers, and particularly his lecture, were written in a style that was more easily understood, in part because he presented his original records. As a result of this ease of understanding, and the elegance of his records and drawings, his experiments rather than the earlier work of Frank are described here.

Starling's apparatus is shown in Fig. 16.1. A part of the apparatus not shown is his method of recording on a **kymograph** or smoked drum. Some authors, such as Winston Churchill, have lamented the passage of the horse from the battlefield and other thoroughfares. Some older physiologists lament the passage of the kymograph, particularly from teaching laboratories. This was a cylinder turned at a constant, slow rate by a clockwork mechanism. A piece of paper was taped around the outer surface of the drum and then covered with soot from an oil lamp. A fine stylus attached to the apparatus scratched a white record in the black soot (see Figs. 16.2 and 16.3). Records were made permanent by shellacking the paper. In spite of its many disadvantages, not the least

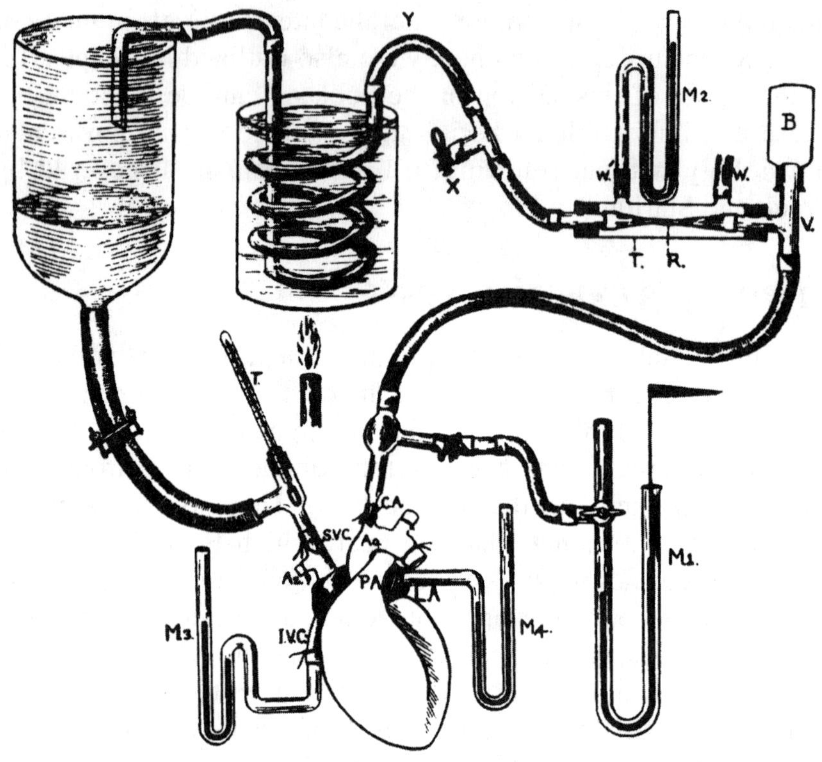

Figure 16.1. *Starling's apparatus. Ao: aorta. PA: pulmonary artery. CA: canula for measuring arterial pressure in inominate artery. LA: left atrium. IVC: inferior vena cava. SVC: superior vena cava. See text for description of apparatus.*

of which was sooty students, it had the great advantage that the preparation and apparatus were connected to the recording device by an obvious mechanical link. There were no obfuscating electronic amplifiers or other mysterious bits of apparatus between the tissue and the record.

The U-shaped manometer labelled "M1" in Fig. 16.1 recorded arterial pressure. The flag shown on the float of this manometer was a stylus that scratched a record of blood pressure on a kymograph. Another part of the apparatus, not shown, was a **plethys-**

mograph, called a "cardiograph" by Starling, that recorded volume changes in the whole heart. This was a rigid jar in which the heart was placed with the edges sealed, so that changes in heart volume changed the pressure within the jar. A manometer with float and stylus recorded these pressure changes on the kymograph as indicators of volume changes. In addition to recording heart volume, this device was a sensitive thermometer, and great care was needed to avoid temperature fluctuations with their confounding effects on the record. Manometers M3 and M4 measured pressures in the right and left atria, respectively.

Starling's apparatus included several clever devices that have since been used by generations of cardiovascular physiologists. One, which became known as a "Starling resistor," was a piece of flexible tubing (R) passing through a rigid outer tubing (T) in which the pressure could be adjusted. This pressure was measured using manometer M2. Arterial resistance, and therefore blood pressure, was adjusted by varying the pressure in the space between the casing and the flexible tubing. Compliance of the arterial circuit was provided by a side arm connection to a closed and partially filled jar, aptly described by its German name *windkessel*, B in Fig. 16.1. Compliance was adjusted by varying the volume of air in the jar. The thermostat consisted of a laboratory assistant who adjusted the height of the flame under the venous reservoir.

In these experiments, mean arterial pressure was adjusted by means of the Starling resistor. The **pulse pressure**, i.e. the difference between the highest, **systolic**, and lowest, **diastolic**, arterial pressures was determined both by the resistance and by the volume of air in the windkessel. The rate of return of blood to the heart was adjusted by the height of the venous reservoir.

The effects of two interventions are shown in Figs. 16.2 and 16.3. In these records, increases in heart volume measured with the plethysmograph (C) are plotted as downward excursions. The other records are arterial pressure (B.P.) and venous pressure (V.P.). The effects of increasing arterial resistance are shown in

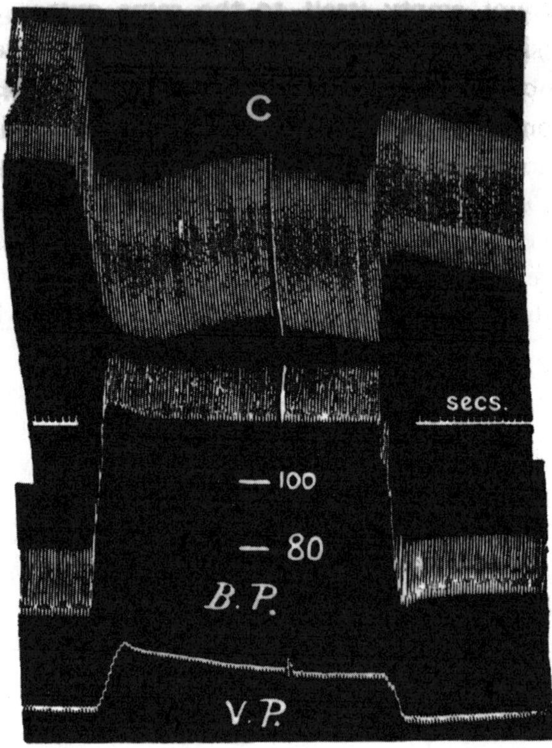

Figure 16.2. *Effect of increased arterial resistance on heart volume and stroke volume. From Patterson et al. (1914).*

Fig. 16.2. Less blood was ejected in each of the first few beats after the pressure increase, so that each succeeding beat began from a larger volume, and venous pressure increased progressively. After several beats, a new steady state was reached in which **stroke volume**, the volume ejected with each beat, and the pulse pressure were at their pre-intervention levels. On the other hand, **stroke work**, the work done during each beat, defined as P·ΔV, was increased because the mean arterial pressure was increased.

The effect of increasing the rate of return of blood to the heart is shown in Fig. 16.3. Since the amount of blood returned during

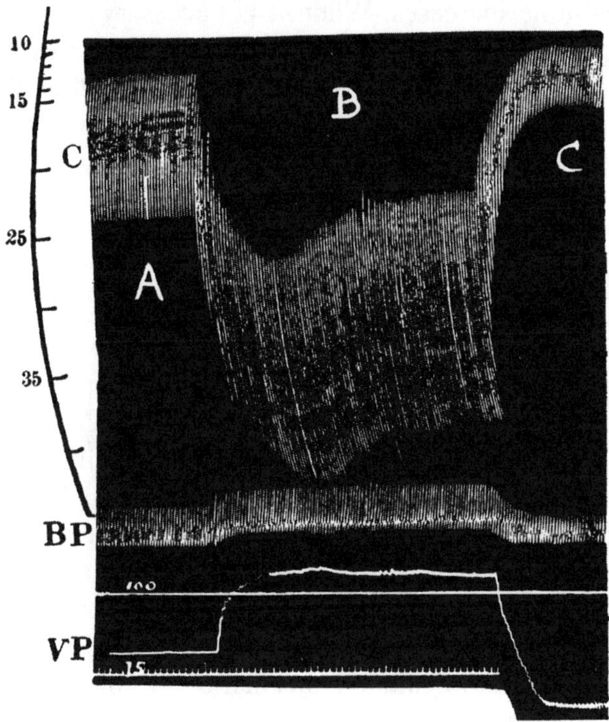

Figure 16.3. *Effect of increasing venous return on heart volume and stroke volume. From Starling (1918).*

diastole was at first greater that the amount ejected during systole, the heart volume increased. The increased volume in turn caused stroke volume to increase, as shown by the increased excursion of the cardiometer tracing. In this case, stroke work was increased, mainly because stroke volume was increased, although there was a slight increase in arterial pressure, as would be expected from the increased flow across an unaltered Starling resistor.

The main point to be made from these experiments is that an increased heart volume elicits an increased stroke work, defined as $P \cdot \Delta V$, and that the way the work increases is determined by the loading conditions. When it is necessary to eject a larger volume, the stroke volume increases. When it is necessary to produce an increased pressure, the pressure increases. From these experiments Starling concluded: "The law of the heart is thus the same as the law of muscular tissue generally, that *the energy of contraction, however measured, is a function of the length of the muscle fibre.*"

REGULATORY MECHANISMS

The importance of the Frank-Starling mechanism is that it is intrinsic to the muscle itself. As explained in Chapter 6, the increased strength of the heat beat at longer muscle lengths appears to result from an increased activation, caused mainly by an increased sensitivity to the myofilaments to calcium. It is thus an "autoregulatory" mechanism, one that compensates for differences in loading conditions without requiring separate sensing mechanisms and feedback pathways outside the heart muscle. The example of a skipped heart beat was used in Chapter 6 to illustrate the need for such autoregulatory mechanics. A change in arterial resistance is another example.

The Frank-Starling Law is not the only regulatory mechanism in the heart. There are other, extrinsic mechanisms that regulate the strength of the heart beat. Because some of his audience came away with the view that cardiac output was regulated entirely by the size of the heart, he gave another prominent lecture published in 1920 entitled "On the circulatory changes associated with exercise," in which he emphasized the importance of other control mechanisms. As he pointed out in the introduction to this lecture, the autoregulation provided by the Frank-Starling mechanism was probably the last line of defense in the body's efforts to provide an adequate cardiac output, being utilized only when the several mechanisms governed by the central nervous system had failed. Much less was known about these other mechanisms then, so that his lecture was somewhat speculative, but it outlined some of the issues to be addressed by future generations of physiologists.

These mechanisms are the subject of Chapter 17. Other regulatory mechanisms are mentioned here because some experiments published after Starling's lecture, particularly those done with hearts *in situ*, suggested that stroke work did not depend upon ventricular volume. Sarnoff and Berglund (1953) subsequently published experiments illustrating the multiplicity of variables that could confound measurements of the relationship between stroke work and ventricular size. Their greatest contribution, however, was the development of a concept for separating the two variables that determine ventricular function.

VENTRICULAR FUNCTION CURVES

Sarnoff and Berglund's concept is illustrated in Fig.16.4, taken from their paper. It shows that the left ventricle has a family of curves relating stroke volume and ventricular size. They called these **ventricular function curves**. The curve describing the ventricle's functions at any moment is determined by its **inotropic state**, i.e. its inherent strength, governed by mechanisms to be described in Chapter 17. The increased inotropic state illustrated in Fig. 16.4 was achieved by the use of adrenaline. The two curves show that the strength of cardiac muscle contraction can be made to vary by two separate pathways, a change in muscle length and a change in inotropic state. A point to be made in Chapter 17 is that substantial confusion over these pathways arose because they were thought to arise from the different mechanisms, when in fact, they use the same basic mechanism. The point to be made here is that defining a heart's inotropic state is equivalent to identifying the function curve on which it is operating.

It is instructive to review the experimental conditions and measurements in Sarnoff and Berglund's experiments because they illustrate issues that will arise later. Unlike Starling's experiments the hearts were studied *in situ* mainly to provide a conveniently oxygenated blood supply to the hearts. These authors made separate measurements for the right and left ventricles, whereas

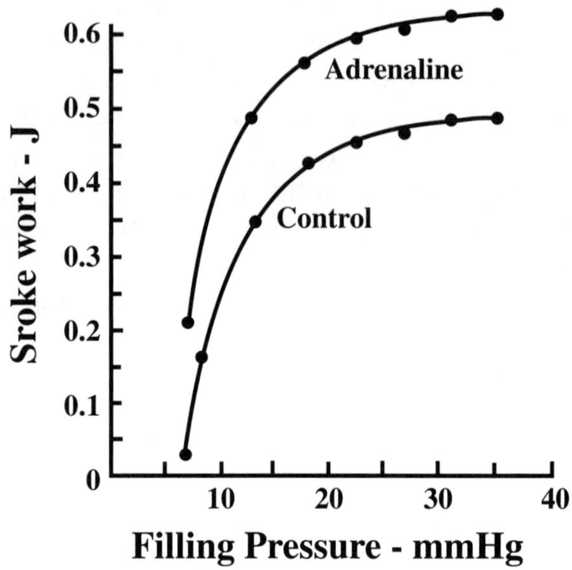

Figure 16.4. *Left ventricular function curves. Upper curve: adrenaline. Lower curves: no adrenaline. Adapted from Sarnoff and Berglund (1953).*

Starling had measured whole heart volumes. In their paper, Sarnoff and Berglund emphasize that some of the earlier studies gave less well defined relationships between ventricular work and size because left ventricular output had been related to right ventricular filling pressures. This importance of relating left ventricular output to left ventricular filling pressures is now reaffirmed on a daily basis around the world by the clinical use of pulmonary artery catheters placed in patients for this purpose. It is relatively simple to assess a patient's right atrial filling pressures, either by direct observation of the height of venous distention or with a **central venous catheter** placed in the right atrium or in the large veins near the right atrium. Placing a pulmonary artery catheter entails the additional risk associated with passing the catheter through the right side of the heart, but these difficulties and risks are readily accepted in return for the added measure of reliability of the data.

Sarnoff and Berglund did not measure ventricular volumes, as Starling and his colleagues had done, but instead inferred volume from pressure changes, as Frank had done. The importance of this difference is illustrated by comparing the nearly straight function curve for the right ventricle in Fig. 16.5 with the curves for the left ventricle in Fig. 16.4. The thick-walled left ventricle is stiffer than the thinner-walled right ventricle, and more passive pressure is required to distend it. The reduced stiffness results in a more linear relationship for the right ventricle than for the left. The relationship for the left ventricle would have been more linear if stroke work had been plotted against ventricular volume rather than filling pressure. The increasing steepness of the passive volume-pres-

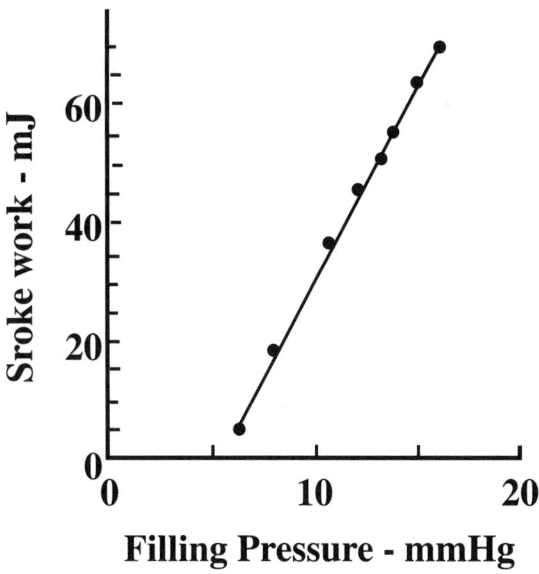

Figure 16.5. *Function curve for the right ventricle. Adapted from Sarnoff and Berglund (1953).*

sure curve causes the function curves of the left ventricle to flatten and give the appearance of approaching a plateau. Such an illusion is not created when work is plotted as function of volume. As described in Chapters 18 and 19, this can be an important consideration in clinical situations where left ventricular filling pressure is being adjusted to obtain the optimum balance between increasing ventricular work and minimizing the venous pressure in the lungs.

Assessments of stroke work

Sarnoff and Berglund emphasized the importance of measuring stroke work as opposed to stroke volume or cardiac output. They presented some examples of how measurements of output alone suggested alterations in inotropic state that had not occurred. They further asserted that some of the earlier work suggesting that Starling's law did not apply might have been due to a failure to consider stroke work as the relevant parameter. While this assertion is undoubtedly correct, it should also be emphasized that there is an important unstated assumption in their measurements of work. The assumption is that the stoke work was measured at the same place on the pressure-work curve, presumably at or near the pressure that would produce maximum work.

A point that is made repeatedly throughout this book is that there is an inverse relationship between muscle shortening and force defining the force-shortening properties. This inverse relationship indicates that muscle work, which is the product of force multiplied by the distance shortened, is zero at each end of the curve and reaches a peak in the middle. Thus, the stroke work can be set to any value between zero and its maximum by the systolic pressure, irrespective of the inotropic state of the muscle. Reliable measurements of stroke work therefore require that sufficient data be obtained to define the peak in the pressure-work curves. There is no evidence in Sarnoff and Berglund's paper that they made any effort to determine the peaks in the work curves. In spite of this shortcoming, the general validity of their experiments, which has been affirmed by much subsequent work, suggests that their measurements were not confounded by a failure to determine the peak of the work curves, but this may have been as much the result of good fortune as of intelligent design.

DESCENDING LIMB OF THE FRANK-STARLING CURVE

Most workers, including Starling and Sarnoff and Berglund, who have studied the effects of ventricular volume on performance have found that stroke work reaches a peak and then declines, producing a descending limb to the curve. The same workers usually conclude also that the falling output is due to regurgitation of blood through a mitral valve made incompetent by the ventricular dilation, or to some other factor such as insufficiency of coronary flow that limits ventricular work. Even though striated muscle possesses a descending limb, it is not clear that the ventricle can be dilated sufficiently to bring the muscle of its wall to the peak of its length-force relationship. The dilation required to reach the falling limb appears to alter the ventricular architecture so severely that measurements of output become unreliable. Katz (1965) reviewed the data suggesting a falling limb to the Starling curve and concluded that it does not occur under any conditions encountered in the body. If the heart ever achieved its falling limb, the consequences would be disastrous. The autoregulatory mechanisms that restore smaller heart volumes would be reversed and the heart would dilate until it reached some elastic limit. It is reasonable to conclude therefore, that natural selection has provided mechanisms that avoid the descending limb.

One reason for discussing the falling limb of the Frank-Starling curve is to dismiss it. It simply does not occur in any *in situ* situation that might be encountered in living animals. A more important reason for this discussion, however, is that it would be extremely useful to know the maximum capability of the heart achieved at the peak of the pressure-work curves. Heart volume could then be expressed as a fraction of its optimum volume, and performance could be related to maximum performance expected at the optimum volume. In the absence of a descending limb, it is not possible to identify a peak in the function curves. Thus, the optimum volumes required to achieve maximum work cannot be defined. This deficiency greatly hampers a comprehensive estimate of inotropic state.

SUGGESTED READING

STARLING, E. H. (1918) The 1915 Linacre Lecture. *The law of the heart.* (Longmans, Green and Co. London). Reproduced in Chapman, C. B. and Mitchell, J. H. (1965) *Starling on the Heart.* (Dawsons of Pall Mall, London)

SARNOFF, S. J. and BERGLUND, E. (1953) Ventricular function I. Starling's law of the heart studied by means of simultaneous right and left ventricular function curves in the dog. *Circulation.* **9**: 706–718.

Chapter 17

INOTROPIC MECHANISMS

The word **inotropic** characterizes agents that affect the strength of the heart beat. It distinguishes them from **chronotropic** effects that increase heart rate. The two effects are not mutually exclusive, and some agents have both inotropic and chronotropic effects. Inotropic agents can be either positive or negative. Negative agents were described in Chapter 6 on fatigue. As discussed there, some negative agents operate by decreasing activation. The level of activation achieved during a twitch is therefore a balance between positive and some negative influences. Other negative agents, such a metabolic inadequacies, have no positive counterpart, i.e., once the negative influence is removed there is no increase in strength to be gained from further optimization.

A major point of this chapter is that probably all positive inotropic agents operate by increasing muscle activation. The heart has a maximum capability determined by the volume and myosin isoform composition of its contractile machinery, but it never achieves this maximum, at least under physiological conditions. In this sense, muscle and an audio amplifier are similar. Both convert one form of energy to another, both have a defined maximum ability, and both normally operate over their low to mid-range of power, regulating output by varying the degree of suppression they achieve. Heart muscle is more complex, however, because the important quantity is not instantaneous power but the amount of work done over time. Instantaneous power is much less important for the bodily function than cardiac work, the time integral of power. The duration and rate of contraction are therefore as important as power in determining the effectiveness of the pump.

The greater complexity of the heart, as compared with the audio amplifier, arises because shortening deactivation lowers power output later in a contraction. Thus, the total work done during a contraction depends on how the heart is loaded. Extracting work early in a contraction may, for example, reduce later contractile ability and total work production. Furthermore, the effects of various inotropic agents on shortening deactivation have not been investigated thoroughly. Thus, the effects of inotropic agents on integrative measures of contractile activity, such as work, are much less well understood than the effects on instantaneous measures, in spite of their importance to cardiac function.

INOTROPIC MECHANISMS

In theory, mechanisms for influencing the contractile strength might be divided into two broad categories, those that alter the number of crossbridges activated and those that increase the contractile strength of the bridges, once they become activated. In reality, there is strong, if not fully unimpeachable, evidence described below that the second mechanism, alterations in the strength of the crossbridges, does not occur acutely through the action of positive inotropic agents. Thus, the strength of the heart's contraction appears to be regulated entirely by changes in activation.

There are several qualifications in these statements. It is important to specify *positive* agents because the removal of the negative influences of metabolic impairments strengthen the heart beat by direct action on crossbridge function, as described in Chapter 6. Another qualification, the word "acutely" for describing inotropic effects, is used because there are different isoforms of the contractile proteins in the heart, and a chronic change in isoform composition will alter shortening velocity and power output of the crossbridges. Finally, there is controversial evidence that muscles with a mixture of myosin isoforms may respond to inotropic agents by selectively activating one isoform more than another. Such a mechanism, if it exists, could alter the contractile performance of the muscle without altering the total level of activation and without altering the contractile properties of the individual crossbridges.

INCREASED CALCIUM ACTIVATION

There are several types of positive inotropic agent, and their effects on isometric twitches are so different as to suggest, incorrectly, that they arise from different mechanisms. The effects of four classes of intervention on the time course of the twitch are shown in Fig. 17.1. The strength of each agent was adjusted to approximately double twitch force. Some pharmacological agents, such as those shown in Fig. 17.1 A&B, increase force without greatly altering the time course of the twitch. **Adrenergic** agents, those which have an adrenaline-like action, decrease twitch duration (Fig. 17.1 C) such that the rate of force development must be increased substantially to achieve the higher force. Other agents, such as caffeine (Fig 17.1 D), prolong the rise of force, so that force rises to a higher and later peak.

The diverse responses of the twitch contraction suggested initially that there might be different mechanisms for increasing the strength of the heart beat. When it was learned that calcium activates muscle, it was easy to interpret the effects of interventions that simply scaled up the twitch (Fig. 17.1 A and B) as being due solely to an increase in the amount of calcium released into the myofilament space. It was also easy to interpret the prolonged rise of force caused by caffeine (Fig. 17.1 D) as being due to an impairment in the re-accumulation of activating calcium. As will be explained, both of these interpretations have been proven correct. A more complicated explanation was required to explain the abbreviated twitch duration caused by adrenergic stimulation (Fig. 17.1 C). It was therefore suggested that adrenergic stimulation

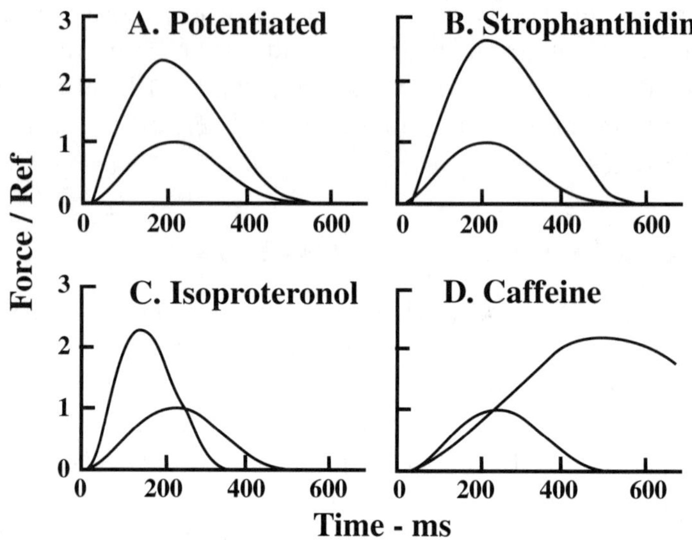

Figure 17.1. *Effects of four classes of inotropic intervention on contraction of isolated heart muscle. The contraction in A was potentiated by interpolating an extrasystole just before the preceding contraction. Adapted from Chiu et al. (1989), with permission.*

might have a direct effect on the contractile machinery independent of any influence on activation. The evidence described below suggests that such a direct effect on crossbridge action does not occur, although this conclusion may not be accepted universally.

ROLE OF CALCIUM ACTIVATION IN INOTROPY

The time courses of calcium in normal and inotropically stimulated contractions are superimposed as solid curves in Fig. 17.2 on the dotted force curves taken from Fig. 17.1. The calcium records were not obtained under exactly identical conditions, and the records have been some interpolated somewhat to make the comparisons. As indicated, some interventions, such as increasing extracellular calcium (Fig. 17.2 A) or the application of acetylstro-

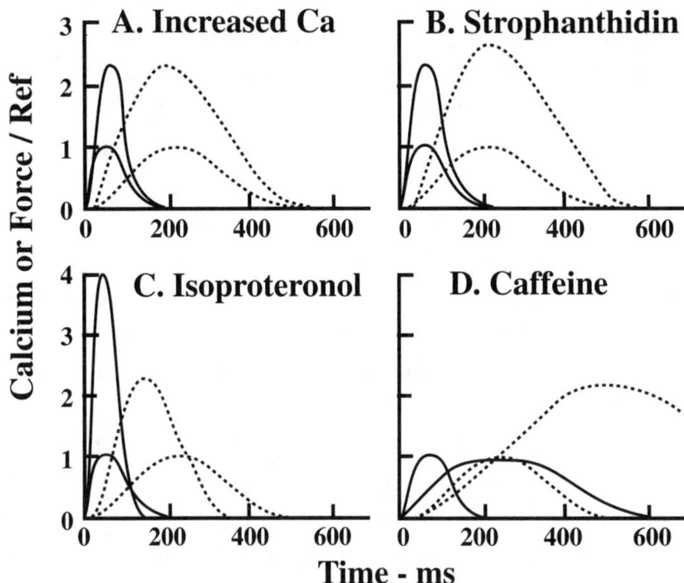

Figure 17.2. *Effects of inotropic stimuli on intracellular calcium transients (solid curves) superimposed on the force records from Fig. 17.1. Curves interpolated from the data of Morgan et al. (1986) to match conditions in Fig. 17.1.*

phanthidin (Fig. 17.2 B) increase calcium and force in approximately the same proportion, without much altering the time course of either. Other agents, most notably those that increase adrenergic tone, abbreviate both the calcium transient and the twitch, while caffeine-like agents prolong both the force and calcium transient.

As might be imagined, calcium activation can be increased by at least two mechanisms, increasing the amount of calcium released and increasing the binding of calcium to thin filaments, once it is released. The first mechanism, increasing free calcium, is show by the effects of increased extracellular calcium (Fig. 17.2 A), post-extrasystolic potentiation (Fig 17.1 A), and the addition of acetylstrophanthidin (Fig. 17.1 B and 17.2 B). All three interventions increase the calcium transient and force in approximately the same proportion.

The second mechanism is demonstrated by the records made with caffeine. This agent slows calcium release (Fig. 17.2 D) and may even decrease the concentration of calcium free in the myoplasm, but it potentiates force by increasing thin filament calcium affinity, so that more calcium binds to the thin filaments.

An inverse effect from this second mechanism is shown in the records made in the presence of isoproterenol (Figs. 17.1 C and 17.2 C). This agent diminishes the sensitivity of the thin filaments to calcium, so that more calcium must be released to achieve the same force level. The reason that such a seeming paradoxical mechanism has evolved is that the diminished calcium affinity is needed to decrease twitch duration.

Activation is also increased by a third mechanism increasing the number of activated thin filament sites for a given amount of calcium bound. Cooperativity of crossbridge binding is an example of such an alteration in calcium sensitivity, and it is likely to be an important determinant of myocardial work. The effects of this cooperativity would not be evident in the isometric records of Figs. 17.1 and 17.2 because they do not become apparent until a length change alters the number of attached crossbridges.

MECHANISMS OF INCREASED CALCIUM ACTIVATION

A small amount of calcium crossing the cell membrane after stimulation causes the release of much more calcium from the sarcoplasmic reticulum, through the calcium-induced calcium release. The amount released depends upon the amount contained in the releasable intracellular pool. The amount of calcium in this pool is, in turn, determined by a balance in prior contractions between the rate of calcium entry into the cell and the rate of its removal. Calcium entry occurs during the plateau of the action potential, when the sarcolemmal calcium channels are open. The amount of calcium crossing the membrane therefore depends on the fraction of the cardiac cycle that the channels are open.

Calcium is removed from the cell against a concentration gradient during diastole by exchange for sodium, with the energy provided by the sodium gradient. This energy depends on the difference between the membrane potential and the equilibrium potential for the sodium gradient. Very little calcium is removed during systole, when the membrane is depolarized, because the membrane potential is very close to the equilibrium potential for the sodium gradient. During diastole, there is a balance between the amount of the released calcium that is re-sequestered by the sarcoplasmic reticulum and the amount that is removed from the cell. This balance is determined by the capacity of the calcium pumps in the reticulum membrane.

Thus, five factors determine the amount of calcium released during activation: the extracellular calcium concentration, the fraction of time the membrane is depolarized, the sodium gradient, the resting membrane potential, and the strength of the sarcoplasmic reticulum calcium pumps. Changes in all these factors occur with the interventions shown.

Extracellular calcium

Twitch force can be made to vary widely simply by changing the concentration of calcium in the extracellular fluid. The effects on the calcium transient are shown in Fig. 17.2 A. It is greatly increased. This increase almost certainly occurs because more calcium flows through the membrane during systole, when the calcium channels are open, and thereby loads the sarcoplasmic reticulum to a greater extent. As indicated, force is increased in approximate proportion to the calcium transient.

Increasing heart rate

The strength of contraction varies widely and inversely with the interval between contractions. This is sometimes called a **positive rate staircase** or **Bowditch effect**. It is explained by the cell membranes being depolarized with open calcium channels for a greater fraction of the cycle at higher heart rates. This increase in the fraction of time spent depolarized occurs because the action potential duration does not shorten in proportion to the reduction in the period between depolarizations. The higher levels of calcium in the myoplasm allow the sarcoplasmic reticulum to accumulate more calcium, and thus increase the pool of releasable calcium. The effect of stimulation rate on force is explained here because post-extrasystolic potentiation is a variation on the same mechanism.

Post-extrasystolic potentiation

The ratio of the duration of systole to the duration of diastole can also be increased by interpolating an extra depolarization between beats. The effect of this intervention on force is shown in Fig. 17.1 A. As indicated there, it increases force without otherwise changing the time course of the twitch. Its effects are very similar to increasing extracellular calcium, shown in Fig. 17.2 A, i.e. it increases force and calcium in approximately the same proportions. It was chosen in the experiments illustrated in Fig. 17.1 because its effects are almost immediate. As explained below, this immediate onset was used in a test for other mechanisms.

Cardiac glycosides

Digitalis is a drug derived from the plant *Digitalis purpura*, known commonly as foxglove. Its use for treating heart failure was discovered by William Withering (1785) who discerned its therapeutic value in a concoction developed by a herbalist known to subsequent generations of students as "the Shropshire sheep woman." Digitalis strengthens the heart beat, and for most of the two centuries that digitalis has been in clinical use, its mechanism of action was unknown. In 1961 Dunham and Glynn showed that the digitalis class of agents, known as **cardiac glycosides**, inhibit **sodium-potassium ATPases**, the membrane pumps that use the energy of ATP hydrolysis to exchange sodium and potassium against their concentration gradients. How this highly specific activity was related to its inotropic effects was a mystery until the sodium-calcium exchange mechanism was discovered by Reuter and Seitz (1968). The discovery of this mechanism suggested to Baker et al. (1969) that digitalis increases intracellular calcium by decreasing the sodium gradient across the sarcolemmal membrane. This lower gradient would, in turn, decrease calcium removal and increase the releasable calcium pool in the sarcoplasmic reticulum. Support for this mechanism derives, in part, from the observation that lowering the sodium gradient by replacing some extracellular sodium with an impermeant ion has an identical inotropic effect (Sheu and Fozzard, 1982).

Digitalis binds tightly and slowly to tissue. Consequently, it requires some time both to load and to remove the drug. It is therefore common to use a less tightly bound cardiac glycoside, **acetylstrophanthidin**, in tissue experiments. The effect of this agent on the time course of the twitch is shown in Fig. 17.1 B and its effect on the time course of the calcium transient is shown in Fig. 17.2 B. As indicated, its effects are identical to those of an extrasystole or increased extracellular calcium; it augments both the twitch and the calcium transient in the same proportion, with little change in their time course. This similarity suggests strongly that its only effect is to increase the amount of calcium released during activation.

Adrenergic stimulation

Adrenaline, also called **epinephrine**, is produced in the adrenal medulla and released into the blood stream to produce the **fight-or-flight responses** required by the body in times of stress. These responses include speeding and strengthening the heart beat to increase cardiac output. There are several analogues of adrenaline, and these **adrenergic agents** have multiple effects that are divided into two broad classes, called α and ß. Inotropic effects in the heart derive from the ß mechanisms, and the records shown in Figs. 17.1 C and 17.2 C were produced with **isoproterenol**, an adrenaline analogue with mostly ß effects. Before describing these mechanisms, some distinctions between chronotropic and inotropic mechanisms should be emphasized because the heart's rhythmic pumping requires a coordination between the two types of effect.

In the normal, intact heart, chronotropic and inotropic effects arise from different tissues. Heart rate is set by the sinoatrial node in the right atrium while the strength of the heartbeat is determined largely by ventricular contraction. This separation of functions is mentioned because chronotropic effects on ventricular tissue are inconsequential, but the duration of systole must be matched to heart rate if sufficient time is to be available for ventricular filling.

The ventricles cannot eject any more blood during systole than is returned to them in diastole. If the duration of systole remained constant as heart rate increased, diastolic filling time would diminish, first decreasing stroke volume, and at higher pulse rates, decreasing cardiac output. To avoid this potential limitation, adrenergic agents also abbreviate the duration of ventricular systole. Finally, they increase the strength of contraction, so that the same volume of blood can be ejected during the abbreviated duration of contraction. As mentioned in Chapter 10, the reduced duration of systole also avoids extreme muscle shortening that might damage the muscle when it is contracting vigorously.

The records in Fig. 17.2 C show that the calcium transient is abbreviated, as expected of the decreased duration of the twitch, and its magnitude is increased to a greater extent than the force. This latter observation suggests that the sensitivity of the myofilaments to calcium is reduced. The reduced sensitivity might at first seem paradoxical until it is realized that such reduced sensitivity is required to abbreviate the duration of the mechanical response.

The calcium records in Fig. 17.2 all show that the free calcium concentration in the myofilament space returns to near baseline levels during the fall of force. Thus, the rate of relaxation appears to be determined not by the rate of removal of free calcium from the myofilament space, but by the rate at which calcium dissociates from thin filaments. Binding affinity between two substances is the ratio of binding rate to dissociation rate. An increased dissociation rate, as is needed to produce more rapid relaxation, will also produce the decreased calcium affinity suggested by the larger ratio of calcium to force in Fig. 17.2 C.

Metabolic consequences of adrenergic stimulation

The inotropic effects of adrenergic stimulation are achieved at the cost of an increased metabolic rate because calcium fluxes are increased out of proportion to the increases in activation. The greater flux increases the energy cost per unit of work done and thereby decreases overall chemomechanical efficiency. While the decreased efficiency is now believed to be due to the greater calcium fluxes, before this mechanism was known, the decreased efficiency was taken as evidence that ß-adrenergic stimulation had a direct effect on the crossbridge cycle. Whatever its cause, the decreased efficiency can be deleterious when myocardial performance is limited by the metabolic supply.

Caffeine

Caffeine has multiple effects in the body, including central nervous stimulation, smooth muscle relaxation, and potentiation of contractile force in striated muscle. In cardiac muscle it appears to have effects opposite to at least two of those produced by adrenergic stimulation. It decreases calcium uptake by the sarcoplasmic reticulum, and it increases calcium binding to the thin filaments. The decreased calcium uptake prolongs the calcium transient, as shown in Fig. 17.2 D. It also diminishes the amount of calcium in the sarcoplasmic reticulum, so that the amount and rate of release are both diminished. The increased calcium binding to thin filaments is demonstrated by the increased twitch force achieved in the presence of the same or lower calcium level at the peak of the calcium transient. Together, the prolonged calcium transient and the decreased rate of calcium dissociation from the thin filaments prolong the rise of twitch force and increase the level of force ultimately achieved. An interesting effect of these agents is to increase the metabolic efficiency of the muscle by decreasing the amount of calcium cycling required to achieve a given level of work.

ADRENERGIC MECHANISMS

The finding that adrenergic agents produce several intracellular alterations raises the further question of the underlying mechanisms. Like many other agents, these hormones achieve their effects by binding to membrane receptors, thereby initiating a cascade of events that includes phosphorylation of proteins. At least three proteins responsible for separate inotropic effects are involved. The effects of phosphorylation on these three proteins, troponin-I, phospholamban, and myosin light chain, are described separately.

Troponin-I phosphorylation

The decreased affinity of the thin filaments for calcium is accomplished by the phosphorylation of troponin. This mechanism was discovered by MacClellan and Winegrad (1978), who developed a "permeabilized" cardiac muscle preparation. This term emphasizes that the cells are permeable to externally applied chemicals, but the membranes retain receptors and the receptor mechanisms. Adrenergic stimulation of the permeabilized preparation had two related effects, troponin-I phosphorylation and a rightward shift of the pCa-force relationship, indicating a decreased calcium sensitivity (Fig. 17.3).

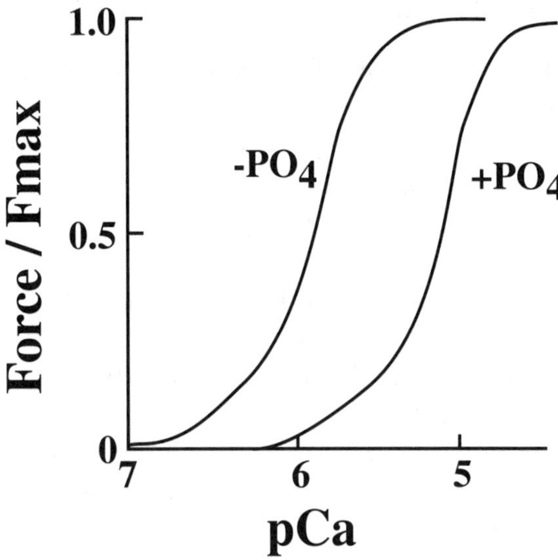

Figure 17.3. *Force vs pCa in the presence (+PO$_4$) and absence (−PO$_4$) of troponin phosphorylation. Adapted from MacClellan and Winegrad (1980) and Winegrad (1983).*

Phospholamban phosphorylation

In 1974 Tada et al. described a sarcoplasmic reticulum protein that regulated the rate of calcium accumulation by the reticulum. The degree of regulation was, in turn, modulated by the degree of phosphorylation of the protein, which they called **phospholamban**, derived from the Greek verb *lamba* meaning *to receive*, i.e phospholamban received phosphate. It has since been discovered that phosphorylation of this membrane protein increased pump activity by de-repression. In the unphosphorylated state, phospholamban binds to the calcium pumps in the sarcoplasmic reticulum membrane and reduces their activity. When it is phosphorylated, the binding is inhibited (Voss et al., 1994), and full pump activity is restored.

Myosin light chain phosphorylation

The regulatory light chain of myosin was one of the first contractile proteins discovered to become phosphorylated (Frearson and Perry, 1975). The discovery led to a series of studies proposing different physiological effects of this biochemical change. Some of the more dramatic of these proposals have not stood the test of time, but one seemingly small effect appears to be genuine and may be more substantial than it first seems. This effect is an increase in the cooperativity of activation.

Metzger et al. (1989) measured steady force produced by skinned skeletal muscle fibers in the presence and absence of light chain phosphorylation. Their results are summarized in the pCa-force curves of Fig. 17.4 A. As indicated by a comparison of the unphosphorylated (dotted) and phosphorylated (solid) curves, the binding of phosphate to the light chain caused a small but definite leftward shift in the curve, without a change in the maximum force at saturating concentrations of calcium. This shift indicates a

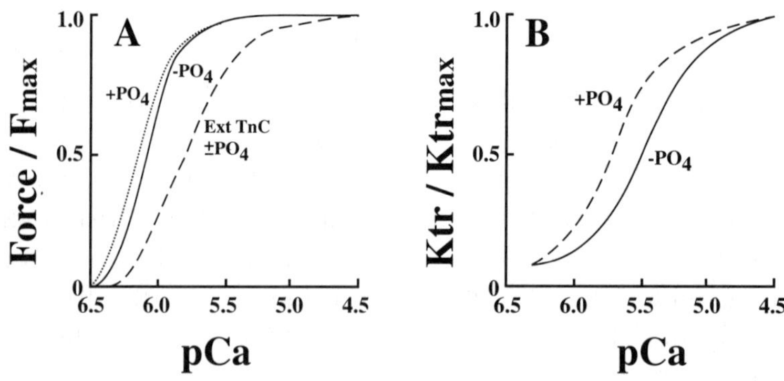

Figure 17.4. *A) Steady force levels at different pCa in the presence (+PO₄) and absence of phosphorylation (-PO₄) and at full and partial (ext TnC) troponin-C. B) Rates of force redevelopment (K_{tr}) in the presence and absence of phosphorylation and full Troponin-C. Adapted from Metzger et al. (1989).*

greater sensitivity of the fiber to calcium, i.e. it produced more force at partial activation.

They also carried out the same experiment after partially extracting troponin-C, an intervention which their laboratory (Moss et al., 1985) and others (Brandt et al., 1984) had previously shown to abolish the cooperativity of activation described at the end of Chapter 5. After this extraction, substantially more calcium was required to activate the fiber (dashed curve), and there was no difference between the phosphorylated and unphosphorylated curves.

These findings suggest that light chain phosphorylation enhances the cooperative binding of calcium and crossbridges to thin filaments. While the effects on the steady force levels plotted in Fig. 17.4 A are relatively small, this is the type of mechanism that is very likely to speed the onset of activation, and in so doing greatly potentiate the transient force development during a twitch. Fig. 17.4 B demonstrates the hastening force development. In these experiments, the rate of force redevelopment, called K_{tr}, is measured after a quick release and restretch that brings isometric force to zero. The perturbation was assumed to detach most of the crossbridges, so the rate of force redevelopment was taken to indicate the rate of reattachment of bridges. It is not known how much of the calcium dissociates from the thin filaments during the time when force is low, and the increase in the speed of force redevelopment indicates only a lower limit to the hastening of force development that might occur at the onset of a twitch. As indicated, the rise of redeveloped force at partial levels of activation is substantially faster when troponin-I is phosphorylated, and the rate could be even faster at the onset of a twitch, if greater increases in cooperativity occur then.

INOTROPIC AGENTS AND CONTRACTILE KINETICS

The foregoing descriptions suggest that the varied effects on twitch force shown in Figs. 17.1 and 17.2 can be explained entirely by effects on activation. A direct alteration of the cross-bridge cycle is not needed to explain the very different effects on the time course of the twitch. Measurements of the changes in the force-velocity properties produced by the four agents further suggest that they do not affect the contractile apparatus.

Post-extrasystolic potentiation was chosen as an inotropic intervention in the experiments of Fig. 17.1 because its effects are so rapid that it seemed unlikely to produce substantial changes in protein phosphorylation. To the extent that this is so (and it might not be), post-extrasystolic potentiation was used as a model for a pure change in activation. Effects of other agents were thus compared to the effects of this potentiation to determine whether a change in contractile kinetics had occurred.

Force-velocity curves obtained at the peaks of unpotentiated and extrasystolically potentiated twitches are shown in Fig. 17.5. As indicated, the main effect of this potentiation is to increase force (Fig. 17.5 A), with only a very small increase in maximum velocity. As explained with Fig. 7.4 in Chapter 7, this result is ex-

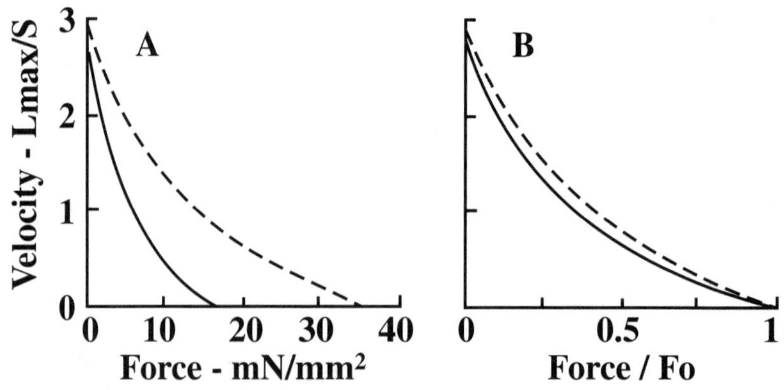

Figure 17.5. *Effect of post-extrasystolic potentiation on the force-velocity relations of rabbit papillary muscle. From Chiu et al. (1987), with permission.*

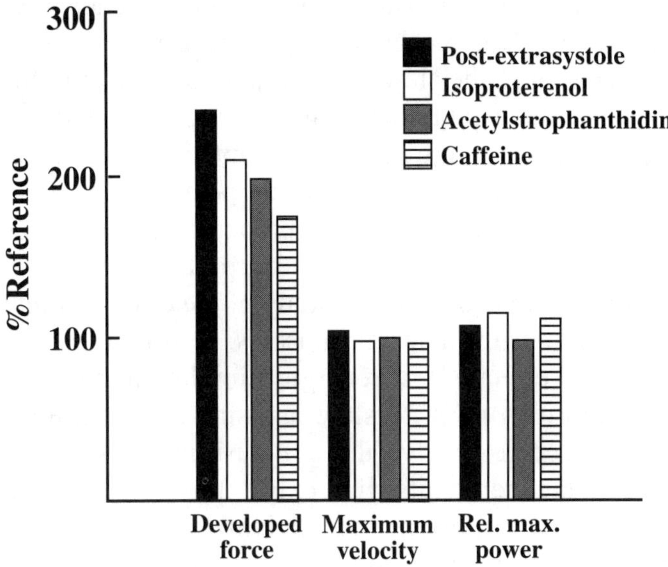

Figure 17.6. *Effects of four inotropic interventions on force-velocity parameters. From Chiu et al. (1987), with permission.*

pected of a doubling of activation in the presence of a small internal load.

Nearly identical changes in the force-velocity curves, summarized in Fig. 17.6, were produced by the other three interventions. None of the other three interventions produced changes that were significantly different from the effects produced by post-extrasystolic potentiation. These results suggest that all of the effects of the four interventions can be explained by changes in activation.

ALTERNATIVE MECHANISMS

Selective activation of myosin isoforms

An intriguing but controversial mechanism described by Winegrad and his colleagues (see Winegrad, 1984) is the selective increase in activation of the fast myosin isoform in the presence of ß-adrenergic stimulation. This mechanism appears to have little effect in muscles having only the slow isoform but increases the force and ATPase in muscles containing some fast isoform. The data supporting this mechanism are not fully accepted, and thus, the mechanism must be regarded as somewhat speculative. Some of the data were obtained in permeabilized preparations, and other workers were unable to reproduce the results, perhaps because of technical differences. In addition, in a study of the effects of maximum shortening velocity in intact rat muscles, deTombe and terKeurs (1991) found no effect of ß-adrenergic stimulation in muscles having different ratios of fast to slow myosin. Thus, further experiments are needed to resolve the conclusions drawn from these very suggestive experiments. If it exists, this selective activation could provide an important mechanism for regulating muscle power in hearts having a mixture of myosin isoforms.

Direct effects on crossbridge cycling

A possibility that has long been considered is that some inotropic intervention might alter the kinetics of crossbridge cycling. Knowing whether such a mechanism exists would be important for designing therapeutic interventions for the failing heart. In addition, such a mechanism would provide a valuable tool for probing the crossbridge cycle. Because of the importance of this issue, a series of experiments testing the possibility are described. This detailed explanation is also given as a heuristic exercise to demonstrate both the complexity of these experiments and the importance of proper attention to detail in all aspects of the work.

While the effects of the four agents described above can all be explained by increasing activation, there is no direct evidence that some of the agents do not affect crossbridge cycling. It is possible, for example, that an alteration in crossbridge function mimics the effects of increased activation with regard to the parameters measured. Thus, it is important to find additional contractile parameters to measure. An example of a different parameter was investigated separately and simultaneously by two laboratories, Hoh et al., (1988) and Berman et al. (1988). While the results from both groups suggested that adrenergic stimulation altered crossbridge cycling, subsequent experiments suggest that this conclusion is incorrect.

Both groups placed muscles in **contracture**, continuous and presumably maximum activation, by substituting barium for calcium in the extracellular solution. These contractures were used on the assumption that activation is constant and fixed in a way that it is not likely to be altered by mechanical perturbations. They then measured the force responses to sinusoidal length oscillations in the presence and absence of adrenergic agents. The effects on the responses to the oscillations suggested that the inotropic agents directly altered the kinetics of crossbridge cycling, but this interpretation is likely to be confounded by other factors.

As explained in Chapter 3, a muscle responds to a sudden length change by an instantaneous change in force that is due to the stiffness of the crossbridges and other structures. This force change is then reversed in several stages. The first stage is due to movement of attached crossbridges in the direction of the length change. Later changes are due to bridge detachment and reattachment. The force response to oscillatory length changes therefore depends on the frequency and amplitude of the applied oscillations. To assess the frequency dependence of the responses, Hoh and Berman and their colleagues measured the **dynamic stiffness**, i.e. the ratio of the force change to the length change at different frequencies. Since the muscles were in a state of contracture, it is safe to conclude that the observed changes were not due to changes of activation. Both groups observed that adrenergic stimulation caused a shift in the frequency at which dynamic stiffness was at a minimum, and that the shift was greater when the muscles contained a higher fraction of the fast myosin isoform. From these observations, they concluded that the adrenergic agents had caused a change in the kinetics of crossbridge movement, and that the agents had a greater effect on the fast isoform. These conclusions are very likely correct, but for reasons other than a direct interaction of the agents with the crossbridges.

The conclusions from this work differ from those of Chiu et al. (1989), described above. Chiu's data were obtained with muscles from adult animals that have a predominance of the slow isoform, so that the effects of seen by Berman and Hoh and their colleagues might have been missed. This possibility suggested that Chiu's experiments should be repeated with muscles containing the fast myosin isoform. The effects of adrenergic stimulation on maximum velocity in rat heart muscles having various ratios of isoforms were thus assessed by deTombe and terKeurs (1991). They found that adrenergic agents had no effect on maximum velocity in their muscles, irrespective of the myosin isoform composition of the muscles. These results suggest that adrenergic agents do not alter the cycling rates of the fast myosin crossbridges. Thus, there is a genuine disparity between the results obtained with shortening muscle and the conclusion derived from oscillatory experiments in contractured muscle.

The cause of the disparity in the two types of study has not yet been identified experimentally. It is possible, for example, that the changes found by Berman and Hoh and their colleagues would not be reflected in maximum shortening velocity. There is, however, one possible source of artifact in the oscillatory experiments that must be considered. As described above, contractile strength increases with increasing the frequency of contraction in cardiac muscle. But this positive rate staircase only occurs up to some optimum frequency. When the muscle is driven to twitch at higher frequencies, there ensues an opposite, **negative rate staircase**, known as the **Anrep effect**. It has further been shown that the frequency optimum is inversely dependent on the diameter of the muscle. This finding suggests that the negative rate staircase results from a metabolic limitation, such as hypoxia at the center of the muscle. The thicker the muscle, the more susceptible it is to hypoxia at its core, and thus it reaches its peak force at lower frequencies. This consideration suggests that a muscle activated continuously and maximally during a contracture would be much more likely to be hypoxic at its core than a similar muscle undergoing cycles of contraction and relaxation. This tendency to hypoxia would be increased in muscles that have a higher metabolic rate because they have more fast myosin. Thus, the results obtained in the contracture experiment might be explained easily as an artifact of the preparation, rather than as an effect of adrenergic stimulation on the crossbridge kinetics. In support of this explanation is a more recent report from Berman's group (Shibata et al., 1990) that barium contracture was associated with depletion of metabolites and increases in the concentrations of the products of contraction. Some of these changes did not occur until after a delay, but others, such as phosphate increases, began almost immediately. While this report does not prove that the metabolic alterations caused the observed mechanical changes in contractured muscle, it is very highly suggestive.

WORK vs POWER

The discussion in this chapter has focused almost entirely on instantaneous measures of cardiac capability, to the exclusion of the very important consideration of how this capability is integrated over time. This exclusion was necessary because little is known about the influence of inotropic agents on mechanisms that abbreviate activation in shortening myocardium. The importance of this shortening inactivation is shown in Fig. 17.7, which compares the instantaneous velocity-force and velocity-power curves for skeletal muscle with the shortening-force and shortening-energy curves integrated over a twitch in cardiac muscle. This comparison shows that while the energy rate increases as the load is lowered, the total energy output over the twitch decreases. The probable reason for the decrease is the abbreviation of the twitch contraction by shortening.

Integrated parameters are much more important for cardiac function than the instantaneous values. For example, the amount

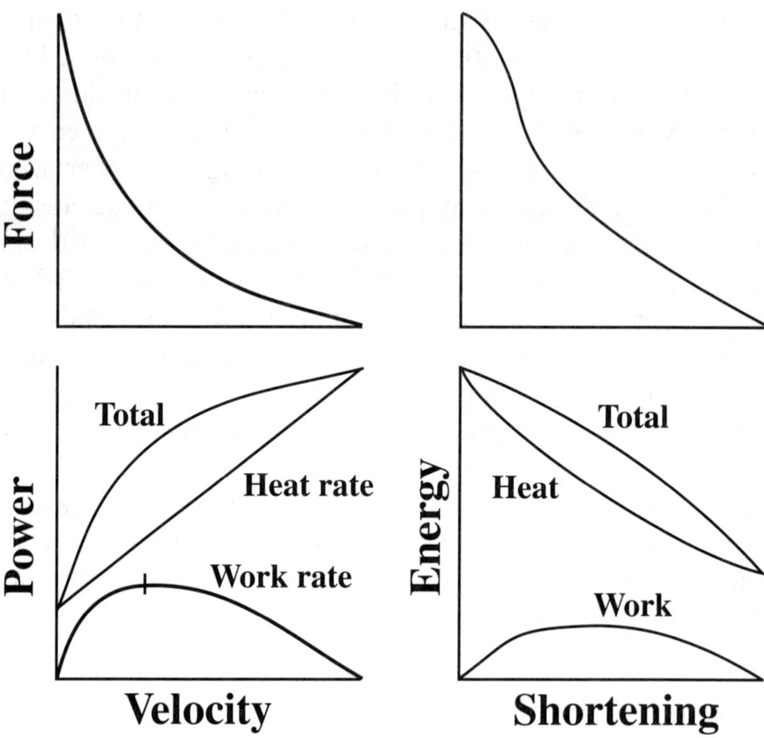

Figure 17.7. *Instantaneous vs. integrated energetic parameters in cardiac muscle. Adapted from Fig 2.9.*

of blood ejected during systole is much more important than the rate at which it is ejected. The effects of shortening inactivation are thus expected to be important, but without data it is difficult to predict how they will influence performance. For example, the greater duration of contraction seen with caffeine would be expected to increase stroke volume to a greater extent than adrenergic stimulation, which abbreviates the twitch. On the other hand, if the twitch fails to shorten sufficiently when pulse rate increases, the reduced time for diastolic filling might greatly limit cardiac reserve. Thus, more data is needed to define the effects of the inotropic agents under the conditions normally found in the body before their full effectiveness can be assessed.

MYOCARDIAL CONTRACTILITY

There are many circumstances where knowing the inotropic state of myocardium would be very useful. These circumstances range from basic scientific need to assess the effect of inotropic agents to the clinical requirement for an accurate assessment of patients' hearts. In spite of this great need, and substantial efforts to develop a reliable measure of contractility, no generally accepted index has emerged. The history of these efforts is not presented in this book because the reasons for the failure lie in technical details whose lengthy description is not justified by the disappointing outcome. There are, however, a few general principles underlying the quest, and a concise understanding of these principles will be presented.

The instantaneous strength of the heart is determined by two separate variables, sarcomere length and inotropic state of the muscle. It was once believed that these two variables achieved their effects through separate mechanisms. The discussion in this chapter and in Chapter 6 suggests otherwise, that both act by varying the level of activation, i.e., by varying the number of thin filament attachment sites for crossbridges. The available evidence suggests that the kinetics of crossbridge activity is not altered by the inotropic interventions or by changes of muscle length. If true, such a conclusion is likely to be quite valuable in the quest for a reliable index of contractility because it indicates that all that is needed is a measure of the level of activation.

An interesting corollary to this conclusion is that the maximum capability of a piece of heart muscle is determined by the number of myosin molecules present and by their isoform composition. If this number could be known, or if the capability of the maximally activated myocardium could be known, a measure of contractile state would be a statement of the fraction of the myosin molecules activated or of the fraction of maximum capability. The complements of these fractions would then be a measure of myocardial reserve. Thus, one reasonable goal in the search for an index of contractility should be a method of assessing the maximum capability of the muscle and the fraction of that maximum achieved at any instant.

A second issue, described briefly above, is that there can be substantial differences between the instantaneous capability of the heart and the capability integrated over the heart beat. Variations in shortening deactivation caused by different inotropic agents may cause large disparities between instantaneous and integrated measures. A complete understanding of contractility therefore requires an understanding of the separate influences of the different inotropic interventions on the instantaneous and integrated measures of performance.

SUGGESTED READING

WINEGRAD, S. (1984) Regulation of cardiac contractile proteins: correlations between physiology and biochemistry. *Circulation Res.* **55**: 565–674.

Chapter 18

PATHOPHYSIOLOGY OF HEART FAILURE

Muscular contraction is so essential for life that the responsible mechanisms have evolved to be extremely robust. Diseases of the contractile machinery or the activation system are very rare because they are so devastating that natural selection has kept their likelihood to a minimum. In spite of this strong negative selection bias, half of all deaths in the United States result from cardiovascular disease, and heart failure is the most common hospital discharge diagnosis in the U.S. Medicare system. The seeming disparity between the health of the machinery and the frequency of cardiac disease occurs because most heart illnesses result from disorders of other tissue and affect the heart only incidently. The most common such disease in the Western World is hardening of the arteries, **atherosclerosis**. Diseased coronary arteries become occluded, causing **myocardial infarction**, death of heart muscle from a lack of blood supply. In the case of heart failure resulting from myocardial infarction, the remaining muscle is normal, and the physiological derangements occur because there is an inadequate amount of muscle left to carry on the work of the heart. This is also true of many other forms of heart failure; the contractile machinery is normal, and there is simply not enough of it or there are disorders of other tissue.

Heart failure is commonly divided into two broad categories called **systolic dysfunction** and **diastolic dysfunction**. Systolic dysfunction results when the contractile machinery lacks the power to meet the circulatory demands of the body. Diastolic dysfunction results when the passive properties of the heart prevent the muscle from being stretched, so that the heart cannot make adequate use of the Frank-Starling mechanism to increase its output. The name is misleading because it implies that the heart has some diastolic function that has been "dyssed", when in fact, the disease results entirely from derangement of passive structures, and in most cases, these structures lie outside the muscle cells.

VENTRICULAR FUNCTION CURVES

The function curves first proposed by Sarnoff and Berglund (1953) and described in Chapter 16 will be developed here to show the interaction between the active and passive properties of the heart. These curves can then be used to explain both systolic and diastolic dysfunction using a common paradigm. The interaction is shown in Fig. 18.1. When a measure of ventricular performance, such stroke work, is plotted against end-diastolic volume, the relationship is likely to be nearly linear, as shown in Fig. 18.1 A. A difficulty arises with this simple relationship, however, because the parameter sensed by the body and measured by the clinician is not the end-diastolic volume but the end-diastolic pressure. When the same measure of ventricular function is plotted against end-diastolic pressure, as in Fig. 18.1 B, the relationship is curved, being less steep at higher filling pressures. The conversion from linear to curved results from the non-linear passive properties of the ventricle, plotted in Fig. 18.1 C. The ventricle is very compliant and easily distended by low pressures at low volumes, but at higher volumes, much greater pressure increases are required to cause an incremental volume increase. In these graphs, and all that follow, the upper limit of normal left-ventricular filling pressures, 19 mm Hg, is plotted as a dotted line extending from the pressure axis. Pressures greater than this cause most patients to experience chest discomfort and shortness of breath.

This simple, non-linear relationship between filling pressure and performance will be used to explain many of the mechanisms in this chapter.

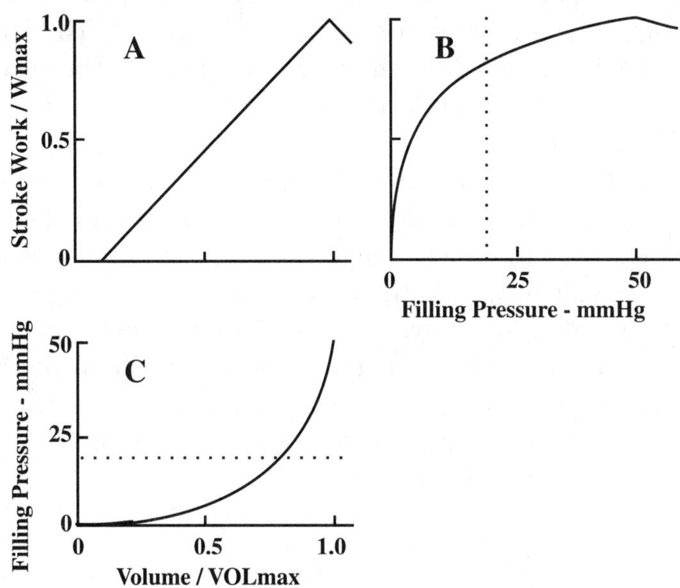

Figure 18.1. *Stroke work, end-diastolic volume, and passive pressure-volume relationships for a normal ventricle.*

Diastolic dysfunction

Fig. 18.2 shows the effect of decreased compliance on ventricular function curves. The curves for the normal ventricle in Fig. 18.1 are plotted as dashed curves, with the properties of a less compliant ventricle superimposed as solid curves. The steeper passive pressure-volume curve for the stiffer ventricle is shown in Fig 18.2 C. The relationship between stroke work and end-diastolic volume, plotted as the solid line in Fig. 18.2 A, is assumed to be the same in both the normal and stiff ventricle. Such a ventricle with normal contractile function but stiff passive components may result from several causes. One such cause is longstanding hypertension. Initially, high blood pressure causes the muscle to hypertrophy, and the muscle thickens. With time, the contractile function diminishes, but the wall remains abnormally thick and stiff, creating the situation represented by the solid curves. The effect of the decreased compliance is shown in Fig. 18.2 B. The stiff ventricle could do as much work as the normal heart if filling pressures were high enough to achieve the necessary volumes.

In general, the ventricle is only infrequently required to distend to the upper limit of its normal filling pressure, and it usually operates over the lower range of filling pressures. Much of the time, the patient with a stiff ventricle has only mildly elevated filling pressures and is asymptomatic. But problems arise when the heart would normally dilate to increase its stroke work. Pulmonary venous pressures increase, the signs and symptoms of pulmonary congestion increase, but the ventricle does not dilate.

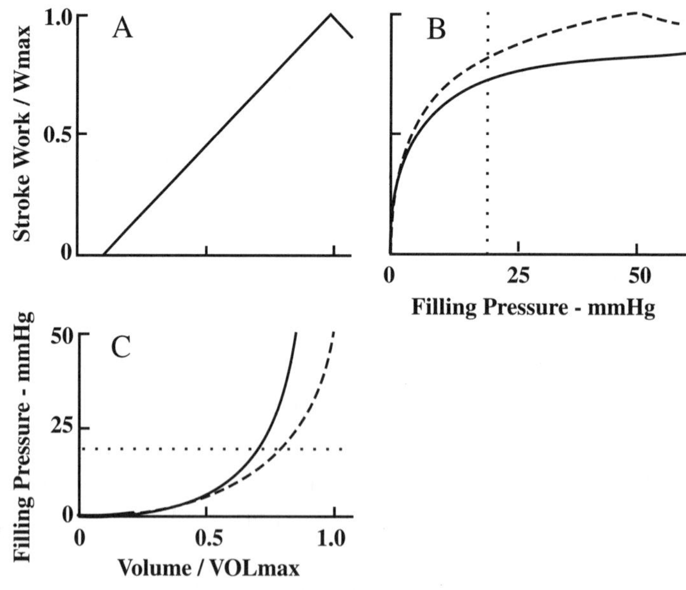

Figure 18.2. *Function curves for a stiff ventricle (solid curves) compared with the normal (dashed curves) from Fig. 18.1.*

Constriction vs restriction

The utility of function curves to explain pathophysiological mechanisms is demonstrated by the contrast between constriction by a stiff pericardium (Fig. 18.3) with restriction due to infiltration of the myocardium (Fig. 18.4).

The heart is situated in a double walled, closed sac, called the **pericardium**. The sac normally contains a small amount of fluid that lubricates the two walls and reduces friction as the heart moves. Sometimes this sac becomes stiff, either because of a resolved inflammation or because of large volumes of fluid accumulated within it. This stiffer pericardium cannot distend when the heart dilates, leading to a condition called **pericardial constriction** or sometimes **constrictive pericarditis** (when the stiffness is caused by inflammation, the suffix **"itis"** is used). The effect of this non-distensibility is shown in Fig. 18.3. When the volume of the heart is less than the volume of the pericardium, the heart behaves entirely normally. When filling pressures are increased to the point where the volume of the heart is equal to the volume of the pericardium, further cardiac dilation cannot occur, and filling pressures rise steeply. At these higher volumes, the heart cannot use the Frank-Starling mechanism, and the signs and symptoms of heart failure develop suddenly and severely. Before the development of ultrasound techniques for imaging the pericardium, these conditions could be difficult to diagnose because the patients functioned normally much of the time but would suddenly develop the signs and symptoms of severe heart failure when their hearts dilated slightly.

There are several conditions where the heart muscle becomes infiltrated with extraneous proteins that encases individual myocytes or groups of myocytes. The two most common such conditions are **amyloidosis,** where the proteins are often fragments of immune globulins, and **hemochromatosis**, where the proteins are hemosiderin and ferritin that store excess iron in the body. Instead of the entire heart being encased in a rigid container, as with peri-

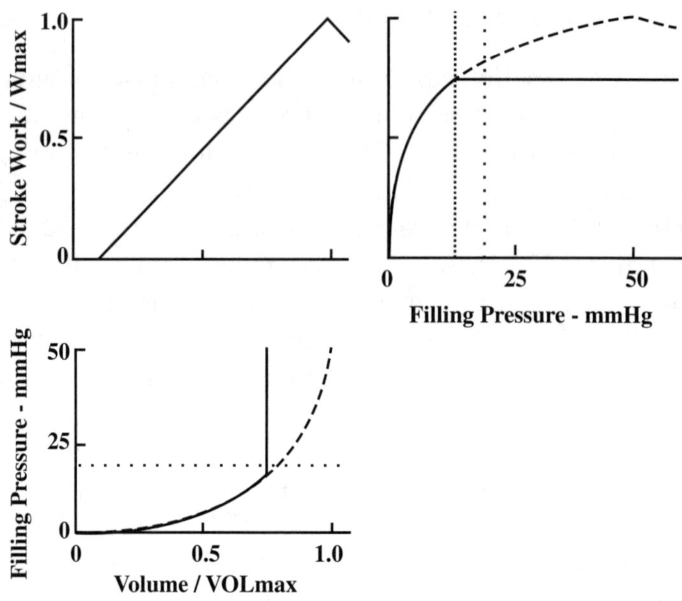

Figure 18.3. *Function curves for a normal heart situated in a stiff pericardium (solid curves) compared with the normal (dashed curves) from Fig. 18.1.*

cardial constriction, individual muscle cells, or groups of cells, are encased. The effects on function are shown in Fig. 18.4. Dilation beyond an upper volume and emptying below a minimum volume are both greatly impeded by the infiltrate. When the upper bound is exceeded, filling pressures rise dramatically. When filling pressures fall below the minimum bound, cardiac output falls precipitously.

Of the two conditions, pericardial constriction is easier to manage because function is normal as long as filling volumes are kept below some critical value. In addition, when medical management becomes difficult, the constricting pericardium can be removed surgically. By contrast, restrictive cardiomyopathy requires that filling volumes and pressures be kept within the narrow range indicated by the vertical cursors in Fig. 18.4 B, and there are no

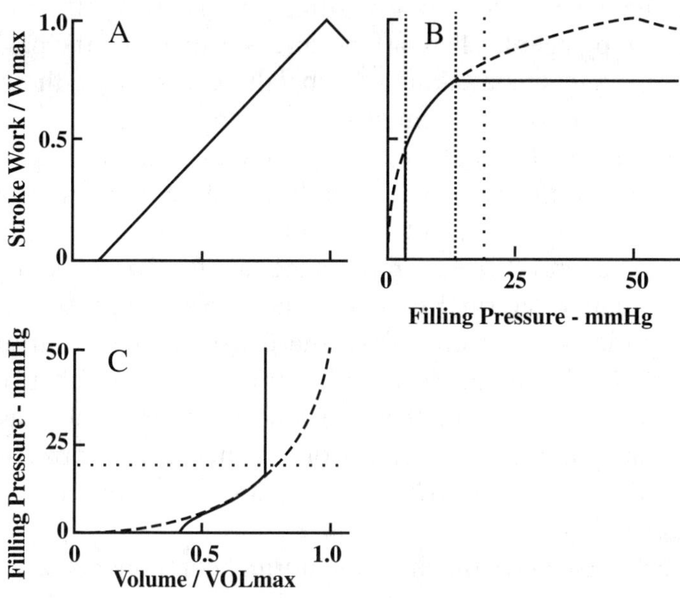

Figure 18.4. *Function curves for restricted heart (solid curves) compared with normal (dashed) curves from Fig. 18.1.*

surgical alternatives. In addition, the infiltrative processes can progress relentlessly, decreasing the working range until adequate function is no longer possible.

SYSTOLIC DYSFUNCTION

In almost all cases where the heart lacks the power to meet the demands of the body, the underlying cause is loss of contractile elements. In addition to the loss of large blocks of myocytes to myocardial infarction, contractile elements may be reduced within cells but without loss of cells. Additionally, individual myocytes may be destroyed randomly throughout the heart. Such diffuse injury may result from viral infections or chemical toxins, such as cobalt, formerly used in brewing, and adriamycin, a cancer chemotherapy agent. There are no diseases in which the basic contractile mechanisms are known to be altered or where the components of the contractile elements are different.

The statement that almost all systolic dysfunction results from loss of contractile elements is likely to be controversial because there are very many reports of altered muscle function in heart failure. In general these alterations are due to variations in the relative amounts of normal components within the muscle and are probably the result, rather than the causes of heart failure. Although these alterations may further impair cardiac function, they are not the root cause of the initial heart failure. The reason for making this point is to dispel any optimism that the correction of a functional impairment will reverse heart failure due to a loss of contractile elements.

The diverse ways in which the normal components of the heart are altered are illustrated by two studies from the same laboratory. In one (Tsutsui et al., 1993) shortening of cells taken from failing hearts was impeded by an overabundance of the normal microtubules. In the other (Tsutsui et al., 1994) cells isolated from failing heart contained a decreased concentration of contractile elements, but the reduction was less if the animals were first treated with an agent that blocked sympathetic tone. One point to be made from these observations is that both alterations result from differences in the expression of normal protein components. Another point is that while reversal of either change will improve function, neither change will correct the original problem. On the other hand, increasing the number of contractile elements in the cells by lowering adrenergic tone is likely to reduce the load on the contractile elements and thereby reduce the stimulus to the alterations that derive from work overload.

Tachycardia-induced heart failure

Increased heart rate, **tachycardia**, maintained continuously for many days, produces reversible heart failure. As the tachycardia becomes established, the atria and ventricles dilate and contract less vigorously. The method of producing the tachycardia is relatively unimportant to the outcome; frequently ventricular failure occurs when the increased heart rate originates in the atrium, either from an atrial dysrhythmia or from an experimental atrial pacemaker. If the heart rate returns to normal, the ventricle returns to its normal size and contracts with its normal vigor.

The mechanisms responsible for tachycardia-induced heart failure are not completely understood, but it is now known that there are at least two separate causes, one due to the tachycardia itself, and the other to increased adrenergic stimulation. Although the two causes are often associated, their separate effects can be distinguished by the use of pacemakers and pharmacological agents that inhibit circulating endogenous adrenergic stimulants.

Tachycardia-induced heart failure is described here because it helps to explain some other forms of failure, as well as the rationale for some forms of treatment.

High-output heart failure

There are some forms of heart failure that probably result from inadequate amounts of contractile elements, even though the number of contractile elements may be higher than normal. One such case is high output failure. Increased cardiac output is produced by volume overload, which can be caused either by increased metabolic demand, or by an increased blood flow, as from ateriovenous shunts or incompetent heart valves. When increased volume demands develop slowly, the heart responds by growing larger in all of its dimensions, as explained in Chapter 14. The increased end-diastolic volume allows an increased stroke volume in the presence of a normal ejection fraction and normal sarcomere shortening. At some point, however, the heart may fail, either because the demands exceed the ability of the heart to grow or for some other, unexplained reason. One possible reason is tachycardia and/or the increased adrenergic tone. This is particularly true of the high-output heart failure that occurs in association with hyperthyroidism, which always increases both adrenergic tone and heart rate.

Effects of systolic dysfunction

The effects of systolic dysfunction are shown in Fig. 18.5. Immediately after an injury to the myocardium, the heart dilates because of reduced emptying. The dilation partially restores cardiac output, but increased adrenergic stimulation, caused by increased secretion of adrenaline and its analogues, strengthens the remaining muscle so that it can operate at lower volumes and filling pressures. Since there must be a stimulus to adrenaline secretion, the

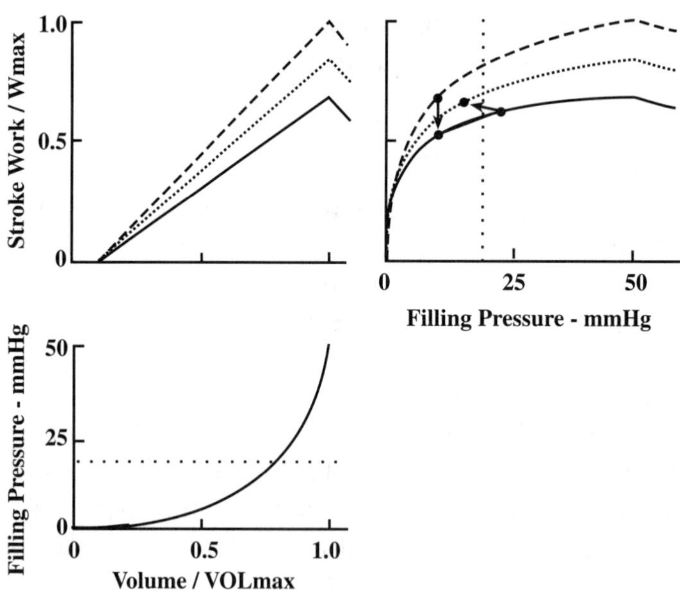

Figure 18.5. *Function curves of a ventricle with systolic dysfunction. Dashed curve: normal ventricle. Solid curve: injured ventricle. Dotted curve: injured ventricle with increased adrenergic tone.*

effects of the myocardial damage are not reversed completely, and cardiac function does not return fully to normal. If it did, the stimulus to increased adrenaline secretion would end, adrenergic tone would return to normal levels, and the heart would not longer be compensated, and it would again dilate.

BIOLOGY OF THE FAILING MYOCARDIUM

While the basic contractile mechanisms of the failing myocardium appear to be normal, there is a distinct abnormality of its biology. Most importantly, it loses its ability to adapt to new loads and to recover from hemodynamic overloads. Within limits, the normal heart will adapt to a hemodynamic overload by growth, as discussed in chapter 15. In addition, normal hearts will regress in size if the overload is removed. Such regression will even occur in athlete's hearts when they stop training (Pellicia, 1991). In a seminal study of this phenomenon, Gaasch et al. (1978) followed the echocardiographic dimensions of hearts in patients undergoing replacement of defective aortic valves. Their results are plotted in Fig. 18.6. As in the original paper, the data are superimposed on the fit used by Ford (1976) to describe normal and overloaded hearts of different animals (Fig. 15.2 in Chapter 15). As shown, the thickness/radius ratios all lay below the line fitted to normal hearts, indicating ventricular dilation and suggesting that none of the hearts were fully adapted to their loads. The data and the patients were, however, separated into two distinct groups, four indicated by closed symbols in the figure who had developed frank heart failure before operation, and the remaining 12, indicated by the open circles, who had not. The thickness/radius ratios for the 12 patients not in failure lay closer to the line than the other 4, and the hearts of these patients regressed in size over many months following surgery. By contrast, the ratios for the patients in failure lay further from the line, indicating greater ventricular dilation and more decompensation. Their hearts did not regress in size, and they did not recover from their heart failure.

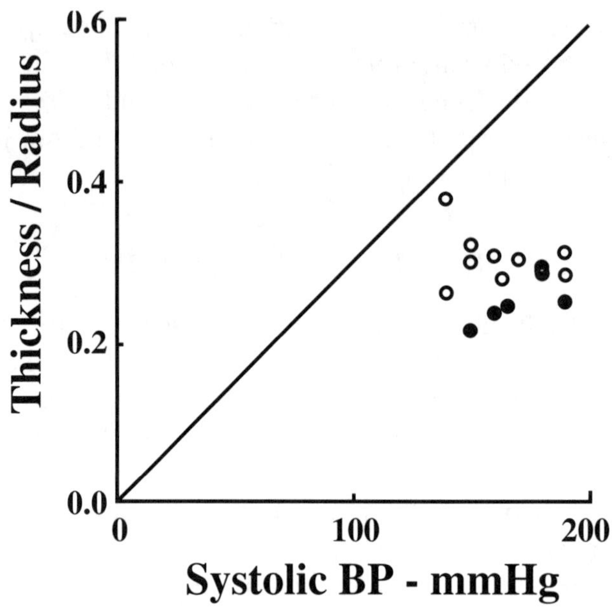

Figure 18.6. *Thickness/radius ratios plotted against chronic ventricular systolic pressure in patients with aortic valvular disease. Closed symbols: patients in overt heart failure. Open symbols: patients not in heart failure. From Gaasch et al. (1978).*

The work of Gaasch and his colleagues has been confirmed by many others and has led to the now well known cardiological conundrum of the optimum time for replacing defective valves. Cardiologists would not like to send their patients for a major operation any earlier than absolutely necessary, but if they wait too long, they risk putting their patients into heart failure from which they may not recover.

This experience with failure resulting from hemodynamic overload suggests that the failing myocardium somehow loses its ability to adapt to changing conditions. A clue to the difference may be found in the experience with beta-blockers in heart failure, described in the next chapter, and in some experimental work by Tsutsui et al. (1994). In this latter study, dogs were induced to develop heart failure by surgically injuring their mitral valves so that some blood flowed back into the left atrium during each heart beat. After heart failure became established, the animals were divided into two groups, one treated with a beta-adrenergic blocking drug and the other an untreated control. The treated group recovered substantially while the untreated controls did not. When the animals' hearts were later examined, the treated group had more contractile elements per cell than the non-treated controls. This study suggests strongly that at least some of the difference in the biology of the failing myocardium is due to an increased adrenergic tone that does not allow the heart to adapt to an increased load placed on the remaining myocytes. As discussed in the next chapter, some, but not all, of this difference may be due to higher heart rates of animals with heart failure.

The true importance of these studies is that they lead to the optimistic conclusions that some of the biological changes in the failing myocardium are reversible, at least partially, and that some recovery through the growth of new contractile elements is possible with proper treatment.

CONSEQUENCES OF HEART FAILURE

The following **symptoms** felt by patients and the **signs** observed by others are often the same whether heart failure results from systolic or diastolic dysfunction.

Edema

The most common sign of heart failure is fluid accumulation in tissue, called **edema**. The accumulation results from two inter-related effects, fluid retention by the kidneys in response to lower arterial pressure and increased capillary pressures resulting from elevated venous pressure. When fluid accumulates in the lungs, it is said to be due to **left-heart failure**. It is said to be due to **right-heart failure** when it accumulates in other sites, most commonly the legs. Usually, both sides of the heart fail together, although one side may fail to a greater degree than the other. Less frequently, fluid accumulates exclusively in the lungs or periphery. It is often said that left-heart failure is the most common cause of right-heart failure, and this is a reason to expect **bi-ventricular failure**, even when the damage is mainly to the left ventricle.

Exertional intolerance

The direct effect of impaired pump function is to reduce the reserve of the heart, so that cardiac output cannot rise sufficiently to meet bodily demands during physical exertion. At rest, cardiac output may be normal, or at least within the normal range, although it may be reduced from the patient's normal value before the onset of disease. To achieve this normal resting function, the heart must use some of its reserve, which leaves less reserve for times of increased activity. Thus, patients with all but the most severe heart failure are comfortable at rest and develop symptoms only when they attempt physical activity. Patients' tolerance to exertion can therefore be used to grade the severity of heart failure. Stair climbing and walking are good measures because they are performed regularly by most people with heart failure. Since energy is expended in transporting the patient's own weight, there is also an inherent correction for body size. The severity of heart failure is graded on the basis of how many stairs they can climb or how far they can walk without stopping. On the other hand, the accuracy of indices of physical exertion in estimating cardiac reserve is confounded by poor physical conditioning in persons made inactive by their disease. Thus, the level of exertional tolerance is only a rough guide to the degree of cardiac dysfunction.

It is also true that the most direct result of heart failure, decreased physical reserve, is often not greatly incapacitating until the heart failure becomes severe because persons in the modern world rarely attempt extreme exertion. This is especially true in developed countries where modern conveniences, such as elevators and automobiles, enable patients to lead nearly normal lives with substantial physical limitations.

Exertional intolerance resulting from decreased cardiac output is generally greater in patients with systolic dysfunction because ventricular filling pressures do not usually rise with moderate degrees of exercise. Under these circumstances, patients with diastolic dysfunction can perform normally because their cardiac outputs are not impaired. On the other hand, when greater exertion occurs, so that diastolic filling pressures rise, the onset of shortness of breath can be sudden and severe. This sudden onset of symptoms results from sudden increases in filling pressure rather than from a sudden decrease in cardiac output.

Dyspnea

Shortness of breath, **dyspnea**, occurs as the result of two separate mechanisms. The more frequent is increased pulmonary venous pressure. High pressure forces fluid out of the capillaries into lung tissue. This fluid impairs the gas exchange between capillaries containing blood and **alveoli** containing air. With higher pressures, edema fluid spills into the alveoli. This fluid further impairs gas exchange by replacing air. Finally, increased pressures themselves may increase the sense of breathlessness through neurological reflexes, even without significant edema fluid.

Less frequently, dyspnea results from decreased cardiac output. Tissues extract oxygen from blood and add carbon dioxide. These chemical changes in the blood stimulate respiration. Tissues of patients with decreased cardiac outputs extract more oxygen and produce more carbon dioxide for each liter of blood pumped by the heart. Thus, patients with severely reduced cardiac output may experience an increase in shortness of breath, even when filling pressures are reduced. The reason for suggesting that this mechanism is less frequent is that increasing dyspnea

is more often treated successfully with diuretics that simultaneously reduce fluid volume, venous pressures, and cardiac output.
Dyspnea is reduced even though cardiac output is also reduced.
The reason for knowing that dyspnea can result simply from diminished cardiac output is demonstrated on the infrequent occa-
sions when patients are diuresed too vigorously. These patients
experience increased dyspnea, particularly with physical exertion, and this symptom is improved when their diuretics are reduced or when they are given salt and water. Thus, dyspnea is increased both when vascular volumes are too large and when they
are too small.

There are several inciting causes of dyspnea. At the mild end
of the spectrum is dyspnea caused by physical exertion, called
dyspnea-on-exertion, sometimes abbreviated **DOE**. The severity
of dyspnea-on-exertion is often graded by the extent of exertion
required to produce it. "One flight DOE" for example would describe a patient who must stop to catch his breath after climbing a
single flight of stairs.

At the other end of the severity spectrum is dyspnea caused
simply by lying down, called **orthopnea**. It occurs because blood
pooled in the veins in the upright position is returned to the central
circulation when the patient becomes recumbent. Because patients
with this condition cannot lie flat, they are frequency forced to
sleep propped up in bed. This type of orthopnea is graded clinically (and somewhat inaccurately) by the number of pillows required maintain comfort at night, e.g. "three-pillow orthopnea."
The inaccuracy arises because the terms do not account for the use
of the headboard. In the more severe cases, patients with orthopnea must sleep sitting upright in a chair.

Between these extremes of severity are patients who accumulate extravascular edema fluid in the lower extremities during the
day, when they are upright, and return the fluid to the circulation
when they lie down at night. In its less severe form, this phenomenon increases urination at night. This **nocturia** is graded by the

frequency with which the patient must get up to void. Accuracy can be confounded by other variables, such as fluid intake in the evening. In addition, insomnia can often send patients to the toilet in the absence of a large volume of urine.

A more severe consequence of the return of lower extremity fluid at night is **paroxysmal nocturnal dyspnea**, sometimes abbreviated **PND**. When this occurs, the patient awakens suddenly with severe shortness of breath about 3–4 hours after going to bed. The symptom occurs when the return of fluid to the circulation is more rapid than the kidney can remove, so that the fluid accumulates in the lungs. Although frightening, this occurrence indicates less severe heart failure than orthopnea that prevents the patient from lying flat at all.

Increased adrenergic tone

Many patients with heart failure show evidence of increased sympathetic stimulation, an increased heart rate, sweating, and arteriolar constriction resulting in cold extremities and poor peripheral circulation. These signs are most prominent with an acute worsening of the heart failure, but some, such as the tachycardia, may be present much of the time. It is now becoming increasingly recognized that this increased adrenergic tone can be a cause of progression of heart failure.

Acute pulmonary edema

There is a group of patients who develop dyspnea and fluid accumulation in the lungs very rapidly, and as a consequence, find themselves in a life-threatening circumstance called **acute pulmonary edema**, or sometimes called **flash pulmonary edema** to emphasize its sudden onset. Many of these patients have diastolic dysfunction, and for many, the symptoms recur with distressing frequency. The condition is described here because it illustrates the importance of the interaction between the heart and circulation and because it represents a form of "heart failure" in which heart function is well above normal.

In many instances, the patients will have been relatively free of symptoms until a few minutes before the sudden onset of dyspnea. When observed, they will be seen to be markedly short of breath, and frequently **cyanotic**, meaning that their skin has a blue tinge due to their arterial blood not being saturated with oxygen. Pulmonary edema is further manifest by the sounds of fluid in the air passage of the lungs, and frequently these sounds are so loud that they are easily heard from a distance. The most striking signs are those due to a large excess of circulating adrenaline.

The symptoms of increased pulmonary vascular pressures cause a vigorous response from the **sympathetic** nervous system responsible for producing the **fight-or-flight** responses needed by animals in times of danger. The sudden increase in circulating adrenaline greatly worsens the initial condition. Adrenaline causes constriction of blood vessels, raising blood pressure and thereby impeding the pump so that left-ventricular filling pressures rise further. The adrenaline also increases heart rate to near maximum values, and the increased rates impede ventricular filling by reducing the fraction of the cardiac cycle spent in diastole.

The increased adrenergic tone causes patients in an attack of acute pulmonary edema to have cold skin and be soaked with perspiration. It may also cause them to have blood pressure and heart rate twice the resting values. Blood pressures of 260/140 mm Hg and heart rates of 150 are not uncommon in patients who otherwise have resting blood pressure of 130/70 and pulse rates of 75. The increased blood pressure and pulse suggest that the heart is not failing as a pump. Cardiac work is sometimes estimated as the product of heart rate multiplied by systolic blood pressure, called the **rate pressure product**. A doubling of both blood pressure and heart rate suggests that cardiac work is increased four-fold. Although the problem may have begun with fluid retention caused by a weak heart, when the heart is working at four times its resting rate, it cannot be said to be "failing." The immediate problem, and its treatment, must be sought elsewhere.

The derangement in acute pulmonary edema is an excessively high mean vascular pressure. This increase in mean pressure is shown by elevated pressures in every part of the circulation. As mentioned, systemic arterial pressure is often twice normal. Systemic venous pressures, as evidenced by distended neck veins, are also often twice normal. Pulmonary venous pressures may exceed normal pulmonary arterial pressures, which now must be elevated substantially to force blood through the pulmonary capillaries into the distended pulmonary veins. This elevation in mean vascular pressure is due to a mismatch between the capacity of the circulation and the volume of blood in it.

Most typically, the problem begins with an increase in blood volume in response to a weak heart. The resulting pulmonary congestion causes an adrenergic response. Although the increased adrenergic tone increases the strength of the heart, it also increases vascular constriction. This constriction worsens the circulatory mismatch, further increasing pulmonary congestion, and causing a greater adrenergic response. Thus, a downward spiral accelerates, with the responses of the body aggravating the circulatory derangement. If not treated immediately, acute pulmonary edema is often fatal, but appropriate treatment can resolve most of the symptoms within minutes. This prompt relief is accomplished by abrupt reductions in mean vascular pressure, with a concomitant reduction in circulating adrenaline and mean vascular pressure. Thirty minutes after an attack, a patient who had been blue, drenched with sweat, and markedly dyspneic will be warm, dry, pink, and resting comfortably, but looking and feeling exhausted from the recent ordeal.

Several very effective treatments of acute pulmonary edema have been developed over the centuries, and all are directed at lowering mean vascular pressure. One of the reasons they are so universally effective is that the heart itself is usually not at risk as long as arrhythmia does not develop. If cardiac output diminishes, either from an abnormal rhythm or from an insufficiency of blood to the heart muscle, blood pressure falls and the patient dies in a state called **cardiogenic shock**, meaning an inadequate blood pressure resulting from an insufficient cardiac output.

CONCLUSION

One major point of this chapter is that several forms of heart failure exist and that most have little to do with the contractile function of the muscle. In many cases, they are lumped together because the signs and symptoms are the same, even though the causes may be very different.

Another major point of the chapter is that systolic dysfunction appears to result from a paucity of normal contractile elements and not from an abnormality of their function. Although there may be alterations in the relative amounts of various cellular components, the components themselves are normal. In addition, the alterations in these relative amounts are usually the result rather than the cause of the failure. The importance of this conclusion is that it suggests a new approach to long-term therapy. This therapy should be directed at providing the environment in which the heart can grow as much new contractile machinery as possible, rather than attempting to increase the work of the remaining elements.

Chapter 19

TREATMENT OF HEART FAILURE

Heart failure is an excellent example of an "experiment designed by nature," providing the medical profession with a wealth of material for study. Although a small organ, normally comprising about 0.5% of body mass, the heart's central role in powering the circulation for the entire body gives it an importance well out of proportion to its size. The prevalence of heart failure makes it readily accessible for study and gives great urgency to the search for improved therapy. Because it is a relatively simple organ, when compared with the brain or the liver, the function of the heart is now fairly well understood. Techniques developed for the study of other tissue, such as nerve and skeletal muscle, have been applied successfully to the heart, and the last half-century has seen enormous advances in both the understanding and treatment of heart failure. A review of the treatment of heart failure is therefore a short course in the history of medical science.

The discussion of therapy for heart failure here will begin with a description of several treatments for acute pulmonary edema. As mentioned in the last chapter, in its fully-developed form, acute pulmonary edema is actually associated with a heart that is performing exceptionally well as a pump. The problem is not with the heart but with the circulation. Therapy is therefore aimed at correcting the circulatory derangements, even though this is not always the recognized purpose. The descriptions of these therapies help to explain the hemodynamic aspects of the treatment of true heart failure.

A point that will become obvious in these descriptions is that many medical therapies were developed long before any understanding of their mechanisms. Treatments evolved through empirical trial and error. Before the era of modern medicine, a therapy found effective in one condition was often applied to another, just to see if it would work. This empirical approach explains how some seemingly irrational remedies came to be used. It would be comforting to state that we now use a more reasoned approach to the development of treatments, but as will also become apparent, this is not the case. The mechanisms underlying the benefits of some of the most promising of the newer therapies for heart failure are as mysterious as the mechanism of action of digitalis when it was first used more than two centuries ago.

TREATMENT OF ACUTE PULMONARY EDEMA

Several very different types of therapy for acute pulmonary edema have evolved, and all are effective, either alone or in combination. The reason for this effectiveness is the robust state of the heart, with signs and symptoms being due to circulatory derangements that are relatively easy to correct. As described in the last chapter, if the heart begins to fail before the hemodynamics are corrected, the condition can become fatal.

Bloodletting

The removal of blood from the circulation was a common medical practice before the advent of effective pharmacological agents. It developed on the theory that it removed "bad humors" from the body and was used in a wide variety of conditions. It was certainly effective in the symptomatic treatment of heart failure. Its use in other conditions, where it mainly caused harm, was probably based on observed benefits to patients who had a component of heart failure contributing to their symptoms. Its major physiological effect is to lower mean vascular pressure immediately. This brings prompt symptomatic relief to persons with pulmonary congestion. The relief can be dramatic in acute pulmonary edema, and it is still an effective means of treating this condition, although it is almost never used today, both because there are easier methods of achieving the same result and because the treatment has the deleterious effect of removing blood components, most notably the red blood cells, that can take weeks to replace.

Rotating tourniquets

As an alternative to bleeding, blood can be made to pool in the extremities by tying tourniquets around the arms and legs. If the tissue pressures beneath the tourniquets are between arterial and venous levels, blood will be pumped out to the extremities but impeded in its return, causing the veins to dilate and increase their capacity. Although this practice is sometimes viewed as primitive by skeptical young doctors, it is highly effective, and was a life-saving practice commonly used a few decades ago, before the more recent intravenous medications were developed. It is very effective in situations where medications and modern equipment are not available; any extensible cord or tubing with the compliance of a surgical drain makes an ideal tourniquet. While the adjective "rotating" is often used in the phrase describing this therapy, the tourniquets are usually removed completely after many minutes and not rotated from one limb to the next.

Morphine

Morphine, particularly when given intravenously, so that its effects are sudden, has been used for generations to treat acute pulmonary edema. This practice is an example of a therapy that developed empirically. Its effectiveness was difficult to explain until the dominant role of the sympathetic nervous system in the sudden onset of symptoms was understood. It almost certainly works by blunting sympathetic responses. A side benefit of its use is that it makes patients more relaxed at the end of the ordeal. As with rotating tourniquets, this inexpensive, rapid, and effective treatment is used less often today. Although morphine is used only for acute pulmonary edema, an analogous treatment in chronic heart failure is beta-adrenergic blockade, which reduces sympathetic tone.

Diuretics

Since acute pulmonary edema produces exaggerated symptoms of fluid accumulation, it might seem logical that diuretics would help to relieve these symptoms. Thus, some of the newer and more potent diuretics are often given intravenously to patients with acute pulmonary edema, with great benefit. The flaw in the logic is that the effects occur over a few minutes, while the diuresis of sufficient fluid to alter hemodynamics takes much longer. The effectiveness of diuretics in treating acute pulmonary is more likely due to their ability to dilate veins. This dilation produces a rapid decrease in filling pressures and a better match between the volume of the blood and the capacity of the circulation.

Afterload reduction

The observation that systemic blood pressure is markedly elevated in acute pulmonary edema led to the postulate that a sudden reduction of blood pressure would reduce left ventricular filling pressures and relieve symptoms (Cohn and Franciosa, 1977). Nitroprusside, a rapidly acting vasodilator given intravenously, is an ideal agent for this purpose because its effects can be terminated immediately, if needed. It has become the drug of choice for treating acute pulmonary edema in situations where blood pressure can be monitored continuously. It is so effective at lowering blood pressure, however, that it may be dangerous in settings where arterial blood pressure cannot be monitored continuously with an arterial catheter.

For historical reasons, therapeutic lowering of blood pressure is called **afterload reduction**. To some extent, it may lower the work the heart must do in ejecting blood, and thereby reduces the high left ventricular filling pressure. It should be emphasized, however, that lowering of systemic vascular resistance can also increase cardiac output and thereby increase cardiac work. It seems likely, therefore, that a major benefit of the reduction in systemic vascular resistance is due to some other mechanism. One likely mechanism is that the decreased resistance shunts blood into systemic veins, in a manner similar to the application of rotating tourniquets. The increase in right heart filling pressures will increase right ventricular work, but since the systemic veins are more compliant than the rest of the circulation, the increase in right heart filling and right ventricular pressure is not as nearly as great as the reduction in pulmonary venous pressure.

Whatever its mechanism, nitroprusside infusion is a highly effective means of treating acute pulmonary edema, and can often produce relief of symptoms within minutes.

Intravenous nitrates

Nitroglycerine, a potent explosive, has been used since the time of Murrel (1879) to relieve **angina pectoris**, pains in the chest due to insufficient coronary blood flow. It is in the same category of drug as nitroprusside, and probably both act through the release of nitric oxide, which relaxes blood vessels. It is known to dilate coronary arteries and to be effective in treating the angina of patients who have coronary artery spasm. However, the best current evidence suggests that its principal effect in most patients with angina, who have pain due to a fixed obstruction in a coronary artery, is to dilate veins and reduce cardiac filling pressures, thereby reducing the work of the heart. It seems to be more selective in dilating the veins than nitroprusside, which greatly dilates arterioles, and thus can be used more safely than nitroprusside to reduce venous pressures without precipitating large decreases in blood pressure.

Although nitroglycerine has been available for much longer than nitroprusside, it has only been widely available for intravenous use since the early 1980s. It is not effective when swallowed, because it is metabolized rapidly in its first pass through the liver, where it is transported by blood leaving the gut. It is therefore usually given either by placing a tablet under the tongue or in an ointment applied to the skin. The dry mouth and sweaty skin caused by increased sympathetic tone in acute pulmonary edema make both of these routes of administration impractical. Thus, nitroprusside came into relatively common use and nitroglycerine has not been considered until recently. Since the development of intravenous preparations of this agent, there have been recent reports of its effectiveness in treating acute pulmonary edema (e.g. Beltrame et al., 1998), and probably more will follow.

Positive pressure breathing

Increasing airway pressure has at least three major benefits; 1) it forces fluid out of the alveoli; 2) by increasing the external pressure on pulmonary capillaries, it increases pulmonary vascular resistance; and 3) by increasing intrathoracic pressure, it impedes the return of blood from systemic veins to the heart. The reduction in alveolar fluid improves gas exchange in the lungs while the increased pressures diminish the return of blood to both sides of the heart. These benefits become apparent to some patients, who can be found breathing out forcibly through pursed lips when their pulmonary congestions becomes acute. The recognition of these benefits also once produced a brief vogue for "pulmonary edema masks," tight fitting face masks with one-way valves that permitted easy inspiration but required increased airway pressure for expiration. The difficulty with these masks was that patients would not wear them voluntarily. Frequently, patients become so anxious that they will not wear even the loose-fitting oxygen masks and will fight all attempts to restrict air flow. Thus, the only circumstance in which the airway pressure can be increased is when the patients are unable to move. Sometimes this immobility occurs because they have become moribund, but more often they are paralyzed deliberately so that an **endotracheal tube** can be placed in their windpipes and a continuous increase in airway pressure can be maintained. In addition to the benefits mentioned, an endotracheal tube also allows the delivery of oxygen directly to the lungs. Such tubes are usually only placed as a last resort, because they can be very uncomfortable. On the other hand, it can be a life-saving maneuver when all else has failed.

DIURETIC THERAPY IN CHRONIC HEART FAILURE

The most obvious and most distressing symptoms of heart failure result from fluid accumulation. These symptoms occur with both systolic and diastolic dysfunction and are usually well treated with diuretics. While doctors have used a variety of agents over the centuries, the first effective diuretics were not developed until the early 1920s, when it was found that some **organic mercurials**, then used to treat syphilis, could cause a diuresis.

The general principle of antibiotics is that they must be more toxic to the infectious agents than to its host. The **therapeutic ratio**, defined as the ratio of lethal dose for host to the lethal dose for organism, may exceed a million with many modern antibiotics. Before the advent of sulfa and penicillin, therapeutic ratios were often in single digits. This was certainly the case for the mercury used to treat syphilis. The infectious agent, *Treponema pallidum*, is exquisitely sensitive to mercury, but unfortunately, the human host is very sensitive as well. Efforts were therefore made to couple mercury to an organic compound that would deliver it more directly to the spirochete. In 1919 the medical house-officers in Vienna used a newly developed organic mercurial to treat a young woman with congenital syphilis. She died a short time later, but not before mounting a vigorous diuresis. This diuresis did not go unnoticed by her observant young doctors, who then tested the agent in several other patients, all of whom diuresed substantial amounts of fluid with no significant side effects (Vogl, 1949). From this small but seminal clinical trial was born a major advance in the therapy of heart failure. Organic mercurials, which had to be given parenterally, became one of the two mainstays of treatment until the 1950s and 60s when less toxic, oral diuretics were developed.

With modern diuretics, it is possible to keep most heart failure patients free of edema, but unfortunately, this reduction in symptoms of fluid accumulation is often achieved at the expense of kidney failure. Thus, the modern heart failure specialist must steer his patient between the Scylla of drowning in edema fluid and the Charybdis of renal insufficiency. While this is usually possible, in severe situations the only alternative is to let the kidneys fail and place the patient on chronic dialysis.

With the advent of effective oral diuretics, physicians began to encounter patients with heart failure who had become sufficiently dehydrated that their cardiac output was reduced because of inadequate return of blood to the heart and low ventricular filling pressures. This experience raises the question of the optimum left ventricular filling pressure. It is answered using ventricular function curves. Fig. 19.1 shows the ventricular function curve derived in the last chapter. When filling pressures are low, raising pressure produces a much larger increase in stroke work than when the filling pressure are high. When pressures are much above the upper limit of normal, indicated by the dotted line in the graph, patients can be severely uncomfortable without deriving much benefit from the higher filling pressure. Thus, the object of diuretic therapy is to bring filling pressures into the high end of the normal range. The issue is to determine when this end-point has been achieved.

In most patients, the dyspnea on exertion caused by low vascular volume is distinguished by the absence of orthopnea. In contrast to patients with fluid overload, who cannot lie flat, persons who have been over-diuresed tolerate the horizontal position more easily, and an improvement in a patient's orthopnea with increasing dyspnea on exertion may be a clue to excessive fluid removal. In addition, accumulation of metabolic waste products in the

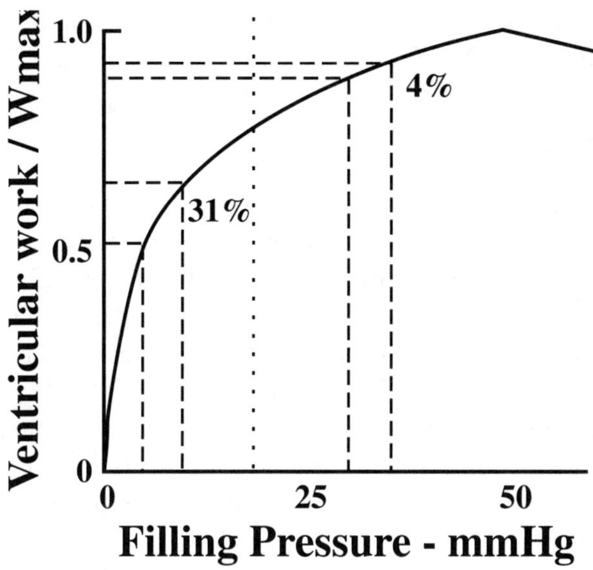

Figure 19.1. *Effect of the same increase in left-ventricular filling pressure starting from low and high values.*

blood may signal a decrease in blood flow to the kidneys resulting from dehydration. In severe cases, the clinician may have to resort to a pulmonary artery catheter to measure the pressure directly. Such catheters are commonly used in intensive care units and are monitored closely to maintain left-ventricular filling pressures in a narrow optimum range when patients are severely unstable.

INOTROPIC THERAPY

If heart failure results from cardiac muscle weakness, it might be supposed that strengthening the heart beat would be an efficacious form of therapy. In fact, this is the oldest form of therapy, developed more than two centuries ago, and it remains a part of the standard treatment of heart failure. But as will also be explained, only some positive inotropic agents are likely to have long-term benefits, and more modern developments have reduced the importance of this treatment.

Digitalis

As described in Chapter 17, Withering (1785) introduced foxglove, *Digitalis purpura*, for the treatment of edema, then called **dropsy**, a contraction of the latin term **hydropsy**. Digitalis was first believed to be a diuretic agent, because it relieved edema. Although it was later recognized to act on the heart, rather than the kidneys, the mechanism of its effect on the heart were not understood until the late 1970s. In spite of the lack of understanding, digitalis was the mainstay of therapy for heart failure for nearly 200 years, and for 145 of those years, it was the only effective medicine available.

In the early 1980s, it was suggested that digitalis be retired from the roster of medications routinely used to treat heart failure. This suggestion was made both because effective diuretics had been developed and because high blood levels of digitalis can produce a variety of cardiac arrhythmias. Predictably, this led to a flurry of studies, all showing that digoxin, a modern variant of digitalis, is effective in reducing morbid events. It seems highly unlikely that a therapy proven to be of such great value over two centuries would suddenly be proved to be harmful. On the other hand, the most comprehensive of these studies (Dig Group Investigators, 1997) also showed that the benefits of digoxin were relatively modest in patients receiving other medications. The reduction in its apparent effectiveness may be due less to the benefits of the newer agents and more to the recent development of the serum digoxin assay. Before the introduction of this assay in the late 1970s, the dose of digoxin was increased either until the desired benefit was achieved or until the patient developed an undesirable side effect. In many cases, the blood levels were undoubtedly much higher than would be tolerated today. While these higher blood levels led to greater therapeutic effects, they also led to increased morbidity and mortality caused by arrhythmia and heart block. Thus, digoxin is not now relied upon in the way that it once was.

A major difference between digitalis analogues and most other inotropic agents is that its mechanism of action is completely different from the adrenergic mechanisms that the body uses to increase cardiac output. As explained below, this may be an important distinction.

Other inotropic agents

Several inotropic agents have been developed for clinical use in the past few decades but these agents have not lived up to their initial promise. In several instances, the medications improved symptoms significantly but also increased mortality. As a consequence, the oral forms of these medicines have been withdrawn from use. Nonetheless, some of these agents are still given intravenously to severely ill patients in a closely monitored environment. This is usually done with the recognition that the improved patient comfort is being achieved at the risk of increased mortality.

NEUROHORMONAL THERAPY

Evidence of the altered biology of the failing myocardium was provided by the seminal study of Gaasch et al. (1979), described in Chapter 18. Patients who underwent replacement of diseased aortic valves did well after operation only if they had not developed overt heart failure prior to surgery. Once heart failure supervened, their hearts did not recover, even though the hemodynamic derangement was corrected. A clue to the nature of the biological alteration was provided by Waagstein et al. (1975), although their work was largely ignored at the time. These physicians in northern Sweden had a group of seven patients with tachycardia and severe heart failure secondary to cardiomyopathy. They recognized that the rapid heart rates were likely to be deleterious and therefore gave the patients the only type of medicine available to slow the heart rate, a beta-adrenergic blocking agent. At the time, it was generally believed that such patients were being kept alive by the increased beta-adrenergic tone in their hearts, and that reducing the tone would be likely to increase the heart failure. In spite of this belief, the patients improved with therapy. One patient's ejection fraction increased from a very low value of 22% to a normal value of 58% over a few months! Additional small studies by the same group showed equal promise (Svedberg et al., 1979, 1980a&b), but other clinicians were slow to follow their lead until the 1990s, mainly because of fear of worsening their patients heart failure with agents that decrease the strength of the heart beat. Much greater enthusiasm for beta-blocker therapy has developed more recently, and the acceptance of this therapy probably developed from the experience with **angiotensin converting enzyme (ACE) inhibitors** combined with the experience that patients with elevated levels of circulating adrenaline do not do well.

ACE inhibitors

An influential editorial by Cohn and Franciosa (1977) led to the widespread and successful use of nitroprusside to treat acute pulmonary edema. From this experience it was concluded that reducing blood pressure would improve hemodynamics for patients with heart failure, and a variety of blood pressure lowering agents have been tested. Most blood pressure lowering agents improve symptoms, but only some prolong life. The first intervention shown to decrease mortality significantly was the combination of hydralazine and nitrates, but the difference between treatment and placebo was not statistically significant (Cohn et al., 1986). A significant and substantial prolongation of survival was found when an ACE inhibitor was used (SOLVD Investigators, 1991). Renin is an enzyme secreted by the kidney at a rate inversely related to blood pressure. It catalyzes cleavage of the 14-amino acid peptide **angiotensinogen** in the plasma to yield an inactive decapeptide,

angiotensin-I. This peptide is further cleaved by ACE to yield the active octa-peptide, **angiotensin-II**. This hormone has several actions that support blood pressure, including inhibition of sodium excretion by the kidney and vasoconstriction of peripheral arterioles. In addition, it stimulates tissue growth in blood vessels. ACE inhibitors block both the conversion of angiotensin-I to angiotensin-II and the degradation of another peptide hormone, **bradykinin**. Through these two actions, these ACE inhibitors lower blood pressure. When given to patients with heart failure, they also increase life expectancy by about 20%. In large clinical trials, mean survival for patients with all but the most severe forms of heart failure is about five years in the absence of ACE inhibitors. This is increased to about six years with the use of ACE inhibitors. Other agents, such as alpha-adrenergic blockers and calcium channel blockers that lower blood pressure to the same extent, do not prolong life. Thus, there appears to be a beneficial effect of the ACE inhibitors that is separate from its effect on blood pressure. This is called the **neurohormonal effect**, although the basis for it is unknown. While the term partially disguises our current ignorance, it is justified on the basis of three related observations: 1) patients with high blood levels of adrenaline-like hormones do not survive as long as patients with lower levels; 2) ACE inhibitors lower the levels of these hormones; and 3) drugs that block beta-adrenergic receptors also prolong life.

It should also be mentioned that several other classes of drugs that lower blood pressure have more recently been shown to increase life expectancy. These classes include drugs that block the receptors for angiotensin-II, the combination of hydralazine and nitrates, and spironolactone, as well as the beta-blockers mentioned above and described in greater detail immediately below.

Beta-blocker therapy

A theme running through this discussion is that chronic beta-adrenergic stimulation is detrimental to the heart. Since the pioneering work of Waagstein and Svedberg and their colleagues, described above, multiple additional studies have confirmed their findings. Some of the deleterious effects of beta-stimulation are due to the associated tachycardia, but this is not the complete explanation. Patients placed on beta-blockers not only live longer, but many have reduced symptoms, increased exertional tolerance, and higher ejection fractions, even when they begin therapy with normal heart rates. One possible benefit was demonstrated by the experiments of Tsutsui et al. (1994), described in the last chapter, showing that dogs made to develop heart failure by damaging their mitral valves had more contractile elements per myocyte when they were treated with a beta-blocker. These observations lead to the conclusion that the myocardium is better able to repair itself and grow new contractile elements when adrenergic tone is reduced. This also is not the complete explanation, as shown by recent experiences with a new generation of beta-blocker.

Carvedilol therapy

Carvedilol is an agent which has three separate effects. In addition to being a beta-blocker, it is also an alpha-adrenergic blocker and an anti-oxidant. For reasons that cannot even be guessed, it has produced very dramatic improvements in survival of patients with heart failure. In one trial of over 1,000 patients, it increased survival by 65% (Packer et al., 1996), and in an earlier and smaller trial, it improved survival by 73% (Bristow et al., 1996). In other trials, it has also increased exercise tolerance and improved ejection fractions. In a trial designed to be identical to a concurrent trial of an older beta-blocker, metoprolol, it had much more beneficial effects on both survival and function (Gilbert et al., 1996).

A problem with trying to answer scientific questions from clinical trials is that it is unethical to withhold treatments likely to be beneficial. In every blinded clinical trial there is a "data and safety monitoring board" behind the scenes that is aware of the clinical outcomes to which the treating physicians are blinded.

Not infrequently, these boards recommend that trials be stopped, either when the tested therapy is found to be harmful or when it is found to be uneqivocally beneficial. The largest trial of carvedilol was stopped for the second of these two reasons; the drug was found to be so efficacious that it appeared improper to withhold it from the patients receiving placebo. As a consequence of these ethics, the very strong evidence supporting the use of carvedilol is based on trials of relatively few patients treated over short periods. Studies are currently in progress to confirm the dramatic benefits, to determine whether the benefits of carvedilol add to the benefits of other agents, and to compare carvedilol with at least one other ß-blocker.

Choice of neurohumoral agents

The preceding discussion indicates that a variety of anti-hypertensive agents improve survival and function in patients with heart failure. This raises the question of whether they achieve their benefit through the same mechanisms, or whether the benefits of the different agents are additive. In addition to the scientific issue, the question has practical implications. Because these agents all lower blood pressure, it is often not possible to use them together in the same patient. When the patient's blood pressure will tolerate only one such agent, a choice must be made between among these different drugs. The greater improvement in survival with beta-blockers in general, and carvedilol in particular, suggests that this should probably be the first agent used. Additional support for this policy comes from the finding that the improvements in survival with beta-blockers are substantially better than with ACE inhibitors. Finally, new and as yet unpublished trials, identified by the acronyms ATLAS and HOPE, suggest an algorithm for selecting the optimum combination of antihypertensive agents. The ATLAS trial suggests that about half the benefit of ACE ihibition is achieved with such small doses of the drug that blood pressure is not lowered. The HOPE trial showed that very low doses of

ACE inhibitor, which lowered blood pressure only slightly, caused substantial and significant reductions in heart attacks and strokes. Both studies indicate that the beneficial effects of these agents do not result from the blood pressure reductions themselves but rather from unexplained tissue effects of these agents. One conclusion to be drawn from these findings is that carvedilol should be prescribed to its full recommended level and that an ACE inhibitor should be added as tolerated, with the reecongition that only a small dose is needed to achieve much of the desired effect. The main point to be derived from this discussion, however, is that the findings supporting these forms of therapy are very new and that clinical practice is likely to evolve and improve as experience increases. It also seems likely that new agents, and newer forms of older agents, will improve therapy further.

SUMMARY

The history of the treatment of heart failure can be divided into two parts. The first part began with Withering's 1785 introduction of digitalis. To this mono-therapy was added parenteral mercurial diuretics in the 1920s, oral diuretics in the 1950s, and oral loop diuretics in the 1960s. The additional use of blood pressure lowering agents completed the "triple therapy" of digitalis, diuretics and afterload reduction in what is sometimes called "hemodynamic therapy" because it improves the circulation without a great improvement in the biology of the failing heart.

The biological phase of a treatment truly began with the findings of Waagstein, Svedberg, and their colleagues in the 1970s that beta-adrenergic blockade improved myocardial performance and survival. The general appreciation of these conclusions date from the later findings in the 1990s that ACE inhibitors improve survival. The very recent improvements in survival and function with beta-blockers in general, and carvedilol in particular, is cause for substantial optimism that progress is ongoing and that the best is yet to come.

REFERENCES

ALEXANDER, R.McN. (1977) Allometry in the limb of antelopes (*Bovids*) *J. Zoo. Lond.* **183**: 125–146.

ALEXANDER, R.McN., JAYES, A.S., MALOIY, G.M.O., and WATHUTA, E.M. (1979) Allometry in the limb bones of mammals from shrew (*Sorex*) to elephant (*Loxodonta*) *J. Zool. Lond.* **189**: 291–300.

ALLEN, D.G. and BLINKS, J.R. (1978) Calcium transients in aequorin injected frog cardiac muscle. *Nature.* **273**: 509–513.

ALLEN, D.G., JEWELL, B.R., and MURRAY, J.W. (1974) The contribution of activation processes to the length-tension relation of cardiac muscle. *Nature* **248**: 606–607.

ALLEN, D.G. and KURIHARA, S. (1982) The effects of muscle length on intracelluar calcium transients in mammalian cardiac muscle. *J. Physiol.* **327**: 79–94.

ALLEN, D.G., MORRIS, P.G., ORCHARD, C.H. and PIROLO, J.S. (1985) A nuclear magnetic resonance study of metabolism in the ferret heart during hypoxia and inhibition of glycoloysis. *J. Physiol.* **361**: 185–204.

ALLEN, D.G. and ORCHARD, C.H. (1983) Intracellular calcium concentration during hypoxia and metabolic inhibition in mammalian ventricular muscle. *J. Physiol.* **339**: 107–122.

ALWAY, S.E., GRUMBT, W.H., STRAY-GUNDERSEN, J., and GONYEA, W.J. (1992) Effects of resistance training on elbow flexors of highly competitive bodybuilders. *J. Appl. Physiol.* **72**: 1512–1521.

ARCILLA, R.A., TSAI, P., THILENIUS, O., and RANNIGER, K. (1971) Angiographic method for evaluation of right and left ventricles. *Chest* **60**: 446–454

ARTS, T. (1979) A model of the mechanics of the left ventricle. *Ann. Biomed. Eng.* **7**: 299–315.

BAGSHAW, C. R., TRENTHAM, D. R,. WOLCOTT, R. G., and BOYER, P. D. (1975) Oxygen exchange in the gamma-phosphoryl group of protein-bound ATP during Mg^{2+}-dependent adenosine triphosphatase activity in myosin. *Proc. Nat'l Acad. Sci. U.S.A.* **72**: 2592–2596.

BAKER, P.F., BLAUSTEIN, M.P., HODGKIN, A.L., and STEINHARDT, R.A. (1969) The influence of calcium on sodium efflux in squid axons. *J. Physiol.* **200**: 431–458.

BAIRD, R.J., MANKTELOW, R.T., SHAH, P.A., and AMEL, F.M. (1970) Intramyocardial pressure: a study of its regional variations and its relation to intraventricular pressure. *J. Thoracic Cardiovasc. Surg.* **59**: 810–823.

BÁRÁNY, M., HEGEDÜS, L., and BÁRÁNY, K. (1994) Dissociation of relaxation and myosin light chain dephosphorylation in smooth muscle. *Biophys. J.* **66**: A139.

BAYLOR, S.M. and OERTLIKER, H. (1975) Birefringence experiments on isolated skeletal muscle fibres suggest a possible signal from the sarcoplasmic reticulum. *Nature* **253**: 97–101.

BEAM, K.G., KNUDSON, C.M., and POWELL, J.A. (1968) A lethal mutation in mice eliminates the slow calcium current in skeletal muscle cells. *Nature.* **320**: 168–170.

BELTRAME, J.F., ZEITZ, C.J. UNGER, S.A., et al. (1998) Nitrate Therapy is an alternative to furosemide/morphine therapy in the management of acute cardiac pulmonary edema. *J. Cardiac Failure* **4**: 271–279

BERMAN, M.R., PETERSON, J.N., YUE, D.T., and HUNTER, W.C. (1988) Effect of isoproterenol on force transient time course and on stiffness spectra in rabbit papillary muscle barium contracture. *J. Mol. Cell. Cardiol.* **20**: 415–426.

BEZANILLA, F. and HOROWICZ, P. (1975) Fluorescence intensity changes associated with contractile activation in frog muscle stained with nile blue A. *J. Physiol.* **246**: 709–735.

BIEWENER, A.A., ALEXANDER, R.McN., and HEGLUND, N.C. (1981) Elastic storage in the hopping kangaroo rat (*Depodomys spectabilis*) *J. Zool. Lond.* **195**: 369–383.

BLINKS, J.R. and ENDOH, M. (1986) Modification of myofibrillar responsiveness to Ca^{++} as an inotropic mechanism. *Circulation.* **73**(Suppl. III): 85–98.

BLIX, M. (1895) Die Länge und die Spannung des Muskels. *Skand. Arch. Physiol.* **5**: 150–206.

BRADY, A.J. (1965) The onset of contractility in cardiac muscle. *J. Physiol.* **184**: 560–580.

BRANDT, P.W., DIAMOND, M.S., and SCHACHAT, F.H. (1984) The thin filament of vertebrate skeletal muscle co-operatively activates as a unit. *J. Mol. Biol.* **180**: 379–384.

BREMEL, R.D. and WEBER, A. (1972) A cooperation within actin filament in vertebrate skeletal muscle. *Nature New Biol.* **238**: 97–101.

BRISTOW, M.R., GILBERT, E.M., ABRAHAM, W.T., et al. (1996) Carvedilol produces dose-related improvements in left ventricular function and survival in subjects with chronic heart failure. *Circulation* **94**: 2807–2816.

BUCHTHAL, F. and STEN-KNUDSEN, O. (1959) Impulse propogation in striated muscle fibers and the role of the internal currents in activation. *Ann. New York Acad. Sci.* **81**: 422–433.

CANNON, S.R., RICHARDS, K.L., and CRAWFORD, M. (1985) Hydraulic estimation of stenotic orifice area: a correction of the Gorlin formula. *Circulation* **71**: 1170–1178.

CECCHI, G., GRIFFITHS, P.J., and TAYLOR, S.R. (1982) Muscular Contraction: kinetics of crossbridge attachment studied by high frequency stiffness measurements. *Science.* **217**: 70–72.

CHANDLER, W.K., RAKOWSKI, R.E., and SCHNEIDER, M.F. (1976a) A non-linear voltage dependent charge movement in frog skeletal skeletal muscle. *J. Physiol.* **254**: 245–283.

CHANDLER, W.K., RAKOWSKI, R.E., and SCHNEIDER, M.F. (1976) Effects of glycerol treatment and maintained depolarization on charge movement in skeletal muscle. *J. Physiol.* **254**: 285–316.

CHAPMAN, C. B. and WASSERMAN E. (1959) *Am. Heart J.* **58**: 282–317, 467–478.

CHIU, Y.-L., BALLOU, E.W., and FORD, L.E. (1987) Force, velocity, and power changes during normal and potentiated contractiions of cat papillary muscle. *Circulation Res.* **60**: 446–458.

CHIU, Y.C., WALLEY, K.R., and FORD, L.E. (1989) Comparison of the effects of different inotropic interventions on force, velocity, and power in rabbit myocardium. *Circulation Res.* **65**: 1161–1171.

CIVAN, M.M. and PODOLSKY, R.J. (1966) Contraction kinetics of striated muscle fibres following quick changes in load. *J. Physiol.* **184**: 511–534.

CLARK, A.J. (1928) *Comparative Physiology of the heart.* Cambridge Univ. Press, New York.

CLOSE, R. (1972) Dynamic properties of mammalian skeletal muscles. *Physiol. Rev.* **52**: 129–197.

COHN, J.N. and FRANCIOSA, J.A. (1971) Vasodilator therapy of heart failure. *New England J. Med.* **297**: 312–316.

COHN, J.N., ARCHIBALD, D.G., ZIESCHE, S., et al. (1986) Effect of vasodilator therapy on mortality in chronic heart failure: results of a veterans administration cooperative study. *New England J. Med.* **314**: 1547–1552

COHN, J.N. and FRANCIOSA, J.A. (1977) Vasodilator therapy of cardiac failure. *New England J. Med.* **297**: 27–31, 254–258.

COOKE, P.H., FAY, F.S., and CRAIG, R. (1990) Myosin filaments isolated from skinned amphibian smooth muscle cells are side polar. *J. Muscle Res. Cell Motil.* **10**: 206–220.

Costantin, L.L. (1974) Contractile activation of frog skeletal muscle. *J. Gen. Physiol.* **63**: 657–674.

Costantin, L.L. and Podolsky, R.J. (1965) Calcium localization and the activation of striated muscle fibers. *Federation Proc.* **24**: 1141–1145.

Costantin, L.L., Franzini-Armstrong, C., and Podolsky, R.J. (1965) Localization of calcium-accumulating structures in striated muscle fibers. *Science.* **147**: 158–160.

Cunningham, J.J. (1980) A reanalysis of the factors influencing basal metaolic rate in normal adults. *Am. J. Clin. Nutr.* **33**: 2372–2374.

deTombe, P.P., and terKeurs, H.E.D.J. (1991) Lack of effect of isoproterenol on unloaded velocity of sarcomere shortening in rat cardiac trabeculae. *Circulation Res.* **68**: 382–391.

Digitalis Investigation Group (1997) The effect of digoxin on mortality and morbidity in patients with heart failure. *New England J. Med.* **336**: 525–533.

Dillon, P.F., Askoy, M.O., Driska, S.P. and Murphy, R.A. (1981) Myosin phosphorylation and the cross-bridge cycle in arterial smooth. *Science.* **211**: 495–497.

Dow, J.W., Levine, H.D., Elkin, M., et al. (1950) Studies of congenital heart disease, IV: uncomplicated pulmonic stenosis. *Circulation* **1**: 267–282.

Dunham, E.T. and Glynn, I.M. (1961) Adenosinetriphosphatase activity and the active movements of alkali metal ions. *J. Physiol.* **156**: 274–293

Ebashi, S. and Ebashi, F. (1964) A new protein component participating in the superprecipitation of myosin. *J. Biochem.* **55**: 604–613.

Ebashi, S. and Kodama, A. (1965) A new protein factor promoting aggregation of tropomyosin. *J. Biochem.* **58**: 107–108.

Ebashi, S. and Lipmann, F. (1962) Adenosine triphosphate linked concentration of calcium ions in a particulate fraction of rabbit msucle. *J. cell Biol.* **14**: 389–400.

Eggleton, P. and Eggleton, G.P. (1927) The inorganic phosphate and labile form of organic phosphate in the gastrocnemius of the frog. *Biochem. J.* **21**: 190–195.

Eisenberg, R.S. and Gage, P.W. (1967) Frog skeletal muscle fibers: changes in electrical properties after disruption of transverse tubular system. *Science.* **158**: 1700–1701.

Endo, M. (1964) Entry of a dye into the sarcotubular system of muscle. *Nature.* **202**: 1115–1116.

ENDO, M. (1966) Entry of fluorescent dyes into the sarcotubular system of frog muscle. *J. Physiol.* **185**: 224–238.

ENDO, M. (1973) Length dependence of activation of skinned muscle fibers by calcium. *1972 Cold Spring Harbor Symp. Quant. Biol.* **37**: 505–510.

ENDO, M., TANAKA, M., and OGOWA, Y. (1970) Calcium induced release of calcium from the sarcoplasmic reticulum of skinned skeletal muscle fibres. *Nature.* **228**: 34–36.

FABIATO, A. and FABIATO, F. (1978) Contraction induced by a calcium-triggered release of calcium from the sarcoplasmic reticulum of skinned cardiac cells. *J. Physiol.* **249**: 497–515.

FABIATO, A and FABIATO, F. (1978) Effects of pH on the myofilaments and sarcomplasmic reticulum of skinned cells from cardiac and skeletal muscles. *J. Physiol.* **276**: 233–255.

FAY, F.S., FUJIWARA, K., REES, D.D., and FOGARTY, K.E. (1983) Distribution of alpha-actinin in single isolated smooth muscle cells. *J. Cell Biol.* **96**: 783–795.

FALLS J. (1977) The Boston marathon. (MacMillan, New York) pp163–165

FELDMAN, T., FORD, L.E., CHIU, Y.C., and CARROLL, J.D. (1992) Changes in valvular resistance, power dissipation and myocardial reserve with aortic valvuloplasty. *J. Heart Valve Dis.* **1**: 55–64

FERENCZI, M.A., GOLDMAN, Y.E., and SIMMONS, R.M. (1984) The dependence of force and shortening velocity on substrate concentration in skinned muscle fibres from Rana temporaria. *J. Physiol.* **350**: 519–543.

FINER, J.T., SIMMONS, R.M., and SPUDICH, J.A. (1994) Single myosin molecule mechanics: piconewton forces and nanometer steps. *Nature.* **37**: 98–99.

FISKE, C.H. and SUBAROW, Y. (1927) The nature of the "inorganic" phosphate in voluntary muscle. *Science.* **65**: 401–403.

FORD, L.E. (1976) Heart Size. *Circulation Res.* **39**: 297–303.

FORD, L.E. (1981) Effect of afterload reduction on myocardial energetics. *Circulation Res.* **46**: 161–166.

FORD, L.E. (1984) Some consequences of body size. *American J. Physiol.* **247**: H495–H507.

FORD, L.E. (1991) Mechanical manifestations of activation in cardiac muscle. *Circulation Res.* **68**: 621–637.

FORD, L.E., DETTERLINE, A.J., HO, K.K., and CAO, W. (2000) Gender and height-related dependent limits of muscle strength in world weightlifting champions. *Appl. Physiol.* **87** (in press).

FORD, L.E., FELDMAN, T., CHIU, Y.C., and CARROLL, J.D. (1990) Hemodynamic resistance as a measure of functional impairment in aortic stenosis. *Circulation Res.* **66**: 1–7.

FORD, L.E. and GILBERT, S.H. (1986) The thermoelastic effect in rigor muscle of the frog. *J. Muscle Res. Cell Motil.* **7**: 35–46.

FORD, L. E., HUXLEY, A.F., and SIMMONS, R.M. (1977) Tension response to sudden length changes in stimulated frog muscle fibres near slack length. *J. Physiol.* 269:441–515.

FORD, L. E., HUXLEY, A.F., and SIMMONS, R.M. (1981) The relation between stiffness and filament overlap in stimulated frog muscle fibres. *J. Physiol.* 311:219–249.

FORD, L.E., HUXLEY, A.F., and SIMMONS, R.M. (1985) Tension transients during steady shortening of frog muscle fibres. *J. Physiol.* **361**: 131–150.

FORD, L.E., HUXLEY, A.F., and SIMMONS, R.M. (1986) Tension transients during the rise of tetanic tension in frog muscle fibres. *J. Physiol.* **372**: 595–609.

FORD, L.E. and PODOLSKY, R.J. (1970) Regenerative calcium release within muscle cells. *Science.* **167**: 58–59.

FORD, L.E. and PODOLSKY, R.J. (1972a) Calcium uptake and force development by skinned muscle fibres in EGTA buffered solutions. *J. Physiol.* **223**: 1–19.

FORD, L.E. and PODOLSKY, R.J. (1972b) Intracellular calcium movements in skinned muscle fibres. *J. Physiol.* **223**: 21–33.

FORMAN, R., FORD, L.E., and SONNENBLICK, E.H. (1972) Effect of muscle length on the force velocity relationship in tetanized cardiac muscle. *Circ. Res.* **31**: 195–206.

FRANK, O. (1895) Zur Dynamic des Herzmuskels. *Z. Biol.* **32**: 370–477. Translated by Chapman, C. B. and Wasserman E. (1959) *Am. Heart J.* **58**: 282–317, 467–478.

FREARSON, N. and PERRY, S.V. (1975) Phosphorylation of light-chain components of myosin from cardiac and red skeletal muscles. *Biochem. J.* **151**: 99–107.

GAASCH, W.H., ANDRIAS, C.W., and LEVINE, H.J. (1978) Chronic aortic regurgitation: the effect of aortic valve replacement on left ventricular volume, mass and function. *Circulation.* **58**: 825–836.

GERTHOFFER, W.T. (1986) Calcium dependence of myosin phosphorylation and airway smooth muscle contraction and relaxation. *Am. J. Physiol.* **250**: C597–C604.

GILBERT, E.M., ABRAHAM, W.T., OLSEN, S., et al. (1996) Comparative hemodynamic, left ventricular functional, and antiadrenergic effects of chronic treatment with metoprolol versus carvedilol in the failing heart. *Circulation* **94**: 2817–2825.

GILLIS, J.M., CAO, M.L. and GODFRAIN-DEBECKER, A. (1988) Density of myosin filaments in the rat anococcygeus muscle, at rest and in contraction. II. *J. Muscle Res. Cell Motil.* **9**: 18–29.

GODFRAIN-DEBECKER, A. and GILLIS, J.M. (1988) Analysis of the birefringence of the smooth muscle anococcygeus of the rate at rest and in contraction. I. *J. Muscle Res. Cell Motil.* **9**: 9–17.

GODT, R. E. (1974) Calcium-activated tension of skinned muscle fibers of the frog. *J. Gen. Physiol.* 63:722–739.

GODT, R.E. and MAUGHAN, D.W. (1976) Swelling of skinned muscle fibers of the frog: experimental observations. *Biophys. J.* **19**: 103–116.

GOLDMAN, Y. E. (1987) Measurement of sarcomere shortening in skinned fibers from frog muscle by white light diffraction. *Biophys. J.* 52:57–68.

GONZALEZ-SERRATOS, H. (1971) Inward spread of activation in vertebrate muscle fibres. *J. Physiol.* **212**: 777–799.

GORDON, A.M., HUXLEY, A.F., and JULIAN, F.J. (1966) The variation in isometric tension with sarcomere length in vertebrate muscle fibres. *J. Physiol.* **184**: 170–192.

GORDON, A.M. and RIDGWAY, E.B. (1978) Calcium transients and relaxation in single muscle fibers. *Eur. J. Physiol.* **7**: 27–34.

GORLIN, R. and GORLIN, S.G. (1951) Hydraulic formula for calculation of the area of the stenotic mitral valve, other cardiac valves and central circulatory shunts. *Am Heart J.* **41**: 1–29.

GRIMM, A.F., LIN, H.-L., and GRIMM, B.R. (1980) Left ventricular free wall and intraventricular pressure-sarcomere length distributions. *Am. J. Physiol.* **239**: H101-H107.

GROSSMAN, W., JONES, D., and McLAURIN, L.P. (1975) Wall stress and patterns of hypertrophy in the human left ventricle. *J. Clin. Invest.* **56**: 56–64.

HALES, S. (1733) Statical essays: containing Hemastatics. (Innys and Manby, London. Reprinted by The Classics of Medicine Library, Birmingam, AL).

HARRIS, J.A. and BENDICT, F.G. (1919) *A biometric study of basal metabolism in man.* Carnegie Inst., Washington, D.C.

HARVEY, W. (1628) Excitatio anatomica de moto cordis et sanguines in anim alibus. Translated 1928 by G. Keynes (Nonesuch press, London. Reprinted 1977 by The Classics of Medicine Library, Birmingham, AL).

HAWKING, S.W. (1988) *A brief history of time.* (Bantam Books, N.Y.)

HEILBRUN, L.V. (1940) The action of calcium on muscle protoplasm. *Physiol. Zool.* **13**: 88–94.

HEILBRUN, L.V. and WIERCINSKI, F.J. (1947) The action of various cations on muscle protoplasm. *J. Cell. Comp. Physiol.* **29**: 15–32.

HELLAM, D.C. and PODOLSKY, R.J. (1969) Force measurements in skinned muscle fibres. *J. Physiol.* **200**: 807–819.

HILL, A.V. (1938) The heat of shortening and the dynamic constants of muscle. *Proc. Royal Soc. Lond. B.* **126**:136–195.

HILL, A.V. (1949a) The abrupt transition from rest to activity in muscle. *Proc. Royal Soc. Lond. B* **136**: 399–420.

HILL, A.V. (1949b) The dimensions of animals and their muscular dynamics. *Proc. Roy. Inst. GB.* **34**: 450–471.

HILL, A.V. (1951) The physical analysis of events in muscular contraction. *Rev. Can. Biol.* **10**: 103–118.

HILL, A.V. (1964) The effect of load on the heat of shortening muscle. *Proc. Royal Soc. Lond. B.* **159**: 297–318.

HILL, L. (1977) A-band length, striation spacing, and tension change on stretch of active muscle. *J. Physiol.* **266**: 677–685.

HODGKIN, A.L., and HOROWICZ, P. (1960) Potassium contractures in single muscle fibres. *J. physiol.* **153**: 386–403.

HOFFMAN, M., PFEIFER, W.A., GUNDLACH, B.L., et al. (1979) Resting metabolic rate in obese and normal weight women. *Int. J. Obesity* **3**: 111–118.

HOH, J.F.Y., ROSSMANITH, G.H., KWAN, L.J., and HAMILTON, A.M. (1987) Adrenaline increases the rate of cycling of crossbridges in rat cardiac muscle as measured by pseudo-random binary noise-modulated perturbation analysis. *Circulation Res.* **62**: 452–461.

HUXLEY, A.F. (1957) Muscle structure and theories of contraction. *Prog. Biophys. Biophys. Chem.* 7:255–318.

HUXLEY, A.F. (1965) Muscle tension on the sliding filament theory. *Proc. XXIIIrd Int. Cong. Physiol. Sci.* 383–387.

HUXLEY, A. F. (1973) A note suggesting that the cross-bridge attachment during muscle contraction may take place in two stages. *Proc. Royal Soc. Lond. B.* 183:83–86.

HUXLEY, A.F. (1980) *Reflections on Muscle.* (Princeton Univ. Press, Princeton, NJ)

HUXLEY, A.F. and NIEDERGERKE, R. (1954) Interference microscopy of living muscle fibers. *Nature.* **173**: 971–973.

HUXLEY, A.F. and SIMMONS, R.M. (1971) Proposed mechanism for force generation in striated muscle. *Nature* **233**: 533–538.

HUXLEY, A.F. and SIMMONS, R.M. (1973) Mechanical transients and the origin of muscular force. *1973 Cold Spring Harbor Symp. Quant. Biol.* 37:669–680.

HUXLEY, A.F. and TAYLOR, R.E. (1955) Function of Krause's membrane. *Nature.* **176**: 1068.

HUXLEY, A.F. and TAYLOR, R.E. (1958) Local activtion of striated muscle fibres. *J. Physiol.* **144**: 426–441.

HUXLEY, H.E. (1964) Evidence for the continuity between the central elements of the triads and extracellular space in frog sartorius muscle. *Nature.* **202**: 1067–1071.

HUXLEY, H.E. (1969) The mechanisms of muscular contraction. *Science* **164**: 1356–1366.

HUXLEY, H.E. (1973) Structural changes in the actin- and myosin-containing filaments during contraction. *1972 Cold Springs Harbor Symp. Quant. Biol.* **37**: 361–376.

HUXLEY, H.E. and HANSEN, J. (1953) Structural basis of the cross-striations in muscle. *Nature.* **172**: 530–532.

HUXLEY, H.E. and HANSEN, J. (1954) Changes in the cross-striations of muscle during contraction and stretch and their structural interpretation. *Nature.* **173**: 973–976.

JEWELL, B.R. and WILKIE, D.R. (1958) An analysis of the mechanical components in frog's striated muscle. *J. Physiol.* **143**:515–540.

JÖBSIS, F.F. and O'CONNOR, M.J. (1966) Calcium release and reabsorption in the sartorius muscle of the toad. *Biochem. Biophys. Res. Commun.* **25**: 246–252.

KATZ, A. M. (1965) The descending limb of the Starling curve and the failing heart. *Circulation.* **32**: 871–875.

KATZ, A. M. (1992) *Physiology of the Heart, 2nd Edition.* (Raven Press, Ltd., NY)

KATZ, A. M. and HECHT, H.A. (1969) The early pump failure of the ischemic heart. *Am. J. Med.* **47**: 497–502.

KATZ, B. (1985) Reminiscences of a physiologist, 50 years after. *J. Physiol.* **370**: 1–12.

KEILEY, W.W. and MEYERHOFF, O. (1948a) A new magnesium- activated adenosinetriphosphatase from muscle. *J. Biol. Chem.* **174**: 387–388.

KEILLEY W.W. and MEYERHOFF, O. (1948b) Studies on adenosine-triphosphatase in muscle. II. A new magnesium activated adenosinetriphosphatase. *J. Biol. Chem.* **176**: 591–601.

KELLY, R.E. and RICE, R.V. (1968) Localization of myosin filaments in smooth muscle. *J. Cell Biol.* **37**: 105–116.

KEYS, A.H., TAYLOR, H.L., and GRANDE, F. (1973) Basal metabolism and age of adult man. *Metabolism* **22**: 579–587.

KLEIBER, M. (1932) Body size and metabolism. *Hilgardia* **6**: 315–353.

KLEIBER, M. (1947) Body size and metabolic rate. *Physiol. Rev.* **27**: 511–541.

KUSHMERICK, M.J and DAVIES, R.E. (1969) The chemical energetics of muscle contraction, II: the chemistry, efficiency, and power of maximally working sartorius muscle. *Proc. Royal Soc. Lond. B*: **174**: 315–353.

LEVY, M.N. (1979) The cardiac and vascular factors that determine systemic blood flow. *Circulation Res.* **44**: 739–747

LIETZKE, M.H. (1956) Relation between weight-lifting totals and body weight. *Science* **124**: 486–487.

LINZBACH, A.J. (1960) Heart failure from the point of view of quantitative anatomy. *Am. J. Cardiol.* **5**: 370–382.

LOHMANN, K. (1929) Uber die Pyrophosphatfraktion im Muskel. *Naturwissensch.* **17**: 624–625.

LOMBARDI, V., PIAZZESI, G., and LINARI, M. (1992) Rapid regeneration of the actin-myosin power stroke in contracting muscle. *Nature.* **355**: 638–641.

LYMN R. W., and E. W. TAYLOR. (1971) Mechanism of adenosine triphosphate hydrolysis by actomyosin. *Biochem.* 10:4617–4624.

MACPHEARSON, L. (1953) A method for determining the force-velocity relation of muscle from two isometric contractions. *J. Physiol.* **122**: 172–177.

MARSH, B.B. (1952) The effects of on the fibre volume of muscle homogenate. *Biochim. Biophys. Acta.* **9**: 247–260.

MATSUBARA, I. and ELLIOTT, G.F. (1972) X-ray diffraction studies on skinned single fibres of frog skeletal muscle. *J. Mol. Biol.* **72**: 657–669.

MCCLARE, C.W.F. (1971) Chemical machines, Maxwell's demon and living cells. *J. Theor. Biol.* **30**: 1–34.

MCCLELLAN G.B. and WINEGRAD, S. (1978) The regulation of the calcium sensitivity of the contractile system in mammalian cardiac muscle. *J. Gen. Physiol.* **72**: 737–764.

MCDANIEL, N.L., CHEN, X.-L., SINGER, H.A., et al. (1992) Nitrovasodilators relax arterial smooth muscle by decreasing $[Ca^{2+}]_i$ and uncoupling stress from myosin phosphorylation. *Am. J. Physiol.* **263**: C461-C467.

MCKUSICK, V.A. (1972) *Heritable disorders of connective tissue, 4th ed.* (C.V. Mosby Co., St. Louis).

MCMAHON, T.A. (1973) Size and shape in biology. *Science* **179**: 1201–1204.

MCMAHON, T.A. (1975a) Allometry and biomechanics: limb bones in adult ungulates. *Am. Naturalist* **109**: 547–563.

McMahon, T.A. (1975b) Using body size to understand the structural design of animals: quadrupedal locomotion. *J. Appl. Physiol.* **39**: 619–627.

McMahon, T.A. (1984) *Muscles, reflexes, and locomotion.* (Princeton Univ. Press., Princeton, N.J.)

Metzger, J.M., Greaser, M.L., and Moss, R.L. (1989) Variations in cross-bridge attachment rate and tension with phosphorylation of myosin in mammalian skinned skeletal muscle fibers. *J. Gen. Physiol.* **93**: 855–883.

Meyerhoff, O and Lohmann, K. (1932) Uber energetische Wechsel-beziehungen zwischen dem Umsatz der Phosphrsaureester im Muskelextrakt. *Biochem. Z.* **253**: 431–461.

Miller, A.T. and Blythe, C.S. (1953) Lean body mass as a metabolic reference standard. *J. Appl. Physiol.* **5**: 311–316.

Mope, L., McClellan, G.B., and Winegrad, S. (1980) Calcium sensitivity of the contractile system and phosphorylation of troponin in hyperpermeable cardiac cells. *J. Gen. Physiol.* **75**: 271–282.

Moss, R.L., Allen, J.D., and Greaser, M.L. (1986) effects of partial extraction of troponin complex upon the tension-pCa relation in rabbit skeletal muscle: further evidence that tension development involves cooperative effects within the thin filament. *J. Gen. Physiol.* **87**: 761–774.

Moss, R.L., Giulian, G.G., and Greaser, M.L. (1985) The effects of partial extraction of TnC upon the tension-pCa relationship in rabbit skinned skeletal muscle fibers. *J. Gen. Physiol.* **86**: 585–600.

Naijer, M.F. and Rowland, M., National Center for Health statistics. (1998) Anthropomorphic reference data and prevalence of overweight, United States, 1976–1980. *Vital and Health Statistics.* Series 11, No. 238. DHHS Pub. No (PHS) 87–1988.

Nathan, D. and Beeler, G.W. (1975) Electrophysiologic correlates of the inotropic effects of isoproteronol in canine myocardium. *J. Mol. Cell. Cardiol.* **7**: 1–15.

Natori, R. (1954) The property and contraction process of isolated myofibrils. *Jikeikai Med. J.* **1**: 119–126.

Needham, D.M. (1971) *Machina Carnis.* (Cambridge University Press, Cambridge, U.K.)

O'Rourke, M.F. (1967) Pressure and flow waves in systemic arteries and the anatomical design of the arterial system. *J. Appl. Physiol.* **23**: 139–149.

Parmley, W.W. and Chuck, L.H. (1973) Length-dependent changes in myocardial contractile state. *Am. J. Physiol.* **224**: 1195–1199.

PATTERSON, S. W., PIPER, H., and STARLING, E. H. (1914) The regulation of the heart beat. *J. Physiol.* **48**: 465–513.

PEACHEY, L.D. (1965) The sarcoplasmic reticulum and transverse tubules of frog sartorius. *J. Cell Biol.* **25**: 209–231.

PELLICCIA, A., MARON, B.J., SPATARO, A., et al. (1991) The upper lilmit of physiologic cardiac hypertrophy in highly trained elite athletes. *New England J. Med.* **324**: 295–301.

PITT, B., SEGAL, R., MARTINEZ, F.A., et al. (1997) Randomized trial of losartan versus captopril in patients over 65 with heart failure. *Lancet* **349**: 747–752.

PODOLIN, R.A. and FORD, L.E. (1986) Influence of partial activation on force-velocity properties of frog skinned muscle fibers in millilmolar magnesium ion. *J. Gen. Physiol.* **87**: 607–631.

PODOLKY, R.J. (1960) Kinetics and muscular contraction: the approach to the steady state. *Nature* **188**: 666–668.

PODOLSKY, R.J. and HUBERT, C.E. (1961) Activation of the contractile mechanism in isolated myofibrils. *Federation Proc.* **20**: 301.

POLINER, L.R., DEHMER, G.J. LEWIS, S.E. et al. (1980) Left ventricular performance in normal subjects: a comparison of responses to exercise in the upright and supine positions. *Circulation* **62**: 528–534.

PRATUSEVICH, V.R., SEOW, C.Y., and FORD, L.E. (1995a) Plasticity in canine airway smooth muscle. *J. Gen. Physiol.* **105**: 73–94.

PRATUSEVICH, V.R., SEOW, C.Y., and FORD, L.E. (1995b) A series-to-parallel transition during the rise of activation of smooth muscle. *Biophys. J.* **68**: A16.

RACK, P.M.H. and WESTBURY, D.R. (1969) The effects of length and stimulus rate on tension in the isometric cat soleus muscle. *J. Physiol.* **204**: 443–460.

RAMSEY, R.W. and STREET, S.F. (1940) The isometric length-tension diagram of isolated skeletal msucle fibers of the frog. *J. Cell. Comp. Phyiol.* **15**: 11–34.

REUTER, H. and SEITZ, N (1968) The dependence of calcium efflux from cardiac muscle on temperature and external ion composition. *J. Physiol.* **195**: 451–470.

RICE, R.V., MOSES, J.A. MCMANUS, G.M., BRADY, A.C., and BLASIK, I.M. (1970) The organization of contractile filaments in a mammalian smooth muscle. *J. Cell Biol.* **47**: 183–196.

RIDGWAY, E.B and ASHLEY, C.C. (1967) Calcium transients in single muscle fibers. *Biochem. Biophys. Res. Comm.* **29**: 229–234.

ROBERTSON, J.D. (1956) Some features of the ultrastructure of reptilian skeletal muscle. *J. Biophys. Biochem. Cytol.* **2**: 369–380.

RÜDEL, R and TAYLOR, S.R. (1971) Striated muscle fibers: fascilitation of contraction at short lengths by caffeine. *Science.* **172**:387–388.

SATO, T.G. (1954) Volume change of a muscle fiber on tetanic contraction. *Annotationes Zoologicae Japonenses* **27**: 165–172.

SARNOFF, S.J. and BERGLUND, E. (1954) Ventricular function. I. Starling's law of the heart studied by means of simultaneous right and left ventricular function curves in the dog. *Circulation.* **9**: 706–718.

SCHNEIDER, M.F. and CHANDLER, W.K. (1973) Voltage depenent charge movement in skeletal muscle: a possible step in excitation-contraction couplilng. *Nature.* **242**: 244–246.

SCHOENBERG, C.F. (1969) An electronmicroscope study of the influence of divalent ions on myosin filament formation in chicken gizzard extracts and homogenate. Tissue Cell 1:83–96.

SEOW C.Y. and FORD, L.E. (1991) Shortening velocity and power output of skinned muscle fibers from mammals having a 25,000-fold range of body mass. *J. Gen. Physiol.* **97**: 541–560.

SEOW C.Y. and FORD, L.E. (1993) High ionic strength and low pH detain activated skinned rabbit skeletal muscle crossbridges in a low force state. *J. Gen. Physiol.* **101**: 487–511.

SHEU, S.S. and FOZZARD, H.A. (1982) Transmembrane Na^+ and Ca^{2+} electrochemical gradients and cardiac muscle and their relationship to force development. *J. Gen. Physiol.* **80**: 325–351.

SHIBATA, T., BERMAN, M.R., HUNTER, W.C., and JACOBUS, W.E. (1990) Metabolic and functional consequences of barium-induced contracture in rabbit myocardium. *Am. J. Physiol.* **259**: H1566-H1574.

SHIMOMURA, O., JOHNSON, F.H., and SAIGA, Y. (1962) Extraction, purification and properties of aequorin, a bioluminescent protein from the luminous hydromedusan, aequorea. *J. Cell. Comp. Physiol.* **59**: 223–239.

SILBER, E.N., PREC, O., GROSSMAN, N, KATZ, L.N. (1951) Dynamics of isolated pulmonary stenosis. *Am. J. Med.* **10**: 21–26.

S0LVD INVESTIGATORS (1991) Effect of enalapril on survival in patients with reduced left ventricular ejection fractions and congestive heart failure. *New England J. Med.* **325**: 293–302.

SOMLYO, A.P. and SOMLYO, A.V. (1982) Smooth muscle structure and function. In *The Heart and Cardiovascular System* (ed. Fozzard, H.A.) 1295–1324 (The Raven Press, N.Y.).

STAHL, W.R. and GUMMERSON, Y.Y. (1967) Systematic allometry in five species of adult primates. *Growth* **31**: 453–460.

STARLING, E. H. (1918) The Linacre Lecture 1915: The law of the heart. (Longmans Greene, London). Reproduced in Chapman C. B. and Mitchell J. H. (1965) *Starling on the heart*. (Dawsons of Pall Mall, London).

STARLING, E. H. (1920) On the circulatory changes associated with exercise. *J. Roy. Army Med. Corps* **34**: 258–272.

SVEDBERG, K., HJALMARSON, A., WAAGSTEIN, F., et al. (1979) Prolongation of survival in congestive cardiomyopathy by beta-receptor blockade. *Lancet* **1**: 1374–1376.

SVEDBERG, K., HJALMARSON, A., WAAGSTEIN, F., et al. (1980) Beneficial effects of long-term beta-blockade in congestive cardiomyopathy. *British Heart J.* **44**: 117–133.

SVEDBERG, K., HJALMARSON, A., WAAGSTEIN, F., et al. (1980) Adverse effects of beta-blockade withdrawal in patients with congestive cardiomyopathy. *British Heart J.* **44**: 134–142.

SZENT-GYÖRGYI, A. (1949) Free-energy relations and contraction of actomyosin. *Biol. Bull.* **96**: 140–161.

TADA, M., KIRCHBERGER, M.A., REPKE, D.I., and KATZ, A.M. (1974) The stimulation of calcium transport in cardiac sarcoplasmic reticulum by adenosine 3':5'-monophosphate-dependent protein kinase. *J. Biol. Chem.* **249**: 6174–6180.

TANABE, T. BEAM, K.G., POWELL, J.A., and NUMA, S. (1988) Restoration of excitation-contraction coupling and slow calcium currents in dysgenic muscle by dihydropyridine receptor complementary DNA. *Nature.* **336**: 134–139.

TANABE, T., BEAM, K., ADAMS, B.A., NIIDOME, T., and NUMA, S. (1990) Regions of the skeletal muscle dihydropyridine receptor critical for excitation-contraction coupling. *Nature.* **346**: 567–569.

TAYLOR, N.A. and WILKINSON, J.G. (1986) Exercised induced skeletal muscle growth. Hypertrophy or hyperplasia? *Sports Med.* **3**: 190–200TAYLOR, S.R. and RÜDEL, R. (1970) Striated muscle fibers: inactivation of contraction induced by shortening. *Science.* **167**: 882–884.

TAYLOR, S.R. and RÜDEL, R. (1970) Striated muscle fibers: inactivation of contraction induced by shortening. *Science.* **167**: 882–884.

TOYOSHIMA, Y.Y., KRON, S.J., MCNALLY, E.M., NIEBLIN, K.A., TOYOSHIMA, C., SPUDICH, J.A. (1987) Myosin subfragment-1 is sufficient to move actin filaments *in vitro*. *Nature.* **328**: 536–539.

TSUTSUI, H., ISHIHARA, K., and COOPER, G., I.V. (1993) Cytoskeletal role in the contractile dysfunction of hypertrophied myocardium. *Science.* **260**: 682–687.

TSUTSUI, H. SPINALE, F.G., NAGATSU, M., et al., (1994) Effects of chronic beta-adrenergic blockade on the left ventricular and cardiocyte abnormalities of chronic canine mitral regurgitation. *J. Clin. Invest.* **93**: 2639–2648.

TZANKOFF, S.P. and NORRIS, A.H. (1977) Effect of muscle mass decrease on age related BMR changes. *J. Appl. Physiol.: Respirat. Environ. Exercise Physiol.* **43**: 1001–1006.

TZANKOFF, S.P. and NORRIS, A.H. (1978) Longitudinal changes in basal metabolism in man. *J. Appl. Physiol.: Respirat. Environ. Exercise Physiol.* **45**: 536–591.

UVELIUS, B. (1976) Isometric and isotonic length-tension relations and variations in cell length in longitudinal smooth muscle from rabbit urinary bladder. *Acta Physiol. Scand.* **97**: 1–12.

VANDENBOOM, R., CLAFLIN, D.R., and JULIAN, F.J. (1998) Effects of rapid shortening on rate of force regeneration and myoplasmic [Ca^{2+}] in intact frog skeletal muscle fibres. *J. Physiol.* **511**: 171–180.

VOGL, A. (1950) The discovery of the organic mercurial diuretics. *Am. Heart. J.* **39**: 881–883.

VOSS, J., JONES, L.R., and THOMAS, D.D. (1994) The physical constants of calcium pump regulation in the heart. *Biophys. J.* **67**: 190–196.

WAAGSTEIN, F., HJALMARSON, A., VARNAUSKAS, E., et al. (1975) Effect of chronic beta-adrenergic receptor blockade in congestive cardiomyopathy. *British Heart J.* **37**: 1022–1036.

WALLEY, K.R. and COOPER, D.J. (1991) Diastolic stiffness impairs left ventricular function during hypovolemic shock in pigs. *Am. J. Physiol.* **260**: H702–H712.

WATANABE, M., TAKEMORI, S., and YAGI, N. (1993) X-ray diffraction study on mammalian smooth muscles. *J. Muscle Res. Cell Motil.* **14**: 469–475.

WEBER, A. (1959) On the role of calicum in the activity of adenosine 5'-triphosphate hydrolysis by actomyosin. *J. Biol. Chem.* **234**: 2764–2769.

WHITE, D.C.S. (1983) The elasticity of relaxed insect fibrillar flight muscle. *J. Physiol.* **343**: 31–57.

WITHERING, W. (1785) An account of the foxgolve, and some of its medical uses. (Robinson, London. Reprinted 1979 by The Classics of Medicine Library, Birmingham, AL).

WINEGRAD, S. (1984) Regulation of cardiac contractile proteins: correlations between Physiology and Biochemistry. *Circulation Res.* **55**: 565–574.

WOLEDGE, R.C, CURTIN, N.A., and HOMSHER, E. (1985) *Energetic aspects of muscle contraction.* (Academic Press, London.)

WOODS, R. H. (1992) A few applications of a physical theorem to membranes in the human body in state of tenstion. *J. Anat. Physiol.* **26**: 362–370.

YOUNG, M.C. (1997) *The guinness book of world records 1998.* (Guinness Publishing Ltd., New York)

YOUNG, T. (1809) The function of the heart and arteries. The Croonian Lecture. *Phil. Trans. Roy. Soc.* **98**: 1–31

INDEX

A-band **9**
Acetylcholine 201
Acetylstrophanthidin 345
Adreneline 325, 345
Acidosis 171–174
Actin, f- and g- **18**, 110
Action potential duration 198
Activation **147–165**
   mechanistic definition, 148
   operational definition **148–149**
Active state 147, 159
   time course 157–158
Adrenaline 201
Adrenergic agents 339, 341, 355
   alpha 345
   beta 345, 354
   blockers 373, 392, 396
Adrenergic mechanisms **348–351**
Aequorin 113, 136, 138
Afterload **123**, **160**
Allen, D.G. 135–138, 173–174
Amyloidosis 365
Angina pectoris 176
Angiotensin converting enzyme (ACE) 392–393
   inhibitors 392–393, 395, 396
Arteriole **273**
Artery 264
   coronary 294–295
Ascending limb 127, 144
Ashley, C.C. 113
Atherosclerosis 361

Atrioventricular node **199, 266, 291**
Atrium **298–299**
Autoregulation **277– 278**
Bagshaw C.R. 83
Banks, B. 83
Bare-zone **11**
Baroreceptor **280**
Basketball 239, 258
Baylor, S.M. 118–119
Beam K. 122–123
Berglund, E. 325, 331–335, 362
Berman, M. 355, 356, 357
Bezanilla, F. 118
Bicycle racing 248–249
Blinks, J.R. 135–137
Blix, M. 160
Body length **28**
Brady, A.J. 124, 164
Brandt, P.W. 126
Bremel, R.D. 123
Boxing 237–238, 258
Cadence, pedaling 248
Caffeine 339, 347
Calcium induced calcium release **118**, 197, 342
Capillary **264, 273**
Carbohydrate loading 176
Cardiac cycle **266–268**
   output **275**
   return 263
   skeleton **292**

Carvedilol 394–396

Cecchi, G. 73,

Chandler, W.K. 120–121

Chapman C.B. 324

Chiu, Y.C. 356

Claudication, intermittent 176

Cohn, J.N. 392

Compliance, arterial 269

Constriction, pericardial 365

Contraction **25**

Contracture 355

Cooper, D.J. 146, 206,

Cooperativity 123–126, 146, 166, 342, 351

Costantin, L.L. 109, 114, 115

Creatine kinase (CK, CPK) 79

Critical closing **279**

Crown **11**

Crossbridge
 head **16**, 17,
 power stroke 64, 67, 84–86
 original theory **52–56**, 155,
 partial cycles 89,

Dantrolene 166,

Davies, R.E. 87–88

Denominator, metabolic 212

Dense bodies **185**

Descending limb 127

deTombe, P.P. 354, 356

Digitalis 340, 390–391, 396

Dihydropyridine receptor 122

Dillon, P.F. 189

Diuretics 384, 396
 organic mercurial 387

Dunham, E.T. 344

Dysfunction
 Diastolic 357, **363–364**
 Systolic 357, **367–371**

Dyspnea **375–377**

Ebashi, F. 110,

Ebashi, S. 103, 105, 107, 109

Edema **373–374**
 pulmonary 373
 acute 276, **377–380**
 treatment **282–387**
 flash 377

Efficiency 43–44

EGTA 115

Eisenberg, R.S. 119

Embolus 297

Endo, M. 109, 118, 139

Endoh, M. 136–137

Enthalpy 43

Entropy 43

Fabiato, A. and F. 122, 173–174

Fatigue **166–177**

Fay, F.S. 185

Fenn effect **35**, 77–78,

Fenn, W.O. 35,

Fiske, C.H. 78

Fight-or-flight response 345, 378

Football 256–257
 running backs 238–239

Ford, L.E. 101, 116, 117, 197, 308, 371

Forman 208

Franciosa, J.A. 392

Frank, O. 271, 324, 325

Frank-Starling mechanism 127–129, 127–129, 265, 309, 314, 320, **324–336**, 362, 365

Gaasch W.H. 371, 372, 391

Gage, P.W. 119,

Gap junction **7**, **203**,

Gating charge 119–120

Gilbert, S.H. 101,

Glycercol shock **119**

Glycogen **169**, 171

Glycolysis **169–170**

Glycoside, cardiac 344

Glynn, I.M. 344

Godt, R.E. 92, 124

Goldman, Y.E. 93,

Gonzalez-Serratos, H. 131–132

Gordon, A.M. 49, 124, 125, 131

Gorlin, R and S.G. 284

Grossman, W. 309

Guyton, A. 287

Gymnastics 238

H-zone

Hale, S. 263

Hansen, J. 47, 91

Harvey W. 290

Heart Failure
    Pathophysiology **361–380**
    Treatment 381–396

Hecht, H.A. 173,

Heilbrun, L.V. 103, 114, 159

Hellam, D.C. 92, 115

Helmholtz, H.L.F. 35

Hemochromatosis 365

Hemodynamics **263**

Hexagonal lattice 12

High-jumping 246–248, 253–255

Hill, A.V. 35, 43, **45–46**, 67, 75, 97,
      126, 158, 161, 162, 163, 211,
      237, 252

Hinge region **18**, 12

Hodgkin, A.L. 47, 119

Hoh, J.F.Y. 355, 356

Horowicz, P. 118, 119

Hubert, C.E. 114

Huxley, A.F. **47–48**, 49, 58, 59, 66,
      67, 71, 73, 85, 93, 97, 102,
      106–109, 119, 131, 155, 162,

Huxley, H.E. 47, 48, 49, 91, 109,
      111

Hypertension, pulmonary 283

Hyperthermia, malignant 166

Hypoxia **171**, 174–175

Infarction, myocardial **320**, **361**

Inositol triphosphate (IP3) 189, 192

Intermediate filaments **185**, 186

Intercalated disks **7**

Internal membranes **14–16**, 104–110

Ischemia 167, **171**, 175–176

Isoproteronol 346

Jewell, B.R. 57, 162

Jöbsis, F.F. 112–113

Johnson, F.H. 110

Julian, F.J. 49, 131

Katz, A.M. 173, 335

Katz, B. 131

Keilley, W.W. 104–105

Klieber, M. 211

Krause's membrane 106

Kurihara, S. 137

Kushmerick, M.J. 87–89

Kymograph **325**

L_{max} **28**, 145

L_0 **28**

Laplace relationship 127, 204

Latch bridge 190

Law
    Laplace 273, **300–304**, 306
    Poiseuille's 273, 284, 285

Lipmann, F. 103, 105, 107, 110

Lohman, K. 78

Lombardi, V. 74–75, 101

Lymn, R.W. 81–82

M-line **10**

MacPhearson, L. 162

Marsh, B.B.105

Maughan, D.W. 92

McClare, C.W.F. 67

McClellan, G.B. 348

McMahon, T.A. 211
Meromyosin
  heavy (HMM) **16**
  light (LMM) **16**
Metzger, J.M. 350
Meyerhoff, O. 104–105
Mitochondria 8
Moss, R.L. 126
Motor unit **200**
Murexide 112
Myofibril **8**, **14**
Myoglobin **5**, 196
Myosin
  head region **16**
  heavy chain **16**
  hinge region 12, **18**
  isoform **20**, 196
  light chain **16**,
    essential **18**,
    phosphorylation 187–188
    regulatory **18**
  rod **16**
  sub-fragment-1 (S-1) **16**, 18
  sub-fragment-2 (S-2) **16**, 18
Natori, R. 91, 114
Neurotransmitter 201
New elastic body theory 77
Nicotinamide-adenine dinuc-
  leotide (NAD) **169–170**
Niedergerke, R. 47, 49
Nitroglycerine 385–386
Nitroprusside 384–386
O'Connor, M.J. 112–113
Oertliker, H. 118
Orchard, C.H. 173–174
Pacemaker 199, **266**
Parallel elastic element **21**, 204,
Pasteur 122

pCa **111**, **115**
Permeabilized muscle **92**
Phases of the transients **59–60**
  interpretation of **60–63**
Plasticity of smooth muscle 180–182
Phospholamban 349
Phosphorylation
  of myosin light chain 350–351
  of phospholamban 349
  of troponin 348 Plethysmograph
    326–327
Podolsky, R.J. 58, 92, 114–117
Post-extrasystolic potentiation
  340–341, 344
Powell, J.A. 122,
Power 281
Pratusevich, V.R. 181, 185
Pre-capillary sphincter **261**, **273**
Preload **160**
Pressure, mean vascular 287–288
Pulmonary circuit **264**
Pulse wave 270–272
Rack, P.M.H. 141–144, 200, 206
Ramsey, R.W. 134
Rate function 54
Recruitment 141, 200
Resistance
  hemodynamic **274–275**
  pulmonary vascular **275**, 282–283
  systemic vascular **275**
  valvular **275**, 283–287
Restoring force 129, 132, 146
Restriction, myocardial 365
Reuter, H. 345
Rice, R.V. 184
Ridgway E.B. 113, 125
Rigor **81–82**, 124–125, 168
Robertson, J.D. 107

Rotatory stimulation 141–143
Rüdel, R. 131–134
Running 249–253
Ryanodine receptor 122
Sarcomere **9**
Sarcoplasmic reticulum 8, **14**, 15, 109–111, 122, 195, 347
Sarnoff, S.J. 325, 331–335, 362
Schoenberg, C.F. 184,
Schneider, M.F. 120,
Schwann 77,
Seitz, N. 345
Seow, C.Y. 197
Series-to-parallel transition **190–191**,
Shimomura, O. 113
Shock, Cardiogenic 380
Shortening deactivation **123**
Shot-putting 246
Simmons, R.M. 49, 58, 59, 66, 67, 73, 74
Sinoatrial node (SA) 199, **266**, **291**, 346
Skinned fiber **22**, 23, **92**
Soccer 259
Spot follower 49, 59
Spudich, J.A. 74
Starling, E. 324, 325
    resistor 327
Street, S.F. 134,
Striations 3
Stroke volume 328
Stroke work 328
Subbarow, Y. 78,
Svedberg, K. 394, 396
Syncytium **7, 203,**
    electrical **8**
Systemic circuit **264**
Szent-Györgyi, A. 91, 92

T_1 **59**,
    curve 63, 64
T_2 **60**,
    curve 63, 64
t-system **14**, 106–109, 134, 161, 195
t-tubule **14**, 15
Tada, M. 349
Tanabe, T. 123
Target zone of thin filaments 13
Taylor, C.R. 167
Taylor, E.W. 81–82, 84
Taylor, R.E. 106–109
Taylor, S.R. 131–134
Tennis 256
terKeurs, H.E.D.J. 354, 356
Terminal cisterns **14**
Tetanus **25**
Thick filament **9**, 17
    evanescence 182–184, 190–191,
    side polar vs. end polar 186,
Thin filament **10**, 17
    native **19**, 110,
    reconstituted **19**, 110
Thrombus, mural 297
Titin **22**
Torricelli principle 285
Transient **56–64**,
    tension **58**
    velocity **58**
Triad **14**, 105
Tricarboxylic acid cycle 169–171
Tropomyosin **19**, 110, 124
Troponin **19**, 110, 124, 195
Tsutsui, H. 376, 398
Twitch **25**
Uvelius, B. 179–180
Valve
    area, calculated 284–285

atrioventricular 293
aortic 268
  bicuspid 293
mitral 293
pulmonic 268
tricuspid 293
Vein 264
Velocity transient **32**, 33
Ventricular function curve 309
Ventriclular remodeling **316**
Venule **273**
Vessel
  capacitance **273**
  conductance **273**
  resistance **273**

Waagstein, F. 391, 394, 396
Walley, K.R. 146, 206
Wasserman, E. 324
Weber A., 104, 124
Weight-height index 234,239, 240, 242, 253
Weightlifting 240–245
Westbury, D.R. 141–144, 200, 206
White, H.D. 84
Wiercinski, F.J. 103, 114, 159
Wilkie, D.R. 57, 162
Winegrad, S. 348, 353
Withering, W. 344, 390, 396
Z-line **8**, 106, 185